Large Marine Ecosystems of the North Atlantic

Changing States and Sustainability

Large Marine Ecosystems

Series Editor: **Kenneth Sherman**
Director, Narragansett Laboratory and Office of
Marine Ecosystem Studies
NOAA-NMFS, Narragansett, Rhode Island, USA
and Adjunct Professor of Oceanography
Graduate School of Oceanography, University of Rhode Island,
Narragansett, Rhode Island, USA

On the cover

The satellite image of North Atlantic Ocean productivity is provided by the SeaWiFS Project, NASA/Goddard Space Flight Center, and ORBIMAGE. The image is a composite of all SeaWiFS data collected between September 1997 and August 2000. The color-enhanced code depicts a shaded gradient of chlorophyll concentration from a high level of 30 mg/m^3 shown in reddish to orange color spectra to lesser concentrations in yellow and light blue down to an ecosystem low value of 0.1 mg/m^3 depicted in the darker blue.

Large Marine Ecosystems of the North Atlantic

Changing States and Sustainability

Edited by

Kenneth Sherman
Director, Narragansett Laboratory and Office of Marine Ecosystem Studies
NOAA-NMFS, Narragansett, Rhode Island, USA
and Adjunct Professor of Oceanography
Graduate School of Oceanography, University of Rhode Island,
Narragansett, Rhode Island, USA

Hein Rune Skjoldal
Senior Scientist and Research Director
Institute of Marine Science, Bergen, Norway
and Chairman, Advisory Committee on Marine Ecosystems
ICES, Copenhagen, Denmark

2002

ELSEVIER

Amsterdam • Boston • London • New York • Oxford • Paris • San Diego
San Francisco • Singapore • Sydney • Tokyo

ELSEVIER SCIENCE B.V.
Sara Burgerhartstraat 25
P.O. Box 211, 1000 AE Amsterdam, The Netherlands

First edition 2002

British Library Cataloguing in Publication Data
A catalogue record from the British Library has been applied for.

Library of Congress Cataloging in Publication Data
A catalog record from the Library of Congress has been applied for.

ISBN: 0 444 51011 7
ISSN: 1570 0461

⊚ The paper used in this publication meets the requirements of ANSI/NISO Z39.48-1992 (Permanence of Paper).
Printed in The Netherlands.

Preface

This volume on changing states of North Atlantic large marine ecosystems (LMEs) is drawn from a symposium sponsored by the World Conservation Union (IUCN); the Institute of Marine Science, Bergen, Norway; U.S. National Oceanic and Atmospheric Administration (NOAA); and the International Council for the Exploration of the Sea (ICES). It represents an assessment during a period of climate change of the ecological state of ten LMEs that span the east-west extent of the North Atlantic from the U.S. Northeast Shelf to the Scotian Shelf and Newfoundland-Labrador Shelf of North America to the West Greenland Shelf, and the Iceland Shelf, Faroe Plateau, Barents Sea, North Sea, Biscay-Celtic Shelf and Iberian Coastal LMEs of Europe.

Using the case-study approach the contributions examine causes and effects of environmental and human-induced changes on LME productivity and sustainability. Scientific and economic considerations are examined for sustaining the health and biomass yields of the principal fish producing areas of the North Atlantic.

The contributions are at the cutting edge of the newly emerging Sustainability Science (*Science* 292: 27, April 2001). The results presented by the contributors are pertinent to one of the core questions of Sustainability Science: "How are long-term trends in environment and development, including consumption and population, reshaping nature-society interactions in ways relevant to sustainability?" The case studies demonstrate the utility of an ecosystem-based approach to the assessment and management of biomass yields and species sustainability. In two LMEs, the U.S. Northeast Shelf and the Iceland Shelf, governance organizations and management actions to promote the recovery of depleted fish stocks have initiated population recovery responses. In both cases, the positive management decisions were based on comprehensive ecosystem-wide assessments of changing ecological conditions, compelling evidence of the extent of human-induced and climatic forcing causing population declines, and enforced management actions to control fishing effort.

Movements toward ecosystem-based management emerge from the case studies of the resources of the Faroe Plateau, the North Sea, and the Barents Sea LMEs. Uncertainties, with regard to environmental and human-generated forcing are addressed in assessment of the states of the Iberian Coastal and Biscay-Celtic LMEs, and in broad-scale studies of the influences at the base of the food chain of climatic variability on the productivity and biodiversity of plankton communities of the North Atlantic. The volume concludes with an insightful perspective on the approaches used and the results reported by the eminent marine scientist and former President of ICES, Professor Gotthilf Hempel.

This is another of the Large Marine Ecosystem volumes in the new series published by Elsevier Science. LME volumes and titles are listed at http://www.edc.uri.edu/lme

Since the initiation of the LME volume series in 1986, a growing number of scientists, resource managers, and national ministries have moved toward the LME as a geographic unit for the assessment and management of coastal environments and resources. With the initiation in 2001 of the Benguela Current, Yellow Sea, Baltic Sea, and Guinea Current Large Marine Ecosystem (LME) projects, 30 countries across the globe in Asia, Africa, and eastern Europe have made ministerial level commitments to ecosystem-based assessment and management practices in support of the global objectives of UNCED. Among the specific project objectives are: (1) the recovery of depleted fish biomass and fisheries to promote greater food security, sustainable productivity and socioeconomic benefits; (2) reduction in pollution and eutrophication levels of coastal waters; and (3) restoration of degraded habitats including corals, mangroves, and wetlands. The biomass recovery and restoration activities encompass whole marine ecosystems from the drainage basin to the outer boundaries of coastal currents. An additional 20 countries are preparing proposals to improve global coastal health and restore depleted biomass yields in west Africa (Canary Current LME), east Africa (Somali Current and Agulhas Current LMEs), Asia (Bay of Bengal LME), and Latin America (Caribbean LME and Gulf of Mexico LME, Humboldt Current LME, and the Pacific Central American Coastal LME).

The coastal ocean restoration and sustainability approach is based on a 5-module assessment and management methodology. The modules are science-based and country driven; they include considerations of ecosystem: (1) productivity, (2) fish and fisheries, (3) pollution and ecosystem health, (4) socioeconomics, and (5) governance. This approach is engaging ministries across traditional sectors (e.g. Environment, Fisheries, Energy, Tourism, and Finance) to an ecosystem-based management improvement campaign that has been made possible through the cooperation of the Global Environment Facility (GEF) and its partner agencies including the World Bank, UNDP, UNEP, UNIDO, FAO, and IOC. Several collaborating agencies including NOAA in the United States, and the International Union for the Conservation of Nature (World Conservation Union) contribute scientific and technical assistance to the countries and the UN support agencies engaged in this effort. The IOC of UNESCO convenes and hosts annual Consultative Meetings on Large Marine Ecosystems and issues Meeting Reports.

Examples of multi-ministerial level commitments for improving the biomass yields, health, and habitats of LMEs can be found incorporated in the Accra Declaration signed in 1998 by six African states (Benin, Cameroon, Ghana, Ivory Coast, Nigeria, and Togo), in the language of the Benguela Current Commission organized in 2001 by Angola, Namibia, and South Africa, and in the project document for the GEF-supported Yellow Sea LME project, to be implemented by China and Korea. Other ecosystem-based projects are presently under consideration by several countries. Support for these activities is provided with a commitment over the next 5 years of $165 million in GEF, national, and donor funding. It is expected that, as other countries participate in the program, the level of support will reach $200 million by 2003.

Acknowledgements

The editors are indebted to the willingness of the contributors to take time out of their busy schedules to prepare the expert syntheses and reviews that collectively nudge forward the linkages between science and more sustainable ecosystem-based resource management. We are pleased to acknowledge the interest and financial support of the National Oceanic and Atmospheric Administration (NOAA) and the National Marine Fisheries Service (NMFS), the Institute of Marine Science, Bergen, Norway, the Norwegian government, and the International Union for the Conservation of Nature (IUCN - World Conservation Union). We gratefully acknowledge the support of the International Council for the Exploration of the Sea (ICES).

This volume would not have been possible without the willingness and capable cooperation of many people who gave unselfishly of their time and effort. We are indebted to Ms. Ann Royko, Charlestown, Rhode Island for her extraordinary dedication, care and expertise in technically editing and preparing the volume in camera-ready format for publication. Also we extend our thanks to Ms. Hetty Verhagen and Ms. Mara Vos-Sarmiento of Elsevier Science for their care in the final production of the volume.

The Editors

Contributors

Olafur S. Astthorsson
Marine Research Institute
Reykjavík, Iceland

Gregory Beaugrand
Sir Alister Hardy Foundation for Ocean
      Science
Plymouth, United Kingdom

Bjarte Bogstad
Institute of Marine Research
Bergen, Norway

Don Bowen
Marine Fish Division
Bedford Institute of Oceanography
Dartmouth, Nova Scotia, Canada

Alida Bundy
Marine Fish Division
Bedford Institute of Oceanography
Dartmouth, Nova Scotia, Canada

Padmini Dalpadado
Institute of Marine Research
Bergen, Norway

Ken Drinkwater
Ocean Sciences Division
Bedford Institute of Oceanography
Dartmouth, Nova Scotia, Canada

Kenneth T. Frank
Marine Fish Division
Bedford Institute of Oceanography
Dartmouth, Nova Scotia, Canada

Eilif Gaard
Faroese Fisheries Laboratory
Nóatún, Tórshavn
Faroe Islands

Harald Gjøsæter
Institute of Marine Research
Bergen, Norway

Bogi Hansen
Faroese Fisheries Laboratory
Nóatún, Tórshavn
Faroe Islands

Gotthilf Hempel
c/o Bremen Marketing GmbH
Bremen, Germany

Joseph Kane
USDOC/NOAA/NMFS/NEFSC
Narragansett Laboratory
Narragansett, Rhode Island USA

Alicia Lavín
Instituto Español de Oceanografía
Centro Oceanográfico de Santander
Santander, Spain

Jacqueline McGlade
Department of Mathematics
University College London
United Kingdom

Sigbjørn Mehl
Institute of Marine Research
Bergen, Norway

Steven Murawski
USDOC/NOAA/NMFS/NEFSC
Woods Hole Laboratory
Woods Hole, Massachusetts USA

Robert N. O'Boyle
Regional Advisory Process Office
Bedford Institute of Oceanography
Dartmouth, Nova Scotia, Canada

Bergur Olsen
Faroese Fisheries Laboratory
Nóatún, Tórshavn
Faroe Islands

William Overholtz
USDOC/NOAA/NMFS/NEFSC
Woods Hole Laboratory
Woods Hole, Massachusetts USA

Søren Anker Pedersen
Greenland Institute of Natural
        Resources
Danish Institute for Fisheries Research
Department for Marine Ecology and
        Aquaculture
Charlottenlund, Denmark

Carmela Porteiro
Instituto Español de Oceanografia
Vigo, Spain

Philip C. Reid
Sir Alister Hardy Foundation for Ocean
        Science
Plymouth, United Kingdom

Jákup Reinert
Faroese Fisheries Laboratory
Nóatún, Tórshavn
Faroe Islands

Jake Rice
CSAS - Science Branch
Department of Fisheries & Oceans
Ottawa, Ontario, Canada

Doug Sameoto
Biological Oceanography Division
Bedford Institute of Oceanography
Dartmouth, Nova Scotia, Canada

Kenneth Sherman
USDOC/NOAA/NMFS/NEFSC
Narragansett Laboratory
Narragansett, Rhode Island USA

Michael Sinclair
Science Branch
Bedford Institute of Oceanography
Dartmouth, Nova Scotia, Canada

Hein Rune Skjoldal
Institute of Marine Research
Bergen, Norway

Andrew Solow
Marine Policy Center
Woods Hole Oceanographic Institution
Woods Hole, Massachusetts USA

Arnold H. Taylor
Plymouth Marine Laboratory
Plymouth, United Kingdom

Also:
Sir Alister Hardy Foundation for Ocean
        Science
Plymouth, United Kingdom

Luis Valdés
Instituto Español de Oceanografia
Centro Oceanográfico de Santander
Santander, Spain

Hjálmar Vilhjálmsson
Marine Research Institute
Reykjavík, Iceland

Tim Wyatt
Instituto de Investigaciones Marinas
Vigo, Spain

Kees C.T. Zwanenburg
Marine Fish Division
Bedford Institute of Oceanography
Dartmouth, Nova Scotia, Canada

Contents

I
North Atlantic Teleconnections

Large Marine Ecosystems of the North Atlantic
K. Sherman and H.R. Skjoldal (Editors)
© 2002 Elsevier Science B.V. All rights reserved.

3

1

North Atlantic Climatic Signals and the Plankton of the European Continental Shelf

Arnold H. Taylor

ABSTRACT

Climatic variability on the European Continental Shelf is dominated by events over the North Atlantic Ocean, and in particular by the North Atlantic Oscillation (NAO). The NAO is essentially a winter phenomenon, and its effects will be felt most strongly by populations for which winter conditions are critical. One example is the copepod *Calanus finmarchicus*, whose northern North Sea populations overwinter at depth in the North Atlantic. Its annual abundance in this region is strongly dependent on water transports at the end of the winter, and hence on the NAO index.

Variations in the NAO give rise to changes in the circulation of the North Atlantic Ocean, with additional perturbations arising from El Niño - Southern Oscillation (ENSO) events in the Pacific, and these changes can be delayed by several years because of the adjustment time of the ocean circulation. One measure of the circulation is the latitude of the north wall of the Gulf Stream (GSNW index). Interannual variations in the plankton of the Shelf Seas show strong correlations with the fluctuations of the GSNW index, which are the result of Atlantic-wide atmospheric processes. These associations imply that the interannual variations are climatically induced rather than due to natural fluctuations of the marine ecosystem, and that the zooplankton populations have not been significantly affected by anthropogenic processes such as nutrient enrichment or fishing pressure.

While the GSNW index represents a response to atmospheric changes over two or more years, the zooplankton populations correlated with it have generation times of a few weeks. The simplest explanation for the associations between the zooplankton and the GSNW index is that the plankton are responding to weather patterns propagating downstream from the Gulf Stream system. It seems that these meteorological processes operate in the spring.

Although it has been suggested that there was a regime shift in the North Sea in the late 1980s, examination of the time-series by the cumulative sum (CUSUM) technique shows that any changes in the zooplankton of the central and northern North Sea are consistent with the background climatic variability. The abundance of total copepods increased during this period but this change does not represent a dramatic change in ecosystem processes. It is possible some change may have occurred at the end of the time-series in the years 1997 and 1998.

INTRODUCTION

The European Continental Shelf is defined in this paper by the boxes shown in Figure 1-1.

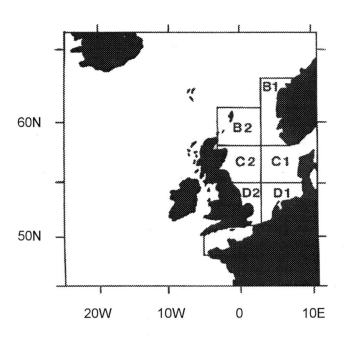

Figure 1-1. Chart of the European Continental Shelf showing Continuous Plankton Recorder Standard Areas that are used in the paper.

It comprises two Large Marine Ecosystems, the North Sea and the Celtic-Biscay Shelf, and part of a third, the Norwegian Shelf. Weather patterns in this region are dominated by events over the North Atlantic. The major source of interannual variability in the atmospheric circulation over the North Atlantic is the North Atlantic Oscillation (NAO), which is associated with changes in the winter surface westerlies across the North Atlantic as far as Europe. Empirical orthogonal function analyses reveal that the NAO is the dominant mode of interannual variability of the surface atmospheric circulation in the

Atlantic and accounts for more than 36% of the variance of the mean December to March sea-level pressure field over the region 20-80°N, 90°W-40°E from 1899 to 1994 (Hurrell 1995). The NAO is present throughout the year in monthly mean data but is most pronounced during the winter (Barnston and Livezey 1987). It takes place between the subtropical high pressures centred on the Azores and the sub-polar low pressures centred on Iceland. The NAO index is the difference in normalised sea level pressures between the Azores (or Portugal) and Iceland during the winter season (The index can be calculated for any period of the year but winter is most often used).

The state of the NAO determines moisture transport and winter temperatures on both sides of the ocean (Hurrell 1995). A high NAO index corresponds to stronger wind circulation than normal in the North Atlantic, high temperatures in western Europe and low temperatures on the east coast of Canada. Conversely, a low NAO index corresponds to the opposite effects. Long-term changes in the NAO have been accompanied by many changes in the North Atlantic and Nordic Seas (Dickson *et al.* 1988; Furnes 1992; Taylor and Stephens 1998).

Several studies have compared time-series of biological populations with variations in the NAO (e.g. Fromentin and Planque 1996; Planque and Taylor 1998; Reid *et al.* 1998b). In addition, the NAO index has shown an increasing trend since the 1980s (Figure 1-2(a)) which has been used in interpretations of population changes occurring at this time (e.g. Reid *et al.* 1998a). Without definite mechanisms these kinds of comparisons can only be suggestive. There is a need to explain why the smooth trend is more significant then the larger year-to-year fluctuations in the NAO index (Figure 1-2(a)). Further, the winter values of the NAO are only weakly correlated with the annual mean values (Figure 1-2(b)), so there is also a need to account for why the populations should be strongly affected by winter conditions.

A second source of climatic variability in the North Atlantic is the Pacific Ocean's El Niño - Southern Oscillation (ENSO) events that have been shown to influence the current system of the North Atlantic (Taylor *et al.* 1998). While these undoubtedly have considerable impact on ecosystems across the tropical Atlantic (e.g. Cane *et al.* 1994), it is unlikely their effects can be demonstrated at northern latitudes.

The present paper discusses some processes by which changes in the North Atlantic influence the European Shelf and considers the consequences for interpreting changes on the Shelf. The paper reviews and synthesises a number of previous studies, in each case extending the original time-series to include more recent years. The results are used to examine the question of whether there have been regime shifts in the North Sea.

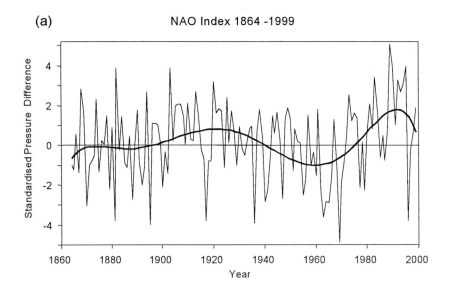

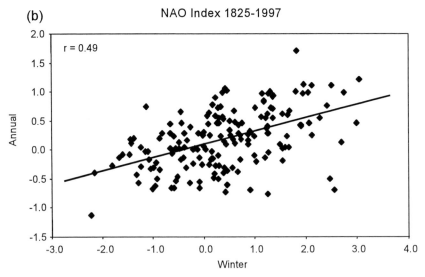

Figure 1-2 (a). The winter NAO index 1864-1999 from Hurrell (1995). A smooth trend calculated by a fourth-order polynomial is plotted through the data. **(b).** Scatter plot between winter and annual values of the NAO index using data from Jones *et. al.* (1997).

MATERIALS AND METHODS

A monthly synoptic survey of the plankton of the North Sea and North-East Atlantic Ocean has been carried out since 1948 using Continuous Plankton Recorders (CPRs) towed by merchant ships and Ocean Weather Ships on regular routes (Glover 1967; Warner and Hays 1994). The CPR is towed at a depth of about 10 m and the seawater is filtered onto a length of 270 μm mesh silk. The zooplankton are counted as individual species and, in addition, the total number of copepods (i.e. the base of the antenna viewed) is counted under x 54 magnification (field of view diameter 2.05 mm) when a traverse is made across the filtering silk in which about 1/40 of the silk is observed. This is the "total copepods" estimate that is used throughout the text.

The latitude of the north wall of the Gulf Stream (GSNW) was calculated from monthly charts of the north wall's position (Taylor and Stephens 1980; Taylor 1995). Six time-series of positions were generated by reading the latitude of the north wall from each monthly chart at the longitudes 79, 75, 72, 70, 67 and 65°W, respectively. An index of monthly position was then calculated as the first principal component of these six series. These data are available at: http://www.pml.ac.uk/gulfstream/.

The zooplankton samples from Lake Windermere (54° 22'N, 2° 56'W) were collected at fortnightly intervals by hauling a net vertically through the water column from a depth of 40 m at a mid-lake station (George and Harris 1985). Biomass was estimated by a simple volumetric technique (George and White 1985). A linear trend was removed from the zooplankton data to correct for the known increasing abundance arising from gradual nutrient enrichment of the lake (George and Harris 1985).

These data are considered in relation to two versions of the North Atlantic Oscillation Index: the widely used index of Hurrell (1995) and that of Jones *et al.* (1997) which extends further back into the early 1800s. Hurrell's index is calculated from the standardised pressure difference between Lisbon and Iceland during winter (December to March), whereas Jones *et al.* calculate a difference between Gibraltar and Iceland for each month of the year.

Three sets of grided atmospheric data covering the North Atlantic are analysed (Taylor 1996): monthly mean sea level pressure, monthly mean height of the 500 hPa pressure level and monthly mean cyclone frequencies. Monthly charts showing tracks of cyclones over the North Atlantic have been published in Mariner's Weather Log. Following Dickson and Namias (1976), the number of tracks crossing each 5° x 5° rectangle was read from each chart (1966-1994) for most of the Atlantic area north of 30°N (more southerly rectangles contained few tracks).

TRANSPORT BY CURRENT FLOWS

Changes in the transport of organisms by current flows can be an important cause of interannual variability in larval fish numbers, e.g. sandeels in the North Sea (Proctor *et al.* 1998). Stephens *et al.* (1998) investigated the effects of fluctuations in North Sea flows on zooplankton abundance by comparing annual abundances observed by the Continuous Plankton Recorder Survey with time-series of transports 1955-1993 from a depth-averaged storm-surge model.   Only one species of zooplankton, *Calanus finmarchicus*, showed a significant dependence on the flows, and then only in the northern regions and on winter flows. In the northern North Sea more than half the variance of this copepod could be accounted for by variations in the winter flows on and off the Shelf edge.   This is consistent with the accepted view that this copepod overwinters in the North Atlantic and enters the North Sea in the early spring. Figure 1-3 shows the comparisons between predictions and observations extended up to 1997. In the northern North Sea area B1 (Figure 1-1) the observed abundances in 1996 and 1997 were considerably less than predicted. However, these discrepancies may still be within the spread of points throughout the time-series so that 1996 and 1997 could still be following similar processes to the rest of the series. In area B2 (Figure 1-1) there is no appreciable discrepancy in these two years.

Fromentin and Planque (1996) have shown that over the period 1958-1995 the abundance of *Calanus finmarchicus* in the NE Atlantic was strongly correlated with the NAO index.   Although other factors such as temperature may have been involved (Planque and Taylor 1998), the strong relationship between flows in winter and the NAO index indicates that variations in flows may be the process leading to Fromentin and Planque's result in the North Sea. Fromentin and Planque report that another dominant copepod, *C. helgolandicus*, which competes with *C. finmarchicus* for food, showed a weaker negative relationship with the NAO. This species overwinters within the North Sea and shows no significant correlation with current flows (Stephens *et al.* 1998).   In 1997, *C. finmarchicus* abundance was also much lower than predicted on the basis of the NAO index, but in this case the 1997 point was well outside the spread of the data points (Reid *et al.* 1998b).

TELECONNECTIONS WITH THE POSITION OF THE GULF STREAM

Plankton Relationships

Apart from *Calanus finmarchicus*, zooplankton species in the North Sea are not strongly dependent on water transports. One indicator of other processes causing interannual variability in zooplankton is the association many species show with the latitude of the Gulf Stream. The measure of Gulf Stream position used is the GSNW index, which is derived from monthly charts of the north wall of the Gulf Stream (Taylor and Stephens

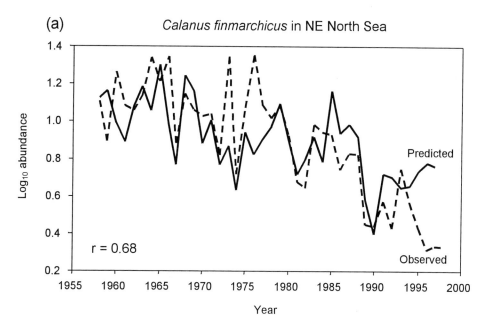

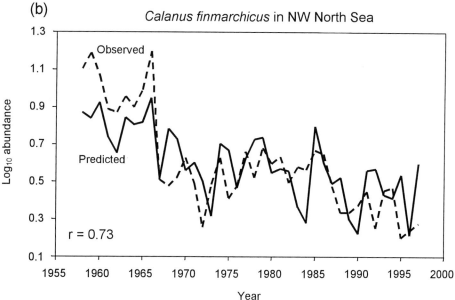

Figure 1-3. Observed (broken line) and predicted (solid line, from a multiple regression analysis of observed annual mean abundance with winter flows across the area boundaries) annual mean abundance of *C. finmarchicus* for CPR Standard areas B1 (NE North Sea) and B2 (NW North Sea). These predictions are from Stephens *et al.* (1998) but extended up to 1998.

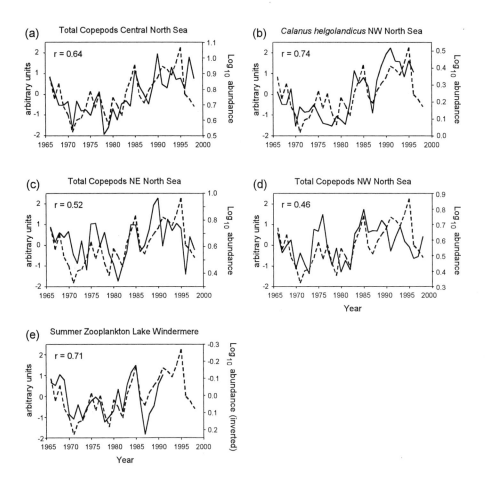

Figure 1-4. The latitude of the north wall of the Gulf Stream 1966-1997 (the GSNW index, broken line) compared with (solid line) the logarithm of: the abundances of copepods (**a, c, d**) and *Calanus helgolandicus* (**b**) measured by the Continuous Plankton Recorder Survey in regions of the central and northern North Sea, and (**e**) the mean summer biomass of zooplankton in Lake Windermere (sedimentation height in cm., from George and Taylor, 1995). A linear trend has been removed from the Windermere data. Central, NW and NE North Sea are CPR standard areas C1, B2 and B1, respectively. All these data are from Taylor *et al.* (in prep.), but updated to 1998.

1980; Taylor 1995). Figures 1-4(a-d) show how closely the abundance of several zooplankton populations follows the trend of the GSNW index. Although these series are only a small subset of the zooplankton species in the North Sea, this phenomenon has been widespread across the zooplankton (Taylor 1995). There are more and larger significant correlation coefficients between individual zooplankton species and the GSNW index than is expected due to chance. An alternative way of demonstrating that associations with the GSNW index permeate the zooplankton, adopted by Taylor *et al.* (in prep.), is to use non-metric Multi-Dimensional Scaling (MDS) as in Figure 1-5.

MDS Plot for east of Orkney and Shetland

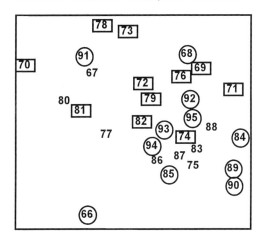

Figure 1-5. Relationships between the log-transformed species abundances in different years expressed in two dimensions by MDS based on Bray-Curtis similarity coefficients for CPR zooplankton from western side of the North Sea east of Orkney and Shetland (stress=0.24), from Taylor *et al.* (in prep.). The positions on the plot are only relative and so have no scales or preferential orientation. Years with a northerly Gulf Stream are marked with circles, years with a southerly Gulf Stream are marked with squares and intermediate years are left unmarked.

The MDS procedure begins by calculating a matrix of similarity coefficients, each of which expresses how similar are a pair of years based on the annual abundance of all the zooplankton species. The coefficient employed is that of Bray and Curtis (1957), which has the advantage that both rare and common species can be included as it is not distorted by the simultaneous occurrence of zero values. Multi-Dimensional Scaling (Shepard 1962; Kruskal and Wish 1978; Clarke and Green 1988) uses the similarity matrix to construct a two-dimensional configuration of the years that attempts to preserve the ranks of the matrix. Thus, if year 1 has higher similarity to year 2 than it does to year 3 then year 1 will be placed closer on the map to year 2 than it is to 3. How

Table 1-1. Percentage probability that the observed difference between groups of years could have arisen by chance (from Taylor *et al.*, in prep.)

Data set	Between north and middle GS years	Between south and middle GS years	Between north and south GS years	Between north and south years after detrending	Between high and low NAO years
CPR zooplankton area 1 east of Orkney and Shetland 1966-1995	28	14	**0.4**	no trend	19
CPR zooplankton area 2 east of Moray Firth to Newcastle 1966-1995	37	36	**1.9**	0.5	12
CPR zooplankton area 3 Dogger Bank 1966-1995	19	74	8.7	no trend	34
CPR zooplankton area 4 Wash/ Humber 1966-1995	15	34	**0.6**	no trend	58
CPR zooplankton area 5 Celtic Sea 1970-1995	0.3	72	**0.0**	no trend	56
CPR summer phytoplankton area 5 Celtic Sea 1970-1995	0.6	43	**0.0**	0.3	33

well the algorithm preserves the rank order in the original matrix is assessed by a stress value (Kruskal's stress formula I). Stress values of 0.2-0.25 are acceptable (Clarke 1993) but smaller values are better.

Figure 1-5 shows such a plot based on logarithmically transformed CPR data from an area to the east of Orkney and Shetland for which there is good data coverage. The figure displays the extent of similarity of the zooplankton populations in different years of the period 1966-1995. It also shows that years with a northward displacement of the north wall of the Gulf Stream at the US coast (circles) tend to be separated from those years in which the wall was displaced southward (squares), the two sets of years being concentrated in the lower right and the upper left part of the figure, respectively. The division between northerly, intermediate, or southerly Gulf Stream years was taken to be at GSNW index values of +0.5 or -0.5, an allocation that gave three approximately equal groups of years. An analysis of similarities (ANOSIM, Clarke 1993) shows that the degree of separation in Figure 1-5 has a 0.4% probability of occurring by chance. Table 1-1 gives the results of this analysis along with those from five other CPR data sets. Only the data from the Dogger Bank region does not show a significant separation between northerly and southerly years. In addition, the table shows that these extreme years are much less separated from the intermediate years. Each of these data sets consists of well over 30 species and the separation between the two groups of years reflects changes spread across a large fraction of these.

Some of the MDS plots display evidence of an overall trend in the populations and, as the GSNW index shows an increase over the 30 years, this could also lead to a separation between northerly and southerly years. To allow for the presence of linear trends, Table 1-1 also includes calculations in which the MDS co-ordinates were detrended by removing a linear trend. The procedure adopted was to construct a three-dimensional MDS configuration of all the years, and then remove a linear trend from any of the co-ordinates in which one was present. In order to avoid de-trending the Gulf Stream series, the linear trend was estimated by means of a multiple regression against time and Gulf Stream position. The detrended co-ordinates were then used as variables to calculate a two-dimensional configuration of the years as in Figure 1-5. Only two of the data sets in Table 1-1, the Moray Firth zooplankton and the Celtic Sea phytoplankton, showed any significant trend. In each of these cases, years with a northerly Gulf Stream remained separate from those with a southerly position even after detrending.

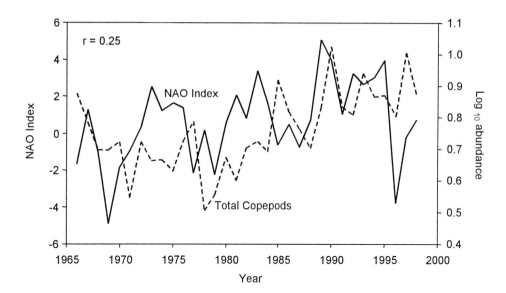

Figure 1-6. The logarithm of the abundances of copepods in CPR Standard Area C1 of the North Sea (broken line) compared with values of the NAO Index (from Hurrell 1995, solid line).

Changes in the position of the Gulf Stream are driven at least in part by the NAO (Taylor and Stephens 1998) and so it is possible that the separations in Figure 1-5 and Table 1-1 could be a direct result of the NAO. To test this possibility, differences between years with a strong or a weak NAO (based on the winter index values in Hurrell 1995) are also shown in Table 1-1. None of the CPR data sets in Table 1-1 shows significantly different zooplankton populations between high and low NAO years. A further example is provided by Figure 1-6, which shows there is little similarity between the NAO index series and the total copepods series from the central North Sea presented in Figure 1-4(a).

Associations with the latitude of the Gulf Stream are widespread throughout the zooplankton community, but they are clearest when estimates of total copepods are used. A possible explanation for this is that individual species have different responses to the climatic forcing. Upon summation these differences cancel out to reveal the underlying common relationship (Taylor 1995).

Taylor *et al.* (1992) have pointed out that the connection between the Gulf Stream and the plankton in the North Sea has to be via the changing weather patterns over the North Atlantic. Confirmation of this is provided by Figure 1-4(e), which shows that zooplankton abundance from a freshwater lake in northern England, Windermere, is also related to the GSNW index. However, in this case northerly positions of the Gulf Stream are associated with less abundant zooplankton instead of the greater abundances observed in the North Sea. This difference may be because an earlier spring bloom and a longer season leads to a mismatch between the zooplankton and its main food supply (George and Harris 1985; George and Taylor 1995).

Importance of spring

Variations in the onset of spring and the start of stratification are likely to be the processes responsible for the association with the latitude of the Gulf Stream. George and Harris (1985) have argued that variations in the timing of spring stratification are the main cause of year-to-year zooplankton abundance fluctuations in Windermere. In addition, CPR estimates of total copepods in the southern North Sea, where no thermal stratification occurs, are not correlated with the position of the Gulf Stream (Figure 1-7). Neither are the *Calanus finmarchicus* populations shown in Figure 1-3 which are driven by water transports rather than stratification. (The Dogger Bank area, which shows no separation in Table 1-1, is also a region with little stratification).

George and Taylor (1995) have constructed an index of the onset of seasonal stratification by analysing temperature profiles at the start of stratification in the beginning of June. The stratification index (SI) was constructed by combining three measures from the early June profiles:

$$SI = G - (R + D).$$

G is the depth of the maximum temperature gradient, R the rate of deepening of the $9°C$ isotherm in the thermocline, and D the difference in depth between the 9 and $10°C$

isotherms in the thermocline. G, R and D were each standardised to zero mean and unit variance.

This index is clearly correlated with the abundance of zooplankton in Windermere (Figure 1-8(a)) and with the Gulf Stream position (Figure 1-8(b)). Indices derived using data from a nearby smaller and shallower lake (Esthwaite Water) and from the northern North Sea show the same general relationship with the position of the Gulf Stream as Windermere. In addition, Figure 1-8(b) shows values calculated from temperature profiles produced by a one-dimensional model of thermal stratification (Mellor and Yamada 1974) when driven by hourly meteorological observations (1966-1994) from Dublin airport (from Taylor *et al.* 1996).

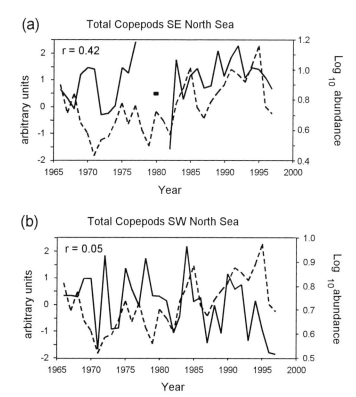

Figure 1-7. The latitude of the north wall of the Gulf Stream 1966-1997 (the GSNW index, broken line) compared with (solid line) the logarithm of the abundance of copepods in CPR southern North Sea Standard Areas **(a)** D1 and **(b)** D2.

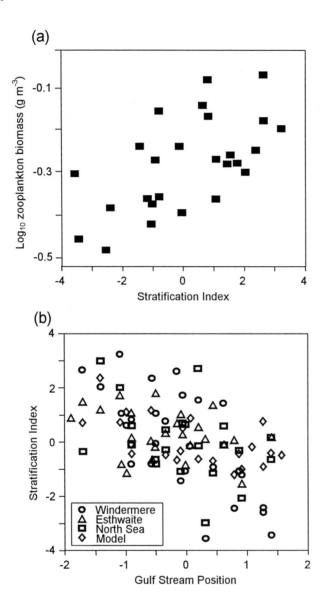

Figure 1-8 (a). Scatter plot relating the logarithm of the mean summer biomass of zooplankton in the north basin of Windermere to the index of the onset of stratification (*SI*) in Windermere (correlation coefficient, r=0.59). A linear trend has been removed from the biomass data to correct for the observed increase in abundance caused by nutrient enrichment of the lake (George and Harris 1985). **(b).** Scatter plot relating the index of stratification *SI* (circles: Lake Windermere; triangles: Esthwaite Water; squares: North Sea; diamonds: Mellor-Yamada model results) to the position of the north wall of the Gulf Stream (r=0.61) (from Taylor *et al.* 1996).

The North Atlantic Oscillation and the Gulf Stream

The GSNW index is statistically related to the NAO index from two years earlier (Taylor and Stephens 1998; Taylor *et al.* 1998), a time-delay that may represent the adjustment time of the ocean circulation (Gangopadhyay *et al.* 1992). Recently, a simple dynamic model incorporating such a time-delay has been developed (Taylor and Gangopadhyay, in press), based on the model of Behringer *et al.* (1979). This predicts the latitude of the Gulf Stream on the basis of variations in the NAO index (Figure 1-9) for all months of the year and on the observed tendency for the wind systems to move north and south as the NAO increases or decreases in strength. The time-lag implicit in the model raises the possibility of forecasting the GSNW index and hence anything correlated with it.

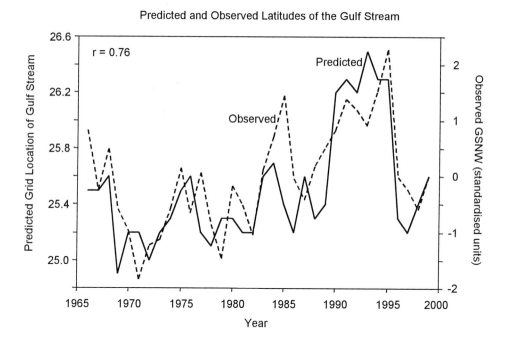

Figure 1-9. Observed latitude of the north wall of the Gulf Stream 1966-1999 (the GSNW index, broken line) compared with (solid line) predicted position from the model of Taylor and Gangopadhyay (in press). *r* is the correlation coefficient between each pair of series.

The atmospheric teleconnection

By applying regression analysis to meteorological data from around the British Isles, Taylor *et al.* (in prep.) have shown that only in the month of April (and perhaps May) is it possible to find year-to-year changes in weather patterns that could connect the GSNW index with the biological series. They have also shown there is a significant difference in the weather types over the UK during April and May between years with a northerly and southerly Gulf Stream. One important factor is the tendency for cloud cover in April to increase during southerly years.

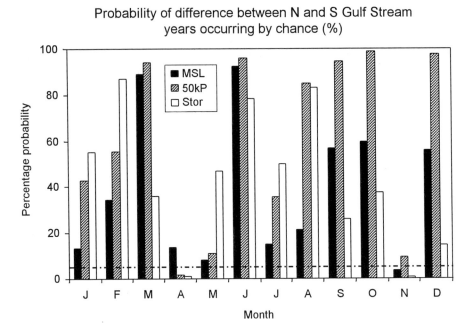

Figure 1-10. Probability that the differences in distributions of storm tracks, mean sea-level pressures and 500 hPa height across the North Atlantic between northerly and southerly GSNW years could have occurred by chance, calculated by the ANOSIM procedure. The broken line shows level of 5% probability.

The association between weather patterns in April (and to a lesser extent in May) and the latitude of the Gulf Stream extends right across the North Atlantic. Multi-Dimensional Scaling was applied to time-series of the number of cyclone tracks on a grid of rectangles spanning the North Atlantic in order to see if the distribution of tracks differed between northerly and southerly years. The similarity measures used were Bray-Curtis coefficients as these allowed rectangles rarely containing tracks to be combined with rectangles having high frequency of tracks. Similar analyses were carried out on mean sea-level pressure and on the mean height of the 500 hPa level, each on a grid covering the North Atlantic. As these measurements are not counts, a normalised euclidean distance was used to measure similarity. Figure 1-10 shows that there appear to be differences between years with a northerly and southerly Gulf Stream in the months of April, May, and November but in no other months, with the April differences in storm tracks and 500 hPa heights each having a probability of less than 5% of arising by chance. These differences in the spring are not a direct result of changes in the NAO. Years with a high value of Hurrell's NAO index showed no statistically significant differences from those with a low NAO index in any month, apart from two of the winter months (December and January) involved in the definition of the index.

The differences between northerly and southerly years in April are shown in Figure 1-11 (a) and (b), the patterns for the number of storm tracks and the mean sea-level pressure matching closely. At this time of the year, southerly Gulf Stream positions have been accompanied by fewer storms and higher pressures south of Iceland than northerly positions, with the opposite changes occurring over Newfoundland. Figure 1-11(c) shows the mean distribution of 500 hPa height during the two sets of years in April. The contours are quite zonal during years with a northerly Gulf Stream. In southerly years, a northward distortion occurs increasing downstream to a maximum south of Iceland. The atmospheric patterns during southerly years appear to show a blocking pattern with cyclones having difficulty reaching the UK during April.

These changes spanning the Atlantic are the source of the variations in weather patterns on the European Shelf discussed in the last section. In April of years with a southerly Gulf Stream, Figure 1-11 shows there is a tendency to higher atmospheric pressures and more northerly winds over the North Sea and the UK. Associated with the northerly winds there is likely to be an increase in cloudiness, which will delay biological growth (Taylor *et al.*, in prep.).

REGIME SHIFTS

Reid *et al.* (1998b, 2001) report that marked changes in phytoplankton colour, fish and the benthos occurred after 1987 in the North Sea. They suggest that these changes signalled the beginning of a regime shift in the ecosystem, which coincided with the rising trend in the NAO shown in Figure 1-2(a). However, the relationships presented in Figures 1-3 and 1-4 indicate that the *Calanus finmarchicus* in the northern North Sea and total copepods and *Calanus helgolandicus* in the central and northern North Sea have shown a uniform response to environmental forcing throughout the period. A sensitive test for any change in level is provided by the cumulative sum (CUSUM) technique, and this can be applied to test for any shift in the relationships shown in Figures 1-3 and 1-4 (Radford and West 1986). The technique was applied to the time-series in Figure 1-4(a) by standardising the plankton and the GSNW series each to zero mean and unit standard deviation, and then forming a running total of the square of the difference between the resulting series. A similar procedure was followed for the data in Figures 1-4(b-d) and for each pair of time-series in Figure 1-3. Figure 1-12 shows the

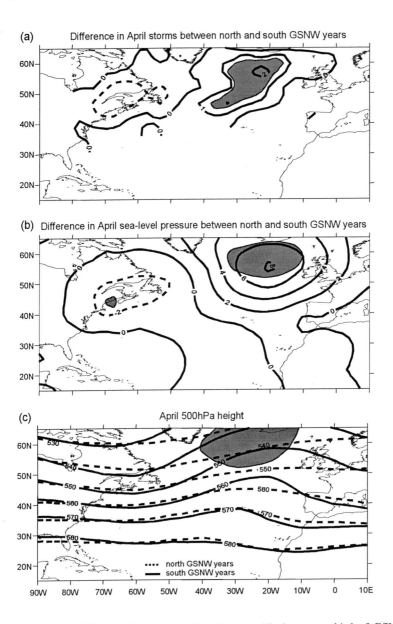

Figure 1-11. Difference between Aprils of years with the upper third of GSNW index values (northerly years) and Aprils with the lower third of values (southerly years) for: **(a)** number of cyclone tracks in 5° x 5° rectangles 1966-1994 and **(b)** mean sea-level pressures (hPa) 1996-1992 on a 5° x 5° grid (signs have been reversed to compare with (a). **(c)** April 500 hPa heights 1966-1992 on a 5° x 5° grid. The solid line is the average of the lower third of GSNW index values, the broken line the average of the upper third. The hatched areas show regions where the difference is significant at the 5% level.

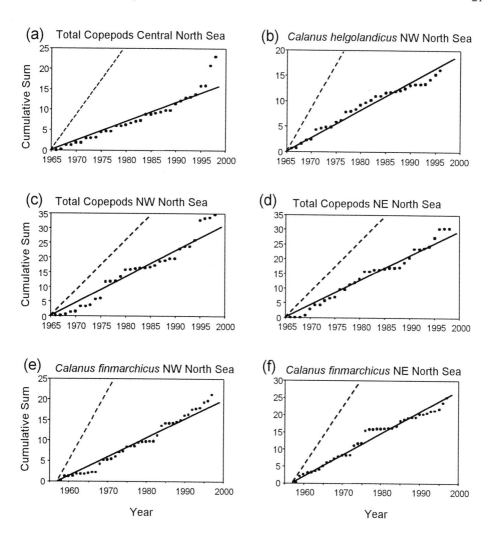

Figure 1-12 (a-d). Cumulative sum of the square of the difference between the abundance of total copepods and the GSNW index, after each time series has been standardised to zero mean and unit standard deviation, calculated from the data in Figures 1-3(a-d). **(e), (f)** Cumulative sum of the square of the difference between the predicted and observed abundance of *Calanus finmarchicus*, after each time series has been standardised to zero mean and unit standard deviation, calculated from the data in Figures 1-2(a), (b). The solid line is a regression fit to the data and the broken line shows the results for an uncorrelated pair of time-series.

resulting cumulative sums. Any change in the relationship between the zooplankton and the GSNW index (Figures 1-12(a-d)), or between *Calanus finmarchicus* and the predictions from the modelled North Sea transports (Figure 1-12(e-f)), will lead to a change in the slope of the plot. A slope obtained from a pair of random time-series is shown for comparison. While the graphs in Figure 1-12 show intermittent variations of slope, some of which approach the random slope, none of the plots shows a marked and permanent change of slope after 1987. The only clear indication of any change is in the total copepods graph from the central North Sea (Figure 1-12(a)) during the years 1997 and 1998. In all the other cases, the changes of slope appear to be within the variability of the time-series.

Although the association with the latitude of the Gulf Stream has continued unchanged from the 1980s to the 1990s for the zooplankton in the central and northern North Sea, this has not been the case for zooplankton in the NE Atlantic. After 1987 the abundance of total copepods declined in this region (Taylor *et al.* 1992) and the correlation with the GSNW index was never restored (Hays *et al.* 1993).

CONCLUSIONS

The processes that have been described in earlier sections are summarised schematically in Figure 1-13. Climatic variability on the European Continental Shelf is dominated by events over the North Atlantic Ocean, and in particular by the North Atlantic Oscillation. The NAO is essentially a winter phenomenon and its effects will be felt most strongly by populations for which winter conditions are critical. One example of this is the copepod *C. finmarchicus* in the northern North Sea, which overwinters at depth in the North Atlantic. Its annual abundance in this region is strongly dependent on water transports at the end of the winter (Figure 1-3), and hence on the NAO index.

Variations in the NAO give rise to changes in the circulation of the North Atlantic Ocean, with additional perturbations arising from ENSO events in the Pacific (Taylor *et al.* 1998). These changes can have delays of several years because of the adjustment time of the ocean circulation. One measure of the circulation is the latitude of the north wall of the Gulf Stream, the GSNW index, which has followed a very similar trend to estimates of the large-scale transport of the Stream (Curry and McCartney in prep.). Interannual variations in the plankton of the Shelf Seas show strong correlations with the fluctuations of the GSNW index (Figure   1-4), which are the result of Atlantic-wide atmospheric processes. These associations (and the relationships between *C. finmarchicus* and water transports shown in Figure 1-3) have two important implications: (i) the interannual variations are climatically induced, and not due to natural fluctuations of the marine ecosystem, and (ii) the zooplankton populations have not been significantly affected by anthropogenic processes such as nutrient enrichment or fishing pressure.

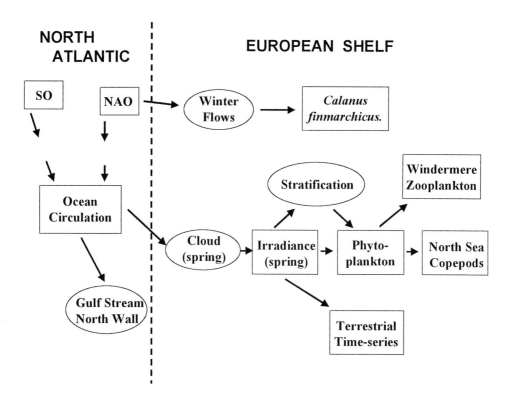

Figure 1-13. Schematic representation of processes connecting climatic events in the North Atlantic to plankton changes on the European Continental Shelf. NAO represents forcing by the North Atlantic Oscillation and SO by the Southern Oscillation in the Pacific Ocean.

While the GSNW index represents a response to atmospheric changes integrated over two or more years, the zooplankton populations correlated with it have generation times of a few weeks with no indications that populations have any memory of events over several years. The simplest explanation for the associations between the plankton and the GSNW index is that the plankton are responding to weather patterns propagating downstream from the Gulf Stream system. Figures 1-10 and 1-11 indicate these weather processes operate in the spring.

The CUSUM analyses in Figure 1-12 show that any changes in the zooplankton of the central and northern North Sea show no evidence for any regime shift in the North Sea during the last half of the 1980s; any changes are consistent with the background climatic variability. Some changes in abundance occurred during this period but they do not indicate a dramatic change in ecosystem processes. A possible exception is total copepods in Standard Area C1 (Figure 1-12(a)) during the years 1997 and 1998.

ACKNOWLEDGEMENTS

I wish to thank the Sir Alister Hardy Foundation for Ocean Science for providing plankton data and, in particular, the late Harry Hunt who gave me assistance over many years. Roger Proctor (Proudman Oceanographic Laboratory) calculated the North Sea flows needed to produce the updated Figure 1-3. J.A. Lindley assembled the data used in the MDS analyses. The illustrations for the paper were prepared by John Stephens. This work forms part of the OC4Z and DYME programmes of Plymouth Marine Laboratory.

REFERENCES

Barnston, A.G. and R.E. Livezey. 1987. Classification, seasonality and persistence of low-frequency atmospheric circulation patterns. Monthly Weather Review, 115: 1083-1126.

Behringer, D., L. Regier, and H. Stommel. 1979. Thermal feedback on wind-stress as a contributing cause of the Gulf Stream. Journal of Marine Research: 699-1979.

Bray, J.R. and J.T. Curtis. 1957. An ordination of the upland forest communities of Southern Wisconsin. Ecological Monographs, 27: 325-349.

Cane, M.A., G. Eshel, and R.W. Buckland. 1994. Forecasting Zimbabwean maize yield using eastern equatorial Pacific sea surface temperature. Nature, 370: 204-205.

Clarke, K.R. 1993. Non-parametric multivariate analyses of changes in community structure. Australian Journal of Ecology, 18: 117-143.

Clarke, K.R. and R.H. Green. 1988. Statistical design and analysis for a 'biological effects' study. Maine Ecology Progress Series, 46: 213-226.

Curry, R.G. and M.S. McCartney. In prep. Ocean gyre circulation changes associated with the North Atlantic Oscillation.

Dickson, R.R. and J. Namias. 1976. North American influences on the circulation and climate of the North Atlantic sector. Monthly Weather Review, 104: 1255-1265.

Dickson, R.R., J. Meincke, S-A. Malmberg, and A.J. Lee. 1988. The "Great Salinity Anomaly" in the northern North Atlantic 1968-1982. Progress in Oceanography, 20: 103-151.

Fromentin, J-M. and B. Planque. 1996. *Calanus* and environment in the eastern North Atlantic. 2. Influence of the North Atlantic Oscillation on *C. finmarchicus* and *C. helgolandicus*. Marine Ecology Progress Series, 134: 111-118.

Furnes, G.K. 1992. Climatic variations of oceanographic processes in the north European seas: a review of the 1970s and 1980s. Continental Shelf Research, 12: 235-256.

Gangopadhyay, A., P. Cornillon, and R.D. Watts. 1992. A test of the Parsons-Veronis hypothesis on the separation of the Gulf Stream. Journal of Physical Oceanography, 22: 1286-1301.

George, D.G. and G.P. Harris. 1985. The effect of climate on long-term changes in the crustacean zooplankton biomass of Lake Windermere, UK. Nature, 316: 536-539.

George, D.G. and A.H. Taylor. 1995. UK lake plankton and the Gulf Stream. Nature, 378: 139.

George, D.G. and N.J. White. 1985. The relationship between settled volume and displacement volume in samples of freshwater zooplankton. Journal of Plankton Research, 7: 411-414.

Glover, R.S. 1967. The Continuous Plankton Recorder Survey of the North Atlantic, Symposia of the Zoological Society of London, 19: 189-210.

Hays, G.C., M.C. Carr, and A.H. Taylor. 1993. The relationship between Gulf Stream position and copepod abundance derived from the Continuous Plankton Recorder Survey: separating biological signal from sampling noise. Journal of Plankton Research, 15: 1359-1373.

Hurrell, J.W. 1995. Decadal trends in the North Atlantic Oscillation: regional temperatures and precipitation. Science, 269: 676-679.

Jones, P.D., T. Jonsson, and D. Wheeler. 1997. Extension to the North Atlantic Oscillation using early instrumental pressure observations from Gibraltar and south-west Iceland. International Journal of Climatology, 17: 1433-1450.

Kruskal, J.B. and M. Wish. 1978. Multidimensional scaling. Sage Publications, Beverly Hills, California.

Mellor, G.L. and T. Yamada. 1974. A hierarchy of turbulence closure models for planetary boundary layers. Journal of Atmospheric Science, 31: 1791-1806.

Planque, B. and A.H. Taylor. 1998. Long-term changes in zooplankton and the climate of the North Atlantic. ICES Journal of Marine Science, 55: 644-654.

Proctor, R., P.J. Wright, and A. Everitt. 1998. Modelling the transport of larval sandeels on the north-west European shelf. Fisheries Oceanography, 7(3/4): 347-354.

Radford, P.J. and J. West. 1986. Models to minimize monitoring. Water Research, 20: 1059-1066.

Reid, P.C., M. de F. Borges, and E. Svendsen. 2001. A regime shift in the North Sea circa 1988 linked to changes in the North Sea horse mackerel fishery. Fisheries Research 50:163-171.

Reid, P.C., M. Edwards, H.G. Hunt, and A.J. Warner. 1998a. Phytoplankton change in the North Atlantic. Nature, 391: 546.

Reid, P.C., B. Planque, and M. Edwards. 1998b. Is observed variability in the long-term results of the Continuous Plankton Recorder Survey a response to climate change? Fisheries Oceanography, 7(3/4): 282-288.

Shepard, R.N. 1962. The analysis of proximities: multidimensional scaling with an unknown distance function. Psychometrika, 27: 125-140.

Stephens, J.A., M.B. Jordan, A.H. Taylor, and R. Proctor. 1998. The effects of fluctuations in North Sea flows on zooplankton abundance. Journal of Plankton Research, 20: 943-956.

Taylor, A.H. 1995. North-south shifts of the Gulf Stream and their climatic connection with the abundance of zooplankton in the UK and its surrounding seas. ICES Journal of Marine Science, 52: 711-721.

Taylor, A.H. 1996. North-south shifts of the Gulf Stream: ocean-atmosphere interactions in the North Atlantic. International Journal of Climatolology, 16: 559-583.

Taylor, A.H., J.M. Colebrook, J.A. Stephens, and N.G. Baker. 1992. Latitudinal displacements of the Gulf Stream and the abundance of plankton in the north-east Atlantic. Journal of the Marine Biological Association, 72: 919-921.

Taylor, A.H. and A. Gangopadhyay. In press. A simple model of interannual displacements of the Gulf Stream. Journal of Geophysical Research.

Taylor, A.H., M.B. Jordan, and J.A. Stephens. 1998. Gulf Stream shifts following ENSO events. Nature, 393: 638.

Taylor, A.H., M.C. Prestidge, and J.I. Allen. 1996. Modelling seasonal and year-to-year changes in the ecosystems of the NE Atlantic Ocean and the European Shelf Seas. Journal of the Advanced Marine Science and Technology Society (Japan), 2: 133-150.

Taylor, A.H. and J.A. Stephens. 1980. Latitudinal displacements of the Gulf Stream (1966 to 1977) and their relation to changes in temperature and zooplankton abundance in the NE Atlantic. Oceanologica Acta, 3(2): 145-149.

Taylor, A.H. and J.A. Stephens. 1998. The North Atlantic Oscillation and the latitude of the Gulf Stream. Tellus, 50 A: 134-142.

Taylor, A.H., J.A. Stephens, M.B. Jordan, E. McKenzie, J.I. Allen, R. Allen, D.J. Beare, R. Clark, N.P. Dunnett, R. Hunt, J.A. Lindley, P.J. Sommerfield, and A.J. Willis. In prep. The latitude of the Gulf Stream as a predictor of the onset of spring in western European ecosystems.

Warner, A.J. and G.C. Hays. 1994. Sampling by the Continuous Plankton Recorder Survey. Progress in Oceanography, 34: 237-256.

Large Marine Ecosystems of the North Atlantic
K. Sherman and H.R. Skjoldal (Editors)
© 2002 Elsevier Science B.V. All rights reserved.

2

Interregional Biological Responses in the North Atlantic to Hydrometeorological Forcing

Philip C. Reid and Gregory Beaugrand

ABSTRACT

Using data from the CPR survey, seven case studies are described that document different spatial and temporal responses in the plankton to hydroclimatic events. Long-term trends in the plankton of the eastern Atlantic and the North Sea over the last five decades are examined. Two of the examples revisit correlations that have been described between copepod abundance in the eastern Atlantic and North Sea and indices of atmospheric variability, the North Atlantic Oscillation index and the Gulf Stream North Wall index. Evidence for an increase in levels of Phytoplankton Colour (a visual index of chlorophyll) on the eastern and western sides of the Atlantic is presented. Changes in three trophic levels and in the hydrodynamics and chemistry of the North Sea circa 1988 are outlined as a regime shift. Two of the case studies emphasise the importance of variability in oceanic advection into shelf seas and the role of western and eastern margin currents at the shelf edge. The plankton appear to be integrating hydrometeorological signals and reflecting basin scale changes in circulation of surface, intermediate and deep waters in part associated with the NAO. The extent to which climatic variability may be contributing to the observed changes in the plankton is discussed with a forecast of potential future ecosystem effects in a climate change scenario.

INTRODUCTION

Changes in the species and biochemical composition of the remains of planktonic organisms through time in oceanic cores have been used as one of the main sources of evidence to support climate change (Mudie and Harland 1996; Kroon *et al.* 1997). Many factors influence the growth of phytoplankton: the degree of mixing/stability of the water column, light intensity, nutrient and trace element concentration, wind strength/direction/frequency, cloudiness, precipitation, and at the mesoscale, eddy formation and other physical structures (e.g. Sverdrup 1953; Margalef 1997). The zooplankton are dependent on the growth of their food and equally influenced by their physical and chemical environment (Verity and Smetacek 1996). The abundance and community structure of the plankton present in a given area of sea is thus likely to be an integration of the recent hydrometeorological history of the area.

Results obtained from the Continuous Plankton Recorder (CPR) survey in the North Atlantic provide support for the above contention at a basin scale. First, Principal Component Analyses by Colebrook in the 1970s, especially in his classic paper of 1978, demonstrated that similar patterns of change in the plankton were evident over very large areas of the Atlantic. The dominant pattern of change in both the oceanic North Atlantic and the North Sea was of a downward trend to approximately 1980; since then there has been a recovery (Reid and Hunt 1998).

Second, Fromentin and Planque (1996) showed that the dominant copepod in the northern Atlantic, *Calanus finmarchicus* was highly correlated with changes in the North Atlantic Oscillation (NAO) in the North Sea and to the northwest of the British Isles. This relationship has broken down since 1996. Other copepod species have also been shown to be correlated with the NAO, with differing regional responses that may reflect their ecological niche.

Third, Taylor and Stephens (1980) showed that small copepods (CPR category: total copepods, 0.3-2mm), especially in the North Sea, were correlated with an index of the position of the North Wall of the Gulf Stream (GSNW index). Because the relationship is between variables on either side of the Atlantic without a lag it must be reflecting a trans-Atlantic atmospheric signal. This relationship also held for an area of the eastern Atlantic from 1958 until 1987. Since 1987 it has broken down.

Fourth, the Arctic indicator copepod *Calanus hyperboreus* has increased substantially in the western Atlantic (Johns *et al.* in press) at the same time as numbers have decreased in the east in an apparent interlinked seasaw relationship. The increased abundance in the western Atlantic is correlated with the thickness of Labrador Sea intermediate water.

There is thus, from a number of sources, good comparative evidence to suggest that changes seen in the plankton on regional scales are reflecting hydrometeorological variability and may act as indicators of climate change. Here we review these earlier studies and document a regime shift in the North Sea circa 1988 that is seen in phytoplankton, zooplankton, fish, benthos, and birds as well as showing up in nutrients, oxygen, temperature, and oceanic inflow. The change at this time appears to be part of a pan North Atlantic event that is especially evident in the Phytoplankton Colour index (a visual index of chlorophyll) measured by the CPR. These events are discussed in relation to documentation of pronounced changes in the deep, intermediate and upper layer circulation, including the eastern and western boundary currents of the northern Atlantic and Nordic Seas.

MATERIALS AND METHODS

Each month of each year since 1946 the Continuous Plankton Recorder survey has deployed high-speed samplers behind voluntary ships of opportunity (SOOP) to sample the upper layer plankton of the northern North Atlantic and adjacent seas (Warner and Hays 1994). Currently more than 100 000 nautical miles of sea are sampled per annum along ~22 routes at an approximate depth of 6.5 m (Hays and Warner 1993). The CPR samples the plankton on a ~15 cm wide band of boulting silk with a mesh of ~270μm that moves across a sampling aperture at a rate that is proportional to the speed of the ship. On return to the laboratory the band is cut into samples of ~10 cm length that are equivalent to ~10 nautical miles of tow and ~3m^3 of seawater filtered. Usually alternate samples are analysed and the phytoplankton and zooplankton counted and identified into up to 400 different species/taxa. Prior to cutting the silk, each 5cm section is allocated a greenness colour into four categories from zero to green by comparison with a colour card. Each of these categories has been given a chlorophyll equivalent value by comparison with fluorometry (Robinson and Hiby 1978). Methods of counting and data processing are described by Warner and Hays (1994).

Principal Component Analyses (PCA) of the CPR data used here have followed the approach of Reid and Hunt (1998), based on Colebrook (1978). A log (x+1) transformation was applied before computing the PCA and the data standardized to zero mean and unit variance.

The North Atlantic Oscillation (NAO) is a basin-scale alternation of atmospheric mass over the North Atlantic between the subtropical and Arctic Atlantic (Hurrell *et al.* 2001). A number of indices of the state of this oscillation have been defined; here we use that of Hurrell based on the winter months December to March. A large positive value of the index indicates reinforced low pressure around Iceland with high pressure over the Azores while a negative value has approximately the reverse pattern (Dickson and Turrell 2000).

The Gulf Stream North Wall (GSNW) is an index of the latitudinal position of the North Wall between 79°W and 65°W (Taylor, this volume). This index is constructed from a Principal Component Analysis of the latitudinal position of the North Wall at different longitudes off the east coast of North America.

LONG-TERM TRENDS

The most consistently sampled area by the CPR survey since 1946 covers the eastern North Atlantic and northwest European shelf. It was this area that the survey retracted to in the late 1980s when it was briefly closed down. In a series of papers, Colebrook used PCA to identify the dominant patterns of year-to-year variation in the plankton of this region (e.g. Colebrook 1978, 1986). He grouped the plankton into approximately 24 species or larger groupings examining their variability in 16 regions (Standard

Areas) that covered the North Sea, western shelf and the open ocean, to the east of 19°W and between approximately 45° and 65°N. His results showed that the dominant pattern of change was a declining trend in more than half of the zooplankton and phytoplankton respectively from 1948 and 1958.

Colebrook's analyses were updated a number of times subsequently, most recently by Reid and Hunt (1998). Here we present reanalyses of this dataset (Figure 2-1) using the approach of the latter authors updated to 1996. It is clear that the minima of the zooplankton decline was reached between 1978 and 1982 in the North Sea and that the decline continued until at least 1996 in the northeast Atlantic. Phytoplankton in the North Sea showed a similar pattern to the zooplankton with a minima in 1978 levelling off in more recent years. The second component of the zooplankton and phytoplankton in the North East Atlantic showed a minima from 1987 and in the North Sea pronounced changes in both the phytoplankton (Component 3) and zooplankton (Component 2) are evident from the mid 1980s and especially after 1987. These changes in the North Sea in the late 1980s reflect, in particular, increases in species of copepods and cladocera such as *Acartia* spp. and *Podon* spp. that lay overwintering resting eggs, and for phytoplankton increases in dinoflagellates.

Planque and Batten (in press) have estimated that the decline in copepods, the dominant group included in the PCA analyses, represented a decline of biomass of approximately 50% in the north-east Atlantic. Colebrook (1978) showed that the pronounced downward trend evident in the first Component occurred in all 16 of the Standard Areas he used and was clearly seen in graphs for many of the individual species. However, soft-bodied plankton such as coelenterates are not well sampled by the CPR and it is not known if the declining trend is representative of all planktonic groups.

THE NAO AND *CALANUS*

The NAO is the dominant influence (>30% of variance) contributing to atmospheric variability in the North Atlantic region. It has been shown to have a modulating effect on a wide range of environmental factors including: temperature, precipitation, wind stress and direction, wave height, general ocean circulation, and the formation of North Atlantic Deep Water and Labrador Sea Intermediate Water (Dickson and Turrell 2000). Through this forcing it strongly influences both marine and terrestrial ecosystems (Ottersen *et al.* in press). In the latter review, the NAO is seen as a proxy for regulating forces on ecosystems with a wide range of ecological responses that are evident at the species, population and community levels.

A strong negative correlation between temporal changes in the copepod *Calanus finmarchicus* (from the North Sea and part of the north-east Atlantic) and the NAO index (Fromentin and Planque 1996) was one of the first biological relationships with the NAO to be discovered. Its congeneric species *C. helgolandicus* was also shown to be correlated, in this case positively and with a lag, to the NAO index. *C. finmarchicus*

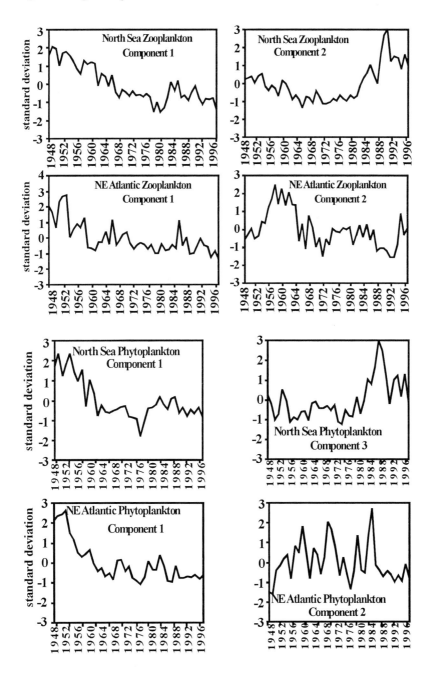

Figure 2-1. Results of a Principal Component Analysis performed on phytoplankton and zooplankton from the CPR survey in the North Sea and the northeastern North Atlantic over the period 1948 to 1997.

is the dominant boreal copepod in the northern Atlantic where it contributes more than half the total biomass and >90% in the Labrador and Norwegian Seas (Planque and Batten, in press). Figure 2-2 shows the North Sea and an area of the eastern Atlantic where abundances of the two *Calanus* species were averaged. These are shown in time series for the period 1958 to 1997 in Figures 2-3 and 2-4. There is considerable interannual variability in abundance in both regions. In the North Sea numbers of *C. finmarchicus* (Figure 2-3) have oscillated until after 1995 when they reduced to the lowest on record. In contrast *C. helgolandicus* has shown an increase to high levels in the last decade. In the 'eastern' Atlantic (Figure 2-4) *C. finmarchicus* is the dominant species and numbers of *C. helgolandicus* are lower than in the North Sea. The relative proportions (*C.fin:C.hel*) of the two species prior to and post 1988 have changed from 80:20 to 38:62 in the North Sea and 85:15 to 80:20 in the 'eastern' Atlantic. Elsewhere in the northern North Atlantic numbers of *C. finmarchicus* have remained high (Planque and Batten, in press; Beare *et al.*, in press).

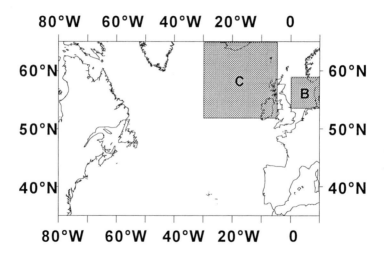

Figure 2-2. Map of the areas for which the abundance of *C. finmarchicus* and *C. helgolandicus* were averaged.

Planque and Fromentin (1996) showed that the relationships with the NAO for the two *Calanus* species was regionally based and differed markedly between the two species with a centre located in the Celtic Sea for *C. helgolandicus* and in the Norwegian Sea for *C. finmarchicus*. Similar regional geographical relationships with the NAO are evident for other species of copepods (Reid and Planque 2000), but differ considerably between species, possibly reflecting their ecological niche and timing in the seasonal succession. The correlations between the NAO and physical variables also show a different regional response. Correlation fields for temperature, precipitation and wind

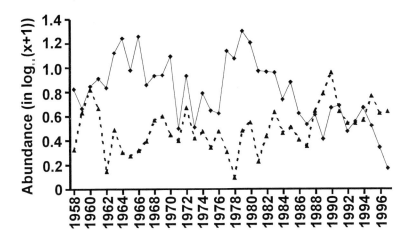

Figure 2-3. Year-to-year changes in the abundance of *C. finmarchicus* (continuous line and diamond) and *C. helgolandicus* (dashed line and triangle) in the North Sea (area B). Day-night and seasonal variability were taken into account in calculating the abundance. A one-way analysis of variance, with replication, was performed for the period 1958-1987 versus 1988-1997 for both *Calanus* species. The difference between the two periods for both species was highly significant (p<0.001). The relative abundance of the two species was CF 80% CH 20% - 1958-87 and CF 38% CH 62% - 1988-97.

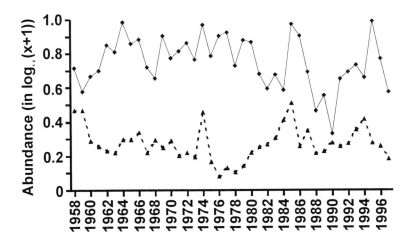

Figure 2-4. Year-to-year changes in the abundance of *C. finmarchicus* (continuous line and diamond) and *C. helgolandicus* (dashed line and triangle) in the northeast North Atlantic (area C). A similar analysis of variance was performed as for Figure 2-3 above, but was not significant (p>0.05). The relative abundance of the two species was CF 85% CH 15% - 1958-87 and CF 79% CH 21% - 1988-97.

all coincide along the west coast of Norway. This was suggested by Ottersen *et al.* (in press) as one of the reasons why NAO ecological linkages are so evident there.

A number of hypotheses have been proposed to explain the switch in the relative abundances of the two *Calanus* species in the North Sea: changes in sea temperature, food availability, inter-species competition, the transport of overwintering populations onto the shelf, the volume of the overwintering habitat in Norwegian Sea deep water, and the flow and temperature of the shelf edge current (Fromentin and Planque 1996; Planque and Taylor 1998; Stephens *et al.* 1998; Heath *et al.* 1999; Reid *et al.*, in press). The true explanation of the changes is likely to be the consequence of a variety of interacting variables. Amongst these sea temperature is likely to be significant, as the two species prefer different temperature regimes. Sea surface temperature is strongly positively correlated with the NAO index in the North Sea (Reid and Planque 2000) and has remained consistently and abnormally high in the recent period of high positive NAO, favouring *C. helgolandicus* and to the disadvantage of *C. finmarchicus*. Using a 2D model, Stephens *et al.* (1998) showed that half the variance of observed changes in *C. finmarchicus* could be accounted for by variations in the flow of water into the northern North Sea (see also Taylor, this volume).

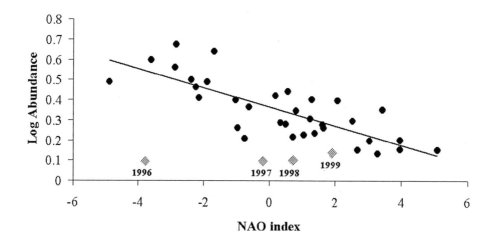

Figure 2-5. Annual log-abundance of *C. finmarchicus* against the North Atlantic Oscillation index. For the period 1958-1995, the NAO explains 58% of the variability in the log-abundance of *C. finmarchicus* (see Planque and Reid 1998).

In the winter of 1995/1996 the NAO index reversed for one winter to one of the most extreme negative states recorded during a period when the index was in a high positive phase immediately prior and subsequent to 1996. In light of the empirical relationship identified by Fromentin and Planque (1996), numbers of *C. finmarchicus* should have increased substantially; they did not do so, but remained at an extremely low level (Planque and Reid 1998) and have remained so since (Figure 2-5). The apparent breakdown in the relationship may indicate that the populations have reduced to such a level that the link with the NAO cannot operate, that the relationship is not consistent over very long periods of time, that there has been a northerly biogeographic shift in the population, or that some factor(s) outside the North Sea are influencing the changes. Alternatively, the atmospheric relationships reflected by the NAO may be changing; there is some evidence to support this as the centre of the Iceland Low pressure cell has moved much further east than normal, to the north of Scandinavia in recent years (Ulbrich and Christoph 1999)

SMALL COPEPODS AND THE GULF STREAM INDEX

The Gulf Stream, where it pulls away from the eastern margin of North America, shows changes from year-to-year in its northerly extent, which may reflect changes in its volume flow or the size and number of eddies. Taylor and Stephens (1980) derived an index of the position of the north wall of the Gulf Stream (GSNW index) and showed that this was highly correlated with the abundance of small copepods (total copepods as measured by the CPR survey) in part of the eastern North Atlantic and North Sea. Subsequent work has confirmed this (see Taylor, this volume), although the correlation between the GSNW index and small copepods in an area to the west of Shetland extending out into the Atlantic broke down in approximately 1987 (Hays *et al.* 1993). The linkage between the Gulf Stream and plankton in the North Sea is believed to be via an atmospheric teleconnection that influences the onset of spring and the initiation of stratification (Taylor, this volume). Part of the relationship may be via the NAO, as the GSNW index of Taylor is correlated with the NAO index two years previously. An alternative index produced by Joyce *et al.* (2000) varies in synchrony with the NAO.

TRANS ATLANTIC CHANGE IN PHYTOPLANKTON COLOUR

Phytoplankton Colour (a visual estimate of chlorophyll measured on CPR silks) was shown by Reid *et al.* (1998a) to have increased substantially in the North Sea and in an area off the shelf to the west of the British Isles after the mid-1980s, whereas the pattern of change was the reverse in the eastern Atlantic to the north of 59°N and south of Iceland (see also Edwards *et al.* 2001). The new higher levels of colour continued in the North Sea until 1999, although there may have been a reversal of the pattern between the two oceanic areas in the eastern Atlantic in 1999 (Figure 2-6). The colour changes in the North Sea form one of the sets of evidence for the regime shift described in the next section. The levels of change were most pronounced in the winter and summer months, with a >90% rise in colour intensity during winter in the North Sea. These results suggest that there has been a substantial increase in levels of chlorophyll, especially in the North Sea. The results must be assessed with care, however, as the colour analysis is based on a simple category system and high levels may reflect

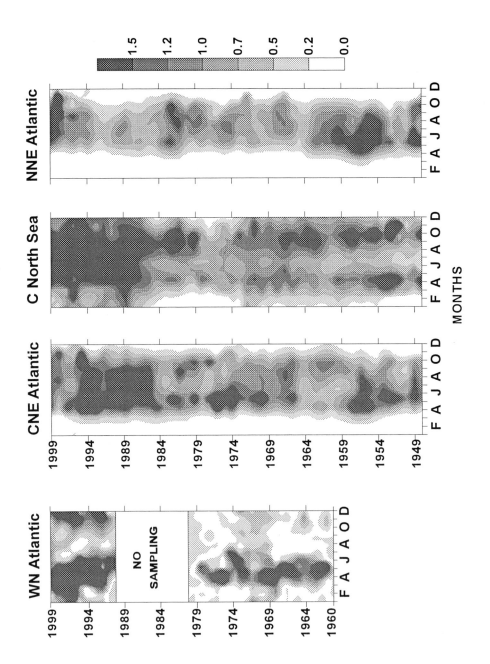

Figure 2-6. Contour plot of mean monthly Phytoplankton Colour during 1948-1999 for the northwest (WN) Atlantic, central (C) North Sea, central northeast (CNE) Atlantic and northern northeast (NNE) Atlantic in part updated from Reid *et al.* (1998a).

clogging of the silk. On face value the colour changes might be considered as evidence for eutrophication. However, they have occurred throughout the North Sea and also to the west of the British Isles (Edwards *et al.* 2001) and must therefore reflect a major hydroclimatic change.

An analysis of changes in Phytoplankton Colour in the 40 standard areas traditionally used for statistical analysis of CPR data (Colebrook 1975) shows that increases occurred after the mid 1980s in all areas of the North Sea, the Celtic Sea, Bay of Biscay, Malin Shelf and Irish Sea and oceanic areas (CPR areas B5 and C5) to the west of the British Isles. A stepwise change is also clearly evident in the western North Atlantic to the east of Newfoundland (CPR areas E8 and E9) over the Grand Banks (Figure 2-6) although the timing of the change is not clear because of a break in sampling between 1980 and 1990. Oceanic areas in between are less well sampled so that patterns of change through time are less clear. The data averaged for the Grand Banks area prior to 1980 show a single spring bloom centred on April and May. Subsequent to 1990 the spring bloom became much earlier and longer and a strong late autumn bloom also occurred. Some indication of higher levels of Phytoplankton Colour on the Scotian Shelf (CPR area E10) is also apparent and is evident between 1991 and 1994 in the plots of Sameoto *et al.* (1996).

NORTH SEA REGIME SHIFT

The term *Regime Shift* has been applied to large decadal scale switches in the abundance and composition of plankton and fish that have been observed in a number of sea areas around the world. Possibly one of the most well defined events of this nature occurred in the North Sea in approximately 1987 (Reid and Edwards, in press). Reid *et al.* (1998b, 2001) outlined this event for the first time, drawing attention to the similarity in the timing circa 1988 of an almost two-fold increase in the Phytoplankton Colour index of the CPR survey, increased sea surface temperature and advection of oceanic water into the North Sea, changes in the abundance of some species of phytoplankton and zooplankton, and a stepwise increase in catches of the horse mackerel (Figure 2-7). It is now clear that a wide range of parameters changed at about the same time including a doubling in the biomass of the benthos off Norderney (Krönke *et al.* 1998) and a pronounced change in the community structure of the benthos off the east coast of the United Kingdom (Warwick and Clarke, pers. comm.). Measurements of nutrients (nitrate, orthophosphate, silicate) and oxygen in deep waters of the Skagerrak down to ~700m show substantial stepwise changes in ~1989 (Dahl and Danielson 1992). As levels of the nutrients increased, concentrations of oxygen decreased. Combined, the above evidence suggests that primary productivity in the North Sea increased after the regime shift and that a substantial part of the phytoplankton was not grazed by the zooplankton but sank to the bottom as detritus to feed the benthos. This hypothesis is further supported by the lag that occurs between the changes seen in the phytoplankton and the benthic, nutrient and oxygen time series. The changes seen in nutrients and oxygen in the Skagerrak are also detected in the

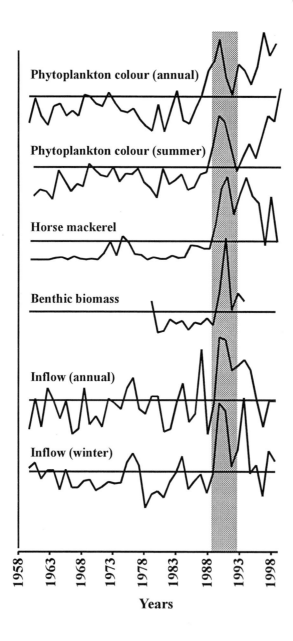

Figure 2-7. Long-term annual means of Phytoplankton Colour (annual means and summer means), horse mackerel catch (tonnes), benthic biomass (and Atlantic inflow into the North Sea (annual means and winter means) highlighting the exceptional peaks seen in the late 1980s. Benthic biomass from 1978-1995 recorded in the Norderney coastal zone (Kröncke *et al.*, 1998). Modelled transport of Atlantic water into the North Sea between Utsira and Orkney Islands (Sverdrup) (Svendsen pers. comm.).

Helgoland Roads dataset in the German Bight (Hickel *et al.* 1996), but approximately two years earlier. Enhanced riverine input of nutrients from the Elbe have been shown to coincide with this change in the German Bight (Radach 1998). It is likely that the hydrometeorological events that took place in 1987/88 have also impacted terrestrial systems leading to an exacerbation of coastal eutrophication effects.

EASTERN BOUNDARY CURRENT

Pingree *et al.* (1999) have shown that the density driven poleward slope current along the European margin "acts as a chute" funnelling water along the slope from at least the south of Ireland to the northern North Sea. To the west of Scotland this shelf edge current is believed to be reinforced by northerly flowing eastern and western branches of a baroclinic flow through the Rockall Trough (Holliday *et al.* 2000). The current appears to continue at the edge of the Norwegian shelf into the Arctic Ocean (Mork and Blindheim 2000). Further south at deeper depths (400-600m), water is carried north in an intermediate current from West Africa and below these depths north of the Strait of Gibralter in the Mediterranean Outflow Water (John *et al.* 1998). Using data taken from a routine cross section across the Rockall Trough at 57°N, Holliday *et al.* (2000) calculated geostrophic transports and showed that the greater portion of the water was directed to the north and along the shelf edge. Holliday and Reid (in press) showed that two periods in early 1989 and early 1998 had flows in the Rockall Trough that were almost double the mean. These two periods coincided with evidence for enhanced oceanic inflow marked by the incursion of large numbers of oceanic lusitanean plankton (Lindley *et al.* 1990; Edwards *et al.* 1999). Large numbers of the doliolid *Doliolum nationalis*, only previously recorded in the North Sea in 1911, were found in the autumn of 1989 with an earlier incursion of oceanic species such as the copepods *Metridia lucens* and *Candacia armata* from 1987. The inflow in 1998 was associated with the presence of the same plankton species (normally found in more southerly latitudes) seen during the late 1980s. These events also coincided with peaks in Phytoplankton Colour especially in the winter and summer months (Figure 2-6). Subsequent work by Reid *et al.* (in press) has shown that the two pulses of oceanic water into the North Sea circa 1988/89 and 1998 also coincided with the presence of anomalously warm water at the edge of the continental shelf. These events are superimposed on the regime shift and provide further evidence to support the suggestion by Reid *et al.* (1998b) that higher flows have occurred in the shelf edge current in the recent period of high positive NAO indices since the regime shift started circa 1988.

CHANGES IN ABUNDANCE OF *CALANUS HYPERBOREUS*

This species is the largest copepod found in the North Atlantic and in the area covered by the CPR survey is largely concentrated in the Labrador and Norwegian Seas where it reflects the presence of Arctic waters. It is typically only found near the surface in the spring and summer (Longhurst 1998) when it undergoes its seasonal ontogenetic migration from intermediate depths that may extend down to 2000m, but may for example, be brought to the surface by upwelling during winter in the entrance to the Skagerrak.

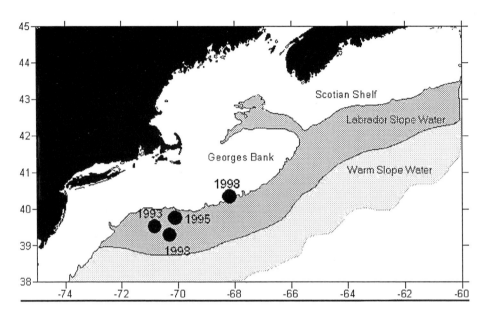

Figure 2-8. Occurrence of *C. hyperboreus* along the shelf edge, south of Georges Bank since 1960 (see Johns *et al.*, in press).

Since the early 1990s there has been a pronounced increase in the abundance of *C. hyperboreus* in the Labrador Sea, with the highest numbers ever recorded in the CPR survey occurring during April and May 1999. In addition, *C. hyperboreus* seems to be spreading further south in the western Atlantic (Johns *et al.*, in press). These findings are consistent with evidence for an increasing southerly penetration of Labrador Sea intermediate water in the 1990s (Pickart *et al.* 1997) that coincides with a high NAO index period. Volume transport in the surface Labrador Current and possibly Labrador slope water appears to be the reverse of what is happening in deeper depths as it is known to be inversely correlated with the NAO (Mayers *et al.* 1989; Colbourne and Foote 2000). In the autumn of 1997 and winter of 1998 along the Scotian Shelf and in the Gulf of Maine, warm slope water was replaced by cold Labrador slope water. Evidence for the presence of this offshore 'slope water' or Labrador Sea intermediate water is confirmed by the most southerly record ever of *C. hyperboreus* in the CPR survey, off the shelf to the southeast of Long Island in 1998 (Figure 2-8). Johns *et al.* (in press) have described these events and shown that the changes in *C. hyperboreus* correlate well with increases in the thickness of Labrador Sea intermediate water. Drinkwater *et al.* (1999) attributed the pronounced southerly extension of Labrador

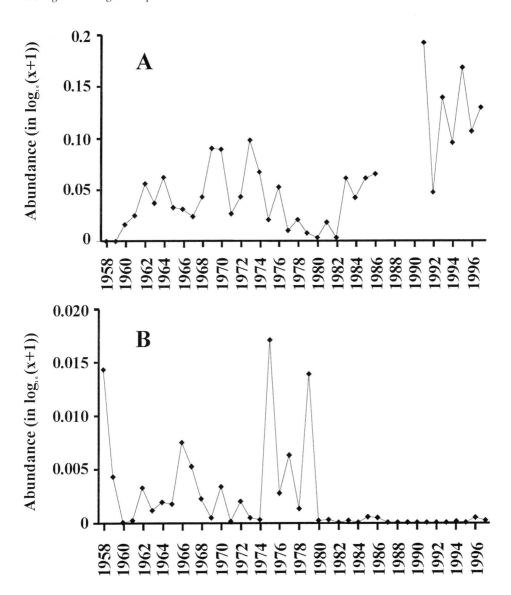

Figure 2-9. a) Year-to-year changes in the abundance of *C. hyperboreus* averaged for an area of the northwestern North Atlantic west of 30°W (80°W, 30°W, 45°N, 63°N). Abundance was estimated taking into account day-night and seasonal variability. **b)** Year-to-year changes in the abundance of *C. hyperboreus* in an area of the northeastern North Atlantic to the east of 30°W (30°W, 10°E, 45°N, 63°N).

slope water in 1998 to a lagged effect from the very low NAO index of the winter of 1995/96. Evidence from the CPR suggests, however, that Labrador water had penetrated well south as early as 1993 and that this southerly extension may instead be a consequence of spreading Labrador Sea intermediate water reflecting the pronounced high NAO situation of the 1990s. The interrelationships between the hydrography of this complex area on the shelf side of the Gulf Stream and the NAO are still far from clear, and the extent to which Labrador slope water is an extension of the surface Labrador Current or an upper layer of Labrador intermediate water in the deep western boundary current has yet to be resolved.

Numbers of *C. hyperboreus* averaged for the western Atlantic to the west of 30°W are shown as a time series in Figure 2-9a. A pronounced increase in the 1990s compared to earlier years is clearly seen. In contrast, in the eastern Atlantic (Figure 2-9b) the species has almost disappeared in CPR samples subsequent to 1979. The data in this plot may be subject to a latitudinal sampling bias as sampling in more northerly latitudes has declined since the early 1980s. However, this period includes two years, 1980 and 1981, when the T route to Ocean Weather Ship Mike was still operating in the Norwegian Sea. In the western and eastern Atlantic therefore, opposite patterns of change appear to have occurred, with higher numbers associated with an increasing volume of Labrador Sea intermediate water in the west and lower numbers that coincide with a pronounced reduction (Østerhus and Gammelsrød 1999) in the thickness of Norwegian Sea deep water in the east.

CONCLUSIONS

This paper started with the hypothesis that plankton can integrate a wide range of different hydrometeorological signals and may act as indicators of the recent climatic history of its enclosing water mass. The primary conclusion from the examined case studies is that both plankton species and communities show systematic, and not chaotic, patterns of change through time over extensive regional areas of the North Atlantic. Some of the changes are highly correlated with proxies of large-scale variability in the ocean climate of the North Atlantic such as the NAO and GSNW indices. The changes seen in the plankton do thus appear to represent responses to widespread and systematic hydrometeorological events. Two of the proven statistical relationships have broken down, however, which suggest that the relationships were an artifact, or that the climate indices and/or the plankton have changed in their characteristics. Establishing the type and causes of these shifts is likely to throw new light on large-scale dynamics of pelagic systems.

While primarily an indicator of winter change, the NAO plays a key role because of its modulation of a wide range of parameters that force hydrographic variability in the North Atlantic, in particular the wind driven circulation, deep convection and sea surface temperature. Reid *et al.* (1998b) proposed two scenarios for changes in the circulation in the North Atlantic in high and low NAO index states. When the NAO index is low, the North Atlantic current is northwesterly directed and strongly penetrates Nordic Seas, deep convection is active in the Greenland Sea and convection to intermediate depths is minimal in the Labrador Sea. In the high NAO state convection is halted in the Greenland Sea leading to a reduction of Norwegian Sea deep water, and is enhanced in the Labrador Sea. At the same time, penetration of the North Atlantic current into Nordic Seas is reduced and warmer waters penetrate all the way to the Arctic in a strong eastern margin current. These alternative scenarios are in part supported by the WOCE observations of Koltermann *et al.* (1999) and are further reinforced by the case studies reported here. Part of the proposed pattern has broken down, however, since approximately 1996 when convection in the Labrador Sea also seems to have reduced (A. Clarke pers. comm.), despite the NAO (with the exception of the winter of 1995/96) being in a moderately high index state.

The changes in the Labrador Sea show temperatures warming during low NAO periods and cooling during high NAO periods (the opposite response to the northeast Atlantic). This seasaw effect between the two areas is reflected in the patterns of change in *C. hyperboreus* and again appears to reflect alternating intensification/weakening of cold Labrador Sea intermediate water and Norwegian Sea deep water. The story is not that simple, however, as the changes seen in both sea areas in the last two decades are unique in living memory and are likely to be a response to larger global changes possibly linked to global warming. In an analysis of early summer zooplankton surveys around Iceland between 1960 and 1996, Beare *et al.* (in press) have noted a pronounced increase in zooplankton of arctic affinity. The species that have increased include *C. hyperboreus*, especially to the northeast of Iceland, and the species has been found further south than usual in the East Icelandic current in the last decade (Beare pers. comm.). A stronger East Icelandic current in positive NAO periods may act as a barrier to the penetration of the North Atlantic current into the Norwegian Sea.

Influx of cold arctic waters into shelf seas may have a pronounced effect on their ecosystems, changing the community structure of the plankton and reducing productivity. This cold dense water, given the right conditions, may flood in at depth gradually replacing the bottom water in the deeper basins. An event of this nature occurred in the Gulf of Maine ecosystem after 1998 (C. Greene pers. comm., Drinkwater *et al.* 1999) and for the North Sea in the period 1978-82 (Edwards *et al.* 2001; Reid and Edwards, in press). In the case of the North Sea, the plankton community became more dominated by boreal species (including *C. hyperboreus*) and the total biomass of the zooplankton appears to have decreased with evidence for a possible 'knock on' effect on fish stocks. The opposite case of an inflow of warm oceanic water may have a similar profound impact on shelf ecosystems, as witnessed by the regime shift in the North Sea.

Phytoplankton Colour, in contrast, has shown similar patterns of change on both sides of the North Atlantic at approximately the same time, suggesting that levels of chlorophyll, and possibly production, have increased substantially. The precise timing of the change in the western Atlantic cannot be determined because sampling with the

CPR was stopped in this area between the mid 1980s and 1990 due to a break in funding support. There is also some suggestion that colour may have increased in the oceanic areas in between. Thus, substantial increases in the Phytoplankton Colour index may have occurred in a trans Atlantic band that stretches in a west north west direction between latitudes 43° to 50°N in the west and 44° to 59-62°N in the east. The mechanism causing these changes is likely to be the same on both sides of the Atlantic through an atmospheric teleconnection, which possibly includes an increase in radiative forcing. Over the same period (1981-1991), the colour results show a remarkable similarity to changes observed by satellites for terrestrial vegetation at similar latitudes of the Northern Hemisphere (Myneni et al. 1997). They reported a pronounced increase in the biological activity of terrestrial plants over a large area of the Northern Hemisphere above 50° N which they attributed, especially in the late winter/spring and summer growing season, to a rising temperature.

Global mean temperatures are projected to rise by at least 1°C over the next 40 years on the basis of the expected rise in CO_2 and other greenhouse gases. This is likely to have a profound effect on northern latitudes leading to melting of the permafrost and Arctic ice. The resulting meltwater is likely to maintain the capping (cessation) of deepwater formation in the Greenland Sea with downstream effects on the "global conveyor belt." Using the same CO_2 scenario, the NAO is also expected to remain positive so that the circulation changes suggested for a high NAO index state are likely to remain in place for the foreseeable future, reinforcing the capping of deep water formation. As the NAO does not seem to be stable, the possibility of a sudden switch to colder weather in Europe as a consequence of sustained closure of deepwater formation in the Greenland Sea cannot be discounted (Rahmstorf 1997). This latter possibility would also require the present warm eastern boundary current to be blocked.

ACKNOWLEDGEMENTS

We wish to thank the contributors to the time series used in this report and in particular the CPR team and the captains and crew of the ships that voluntarily tow CPRs. We are indebted to Darren Stevens and Marion Smith for assistance in accessing the CPR database and preparing the manuscript respectively. The CPR survey is funded by a consortium that has recently been comprised of agencies from Canada, Denmark, Faeroes, France, Iceland, Ireland, Netherlands, UK, USA, the European Union, UNIDO and IOC.

REFERENCES

Beare D.J., A. Gislason, O.S. Astthorsson, and E. McKensie. In press. Assessing long-term changes in early summer zooplankton communities around Iceland. ICES Journal of Marine Science.

Colbourne, E.B. and K.D. Foote. 2000. Variability of the stratification and Circulation on the Flemish Cap during the decades of the 1950s-1990s. Journal of Northwest Atlantic Fishery Science, 26: 103-122.

Colebrook, J.M. 1975. The Continuous Plankton Recorder survey: automatic data processing methods. Bulletins of Marine Ecology, 8: 123-142.

Colebrook, J.M. 1978. Continuous Plankton Records: zooplankton and environment, north-east Atlantic and North Sea, 1948-1975. Oceanologica Acta, 1: 9-23.

Colebrook, J.M. 1986. Environmental influences on long-term variability in marine plankton. Hydrobiologia, 142: 309-325.

Dahl, E. and D.S. Daniellsen. 1992. Long term observations of oxygen in the Skagerrak. ICES Marine Science Symposium, 195: 455-461.

Dickson, R.R. and W.R. Turrell. 2000. The NAO: the dominant atmospheric process affecting oceanic variability in home, middle and distant waters of European Atlantic salmon. In: D. Mills (ed.). The ocean life of Atlantic salmon. Environmental and biological factors influencing survival. Fishing News Books, Bodmin, 92-115.

Drinkwater, K.F., D.B. Mountain, and A. Herman. 1999. Variability in the Slope Water Properties off Eastern North America and their Effects on the Adjacent Shelves. ICES Annual Science Conference Theme Session 0:08.

Edwards, M., A.W.G. John, H.G. Hunt, and J.A. Lindley. 1999. Exceptional influx of oceanic species into the North Sea late 1997. Journal of the Marine Biological Association UK, 79: 737-739.

Edwards, M., P.C. Reid, and B. Planque. 2001. Long-term and regional variability of phytoplankton biomass in the Northeast Atlantic (1960-1995). ICES Journal of Marine Science, 58: 39-49.

Fromentin, J-M. and B. Planque. 1996. *Calanus* and environment in the eastern North Atlantic. II. Influence of the North Atlantic Oscillation on *C. finmarchicus* and *C. helgolandicus*. Marine Ecology Progress Series, 134: 111-118.

Hays, G.C. and A.J. Warner. 1993. Consistency of towing speed and sampling depth for the Continuous Plankton Recorder. Journal of the Marine Biological Association of the UK, 73: 967-970.

Hays, G.C., M.R. Carr, and A.H. Taylor. 1993. The relationship between Gulf Stream position and copepod abundance from the Continuous Plankton Recorder Survey: separating biological signal from sampling noise. Journal of Plankton Research, 15: 1359-1373.

Heath, M.R., J.O. Backhaus, K. Richardson, E. McKenzie, D. Slagstad, D. Beare, J. Dunn, J.G. Fraser, A. Gallego, D. Hainbucher, S. Hay, S. Jónasdóttir, H. Madden, J. Mardaljevic, and A. Schacht, 1999. Climate fluctuations and the spring invasion of the North Sea by *Calanus finmarchicus*. Fisheries Oceanography, 8: 163–176.

Hickel, W., M. Eickhoff, H. Splindler, J. Berg, T. Raabe, and R. Muller. 1996. Auswertungen von Langzeit-Untersuchungen von Nahrstoffen und phytoplankton in der Deutschen Bucht. – 213 p.; Berlin (Umweltbundesamt).

Holliday, N.P., R.T. Pollard, J.F. Read, and H. Leach. 2000. Water mass properties and fluxes in the Rockall Trough 1975-1998. Deep Sea Res, Part I, 47:1303-1332.

Holliday, N.P. and P.C. Reid. In press. Is there a connection between high transport of water through the Rockall Trough and ecological changes in the North Sea? ICES Journal of Marine Science.

Hurrell, J.W., Y. Kushnir, and M. Visbeck. 2001. The North Atlantic Oscillation. Science, 291: 603-605.

John, H-C., E. Mittelstaedt, and K. Schulz. 1998. The boundary circulation along the European continental slope as transport vehicle for two calanoid copepods in the Bay of Biscay. Oceanologica Acta, 21(2): 307-318.

Johns, D.G., M. Edwards, and S.D. Batten. In press. Arctic boreal plankton species in the Northwest Atlantic. Canadian Journal of Fisheries and Aquatic Sciences.

Joyce, T.M., C. Deser, and M. Spall. 2000. On the relation between decadal variability of subtropical mode water and the North Atlantic Oscillation. Journal of Climate, 13: 2550-2569.

Koltermann, K.P., A.V. Sokov, V.P. Tereschenkov, S.A. Dobroliubov, Lorbacher, K and Sy 1999. Decadal changes in the thermohaline circulation of the North Atlantic. Deep-Sea Research, Part 2, 46: 109-138.

Kröncke, I., J.W. Dippner, H. Heyen, and B. Zeiss. 1998. Long-term changes in macrofaunal communities off Norderney (East Frisia, Germany) in relation to climate variability. Marine Ecology Progress Series, 167: 25-36.

Kroon, D., W.E.N. Austin, M.R. Chapman, and G.M. Gassen. 1997. Deglacial surface circulation changes in the north eastern Atlantic: temperature and salinity records off North West Scotland on a century scale. Palaeoceanography, 12: 755-763.

Lindley, J.A., J. Roskell, A.J. Warner, N.C. Halliday, H.G. Hunt, A.W.G. John, and T.D. Jonas. 1990. Doliolids in the German Bight in 1989 – Evidence for exceptional inflow into the North Sea. Journal of the Marine Biological Association of the UK, 70: 679-682.

Longhurst, A., 1998. Ecological Geography of the Sea. Academic Press, London. 398 p.

Margalef, R. 1997. Turbulence and marine life. Scientia Marina, 61(supplement 1). 109-123.

Mayers, R.A., J. Helbig, and D. Holland. 1989. Seasonal and interannual variability of the Labrador Current and West Greenland Current. ICES C.M. 1989/C:16, 18p.

Mork, A.M. and J. Blindheim. 2000. Variations in the Atlantic inflow to the Nordic Seas, 1955-1996. Deep Sea Research, Part I, 47:1035-1057.

Mudie, P.J. and R. Harland. 1996. Chapter 21. Aquatic Quaternary. In: J. Jansonius and D.C. McGregory (eds.). Palynology: principles and applications; American Association of Stratigraphic Palynologists Foundation, 2: 843-877.

Myneni, R.B., C.D. Keeling, C.J. Tucker, G. Asrar, and R.R. Nemani. 1997. Increased plant growth in the northern high latitudes from 1981 to 1991. Nature, 386: 698-702.

Østerhus, S. and T. Gammelsrød. 1999. The Abyss of the Nordic Seas is warming. Journal of Climate, 12: 3297-3304.

Ottersen, G., B. Planque, A. Belgrano, E. Post, P.C. Reid, and N.C. Stenseth. In press. Ecological effects of the North Atlantic Oscillation.

Pickart, R.S., M.A. Spall, and J.R.N. Lazier. 1997. Mid-depth ventilation in the western boundary current system of the sub-polar gyre. Deep Sea Research, Part I, 44:1025-1054.

Pingree, R.D., B. Sinha, B. and C.R. Griffiths. 1999. Seasonality of the European slope current (Goban Spur) and ocean margin exchange. Continental Shelf Research, 19: 929-975.

Planque, B. and S.D. Batten. In press. *Calanus finmarchicus* in the North Atlantic: the year of *Calanus* in the context of interdecadal change. ICES Journal of Marine Science.

Planque, B. and J-M. Fromentin. 1996. *Calanus* and environment in the eastern North Atlantic. I. Spatial and temporal patterns of *C. finmarchicus* and *C. helgolandicus*. Marine Ecology Progress Series, 134: 101-109.

Planque, B. and P.C. Reid. 1998. Predicting *Calanus finmarchicus* abundance from climatic signal. Journal of the Marine Biological Association of the UK, 78: 1015-1018.

Planque, B. and A.H. Taylor. 1998. Long-term changes in zooplankton and the climate of the North Atlantic. ICES Journal of Marine Science, 55: 644-654.

Radach, G. 1998. Quantification of long-term changes in the German Bight using an ecological development index. ICES Journal of Marine Science, 55 (4): 587-599.

Rahmstorf, S. 1997. Risk of sea change in the Atlantic. Nature, 388: 825-826.

Reid, P.C. and M. Edwards. In press. Long-term changes in the pelagos, benthos and fishery of the North Sea.

Reid, P.C. and H.G. Hunt. 1998. Are observed changes in the plankton of the North Atlantic and North Sea linked to climate change? In: A.C. Pierrot-Bults and S. van der Spoel (eds.). Pelagic Biogeography, IcoPB II, Proceedings of the 2nd International Conference. IOC/UNESCO, Paris. Workshop Report No. 142: 310-315.

Reid, P.C. and B. Planque. 2000. Long-term planktonic variations and the climate of the North Atlantic. In: D. Mills (ed.). The ocean life of Atlantic salmon. Environmental and biological factors influencing survival. Fishing News Books, Bodmin, 153-169.

Reid, P.C., M.E. Edwards, H.G. Hunt, and A.J. Warner. 1998a. Phytoplankton change in the North Atlantic. Nature 391: 546.

Reid, P.C., B. Planque, and M. Edwards. 1998b. Is variability in the long-term results of the Continuous Plankton Recorder survey a response to climate change? Fisheries Oceanography, 7:282-288.

Reid, P.C., N.P. Holliday, and T.J. Smyth. In press. Pulses in the eastern margin current and warmer water off the north west European shelf linked to North Sea ecosystem changes. Marine Ecology Progress Series.

Reid, P.C., M. de F. Borges, and E. Svendsen, E. 2001. A regime shift in the North Sea circa 1988 linked to changes in the North Sea horse mackerel fishery. Fisheries Research, 50:163-171.

Robinson, G.A. and A.R. Hiby. 1978. The Continuous Plankton Recorder survey. In: A. Sournia (ed.). Phytoplankton manual. 59-63, UNESCO, Paris.

Sameoto, D.D., M.K. Kennedy, and B. Petrie. 1996. SW Grand Banks, Scotian Shelf and Gulf of Maine: zooplankton and phytoplankton measured by the Continuous Plankton Recorder 1961-1993. Canadian Technical Reports on Fish and Aquatic Science, 2116, xi: 219 p.

Stephens, J.A., M.B. Jordan, A.H. Taylor, and R. Proctor. 1998. The effects of fluctuations in North sea flows on zooplankton abundance. Journal of Plankton Research, 20: 943-956.

Sverdrup, H.U. 1953. On conditions for the vernal blooming of phytoplankton. Journal du Conseil Permanent International pour l'Exploitation de la Mer, 18: 287-295.

Taylor, A.H. (this volume). North Atlantic climatic signals and the plankton of the European continental shelf.

Taylor, A.H. and J.A. Stephens. 1980. Latitudinal displacements of the Gulf Stream (1966-1977) and their relation to changes in temperature and zooplankton abundance in the NE Atlantic. Oceanologica Acta, 3: 145-149.

Ulbrich, U. and M. Christoph. 1999. A Shift of the NAO and increasing storm track activity over Europe due to anthropogenic greenhouse gas forcing. Climate Dynamics 15: 551-559.

Verity, P. G. and V. Smetacek. 1996. Organism life cycles, predation, and the structure of marine pelagic ecosystem. Marine Ecology Progress Series, 130: 277-293.

Warner, A.J. and G.C. Hays. 1994. Sampling by the Continuous Plankton Recorder survey. Progress in Oceanography, 34: 237-256.

II
Northwest Atlantic Large Marine Ecosystems

Large Marine Ecosystems of the North Atlantic
K. Sherman and H.R. Skjoldal (Editors)
© 2002 Elsevier Science B.V. All rights reserved.

3

Changes to the Large Marine Ecosystem of the Newfoundland-Labrador Shelf

Jake Rice

INTRODUCTION

The old years

The Grand Banks of Newfoundland and Labrador have been fished since the 1400s, with fleets coming annually from many of the fishing nations of Europe, and by the 1600s, from North America. The Banks and coastal areas were rich and productive, and formed the basis for settlement of coastal areas of Newfoundland and Labrador, despite policies that were intended to deter settlement (Gough 1993; Parsons 1993; Lear and Parsons 1993). For most of this history, fishing meant fishing for Atlantic cod (*Gadus morhua*). Export records of the British Colonial Office have allowed a historic record of catches of Atlantic cod to be reconstructed back to 1677 (Figure 3-1; Forsey and Lear 1987). These records, and other investigations of journals and correspondence from those centuries, document clearly that instability and variation characterized the entire history of exploitation of fisheries off the coast of Newfoundland and Labrador.

Lear (1998) describes how, although periods of war and peace between France and England unquestionably had impacts on the population and the fisheries of Newfoundland, large stretches of the coast showed synchronous booms or collapses of the cod fisheries, extending over several years. There is convincing evidence that both overfishing and environmental stresses contributed to historic collapses. For example, Head (1976) quotes from the Colonial Office Records of 1683 in the Public Records Office, London "… there is not fishing ground or can constantly be Fish enough for so many Boates as they have kept, … whereas were there half so many Boates fish there, they could not make so great a Destruction One Year as to prejudice the next years fishery". However, Munn (1938) reports, "For some reasons, never explained, the salt water around our country became hostile to fish life. The old fishermen still tell us the water was perfectly clear, and you could see objects on the bottom in twenty fathoms of water. The nets moored in the water would become filthy with slime. The codfish could not live in it, and the spawning or reproduction must have been brought to a stand still. … This became acute in 1862, and got worse and worse during the next five years."

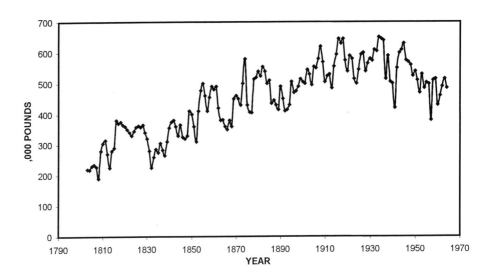

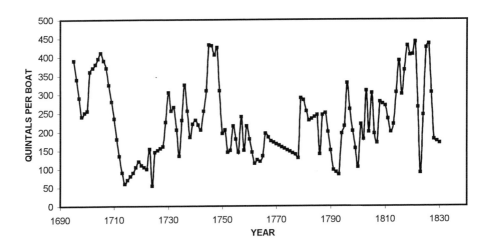

Figure 3-1. Time series of cod catches in Newfoundland. Top panel - Total coastal catches from 1800 to period of expansion of foreign fleets offshore, in traditional units (left axis) and thousands of metric tonnes (right axis). Lower panel - Catch (quintals are traditional units for recording catches) per unit of effort for 150 years reported from communities around Conception Bay, Newfoundland. Data from Lear and Forsey (1986) and Lear (1998).

(The latter quote is a clear description of a period of intense *Okiopleura* bloom, to the present characteristic of influxes of exceptionally cold Arctic conditions to coastal waters.)

With such a long history of variation in the cod fishery, it would be a laudable goal to evaluate scientifically the patterns, causes, and consequences of four centuries of changing states of the large marine ecosystem of the Newfoundland and Labrador Banks. Regrettably, that goal is unattainable. Records of cod exports alone are inadequate to address the dynamics of a large marine ecosystem, although they are sufficient to underscore the inevitability of change. Change is similarly apparent in the other fisheries that developed in the 19th century; for salmon (*Salmo salar*), seals (primarily *Phoca groenlandica*), and lobster (*Homarus americanus)* (Figure 3-2). Moreover, all large cetaceans and several seabird populations, most notably the Great auk (*Pinguinus impennis*) populations had been severely depleted (Mitchell and Reeves 1983; Bengtson 1984). Insights into the role of these top predators in the ecosystem remain the domain of hypotheses and speculative modelling.

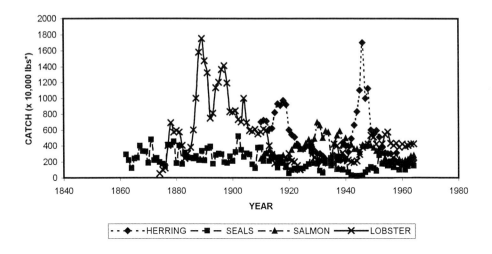

Figure 3-2. Time series of Newfoundland catches going back to beginning of century for herring, seals, salmon, and lobster, in units of export reports. All data from Templeman (1966).

In all cases some, but not all, of the fluctuations have been attributed to over-expansion and over-harvesting by the fisheries. For example, Templeman (1966) conjectured that haddock (*Melanogrammus aeglefinus*) on the southern Grand Banks had increased explosively in a warm period during the 1920s. The state of haddock stocks on the southern Grand Banks prior to the 1920s is unknown, but Templeman's conjecture coincides with other indications that the 1920s were characterized by warm oceanographic conditions that had widespread effects on fish communities. In that decade a large cod stock appeared suddenly off Greenland (Buch *et al*. 1994; Pedersen & Rice, this volume), and the inshore cod fishery off Newfoundland and Labrador relocated much further up the coast, reaching northern Labrador but abandoning many bays of southeastern Newfoundland (Lear 1998).

Unconfounding the effects of fishing and environmental forcing requires more consistent time series data on species' abundances, and on more parts of the ecosystem than are present in catch series. Researchers at the old Newfoundland Fisheries Institute undertook biological investigations on components of the marine ecosystem from the 1930s onward. However, much of this work focused on exploratory surveys to find out what fishing opportunities existed, in order to augment the declining traditional fisheries (Templeman 1966). Spatial coverage was opportunistic and time series were not developed. Systematic data sets on most components of the ecosystem start no earlier than the 1950s for major fish and ocean temperature data, and much later for other parts of the ecosystem.

Notwithstanding the long history of change, the two most marked fluctuations in biological resources of the Newfoundland Shelf have occurred in the past half century. The first was almost certainly driven by the rapid expansion of distant water fleets during the late 1950s, and the nearly two decades of intensive fishing effort those fleets inflicted on the major fish stocks of the Shelf (Murawski *et al*. 1997). After rarely exceeding 300,000t per year, cod catches increased between 1958 and 1968 to over 800,000 t, and then steadily declined to under 150,000 t by 1977. Comparable explosive increases and then collapses were observed in catches of other targeted species such as haddock, American plaice (*Hippoglossoides platessoides*), and redfish (*Sebastes* sp.), and even species taken mainly as bycatch, such as witch flounder (*Glyptocephalus cynoglossus*). All these fisheries had barely begun in 1950, but were in states of collapse 25 years later.

For the flatfish and redfish, it is probable that the stocks had always been present, but inadequate harvesting and processing methods prevented directed fishing. Haddock were possibly uncharacteristically abundant during the early years of the foreign fishery, having produced several exceptional year-classes between 1949 and 1965 (Templeman 1966). Once the fisheries were underway, however, overfishing and likely high rates of discarding undersized fish allowed the fisheries to play a major, if not exclusive role in the reduction or collapse of all the major fish stocks targeted by the distant-water fisheries (Lear and Parsons 1993).

The second great fluctuation was the collapse of many groundfish stocks on the Newfoundland Shelf at the end of the 1980s and beginning of the 1990s. During this collapse there were systematic multispecies research surveys and consistent oceanography monitoring, covering the rebuilding period following extension of jurisdiction, the collapse, and the post-collapse period, when almost all groundfish fisheries were either under moratoria or reduced to very low effort. This overview will describe major changes in selected ecosystem attributes that have been monitored since the 1950s, and report present understanding of contributions of environmental forcing and human activities to account for these changes. Note that "present understanding" is a dynamic factor – different explanations for changes in the dominant fish stocks have been promoted by different researchers at different times, and new relationships continue to emerge as more data sets are brought together. This Symposium on large marine ecosystems in the north Atlantic is welcome, presenting both a chance to critically evaluate our current understanding of the north Atlantic ecosystems and an opportunity to set the course for future integration of the diverse lines of research on-going at present.

Limitations on Application of Analytical Approaches

The ability to explain the dynamics of the large marine ecosystem of the Newfoundland Shelf is severely limited by the lack of time series of data on living components of the system except for a few species of fishes and seals. Even for the fishes, the series of consistent surveys extend back only to 1977 for the full area, and 1951 for parts of the southern Grand Banks and St. Pierre Bank. Moreover, vessels changed several times, and the survey gear was altered in the early 1990s, to increase the capture of small fish (< 12 –15 cm). Comparative fishing was conducted to calibrate catchability of the important commercial species (Stansbury 1997; Warren 1997a; Morgan *et al.* 1998; Walsh *et al.* 1998a), but the surveys have not been calibrated for the majority of the fish species. Furthermore, many patterns are inferred from commercial catches, where variability due to factors associated with harvesting technologies and markets may be important, and may disguise additional effects of environment or previous spawning biomass on population sizes.

The paper intentionally does not focus on summary indices of ecosystem status, such as indices of diversity, richness, slope of size spectra, etc. for two reasons. The first is the lack of calibration of time series of survey catches for most fish species. Casey and Myers (1998) demonstrate that with sufficient attention to detail, it is possible to make progress on these problems, but they also call justified attention to the dangers of not taking seriously the need to account for changes in catchability when working with survey data sets. Hence the documentation of trends in the LME as a whole is far less quantitative than is desirable. The second is the questionable information content in many of these indices (review in Rice 2000). Although indices emphasising richness and size composition of survey catches may contain useful information, the presence and length frequencies of non-commercial species were not recorded with equal vigilance over the survey series for this area.

Nonetheless, the changes in states are of such large magnitude, and the differences between the two periods of recent collapses so marked, that many important aspects of the story emerge directly. If reliable aggregate and summary indices finally become available, they may add even more insights to the story. However, even without the indices, we see some clear patterns and can infer some important messages about processes and perturbations.

With regard to plankton dynamics, there have been several local studies, generally at sites adjacent to St. John's, on the Avalon Peninsula or on the southern Grand Banks (Anderson and Gardner 1986; Prasad and Haedrich 1993; Prasad *et al.* 1992; Therrieault *et al.* 1998). However, studies have shown convincingly that oceanographic signals are decoupled on scales of 200-300 km (deYoung and Tang 1989; Mertz *et al.* 1994), so it is inappropriate to generalize patterns observed at one site to either the coastal area as a whole, or to the main Shelf portion of the LME. There was a continuous plankton recorder (CPR) transect through this area in the 1950s to 1970s, but this transect was discontinued in the early 1970s. Data from the earlier years were presented in Robinson *et al.* (1973), and analyzed by Mertz and Myers (1994). They provide some picture of the magnitude of variation in primary producers and some small invertebrates, but the series ends prior to the more consistent and comprehensive surveys of upper trophic levels began.   Again, although it would be highly desirable to incorporate dynamics of the primary producers and small zooplankton into a grand synthesis, the data simply do not allow that step to be taken.

THE LARGE MARINE ECOSYSTEM OF THE NEWFOUNDLAND & LABRADOR SHELVES

Bathymetry

The seabed of the Newfoundland Shelf is structurally complex. A trough runs along most of the coast of Newfoundland and Labrador, often reaching depths of over 250 m. The coastline is pierced by numerous fiords, some with maximum depths greater than 300 m. Beyond the coastal trough is a series of large plateaus, collectively known as the Grand Banks of Newfoundland, which rise to less than 150-200 m, and as little as 55 m on the Southeast Shoal of the Grand Bank and near surface at the Virgin Rocks. The Grand Banks are actually a complex of multiple banks; Grand Bank, Funk Island Bank, Belle Isle Bank, and Hamilton Bank, from south to north, separated by deeper channels, or saddles, running generally northeast by southwest (Figure 3-3). Beyond the plateaus – which may extend as much as 350 km from the coast on the Nose and Tail of the Grand Bank, but more typically extend about 150 km offshore, is the continental slope. Along the slope the continental margin drops precipitously to more than 1500 m.

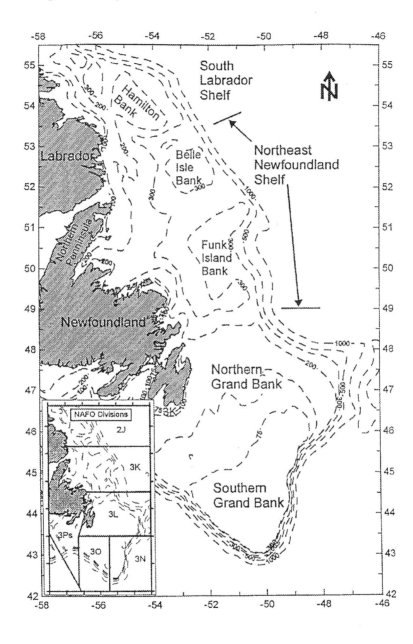

Figure 3-3. Bathymetry of Newfoundland-Labrador Shelf LME. Inset panel shows NAFO Divisions which are used to name fish stocks. Station 27 is 57°25'N, 52°37'W.

Physical oceanography – Features

The dominant physical oceanographic feature in this system is the Labrador Current complex (about 5-6 SV). It brings cold, fresh boreal water down from Davis Strait, constituting the southernmost penetration of polar waters (T <0°C, S < 34 psu) in the northern hemisphere (Helbig *et al.* 1992). The Labrador Current bifurcates around the Grand Banks with inshore (western) and much larger offshore (eastern) arms, in the coastal trough and as a jet along the outer slope respectively.

The inshore branch can bring extremely cold water very close inshore in the fiords and Avalon Channel, where upwelling winds can bring this water in contact with the shoreline (Mertz *et al.* 1994). Labrador Current water also intrudes onto the Grand Banks, contributing to the Cold Intermediate Layer (CIL). The CIL, defined as waters below 0 °C, but with an annual average often below −1.4 °C, is a major feature on the Banks and Avalon Channel (Petrie *et al.* 1988). It varies in both depth and areal extent across the Shelf, in some years having extensive contact with the sea floor on the plateaus and abutting the coastline.

The salinity and strength of the Labrador Current varies inter-annually, and is strongly influenced by freshwater runoff and ice melt to the north (Petrie *et al.* 1992; Davidson and deYoung 1995). A strong Labrador Current is also associated with more extensive ice coverage in winter and spring across this system. The extent of ice coverage has varied interannually by more than 5° of latitude, and more than 250 km seaward from the coast. The melting of this winter ice contributes to the shallow (~25-50 m) layer of fresh surface water that warms through the summer, and strengthens the stratification relative to the CIL below it. Typically warmer (~2-4°C) and more saline (34-36 psu) slope waters underlay the CIL across large areas of the Shelf.

The extent to which the bottom portion of the water column and the seafloor are covered with slope versus CIL waters has many ramifications for biological components of the ecosystem. In summer, surface warming to 8-12 degrees and sometimes more, particularly in the fiords and bays, is typically down to 50 m. Its timing and intensity are an important signal in the ecosystem. On the southern Grand Bank and St. Pierre Bank the entire water column may be somewhat warmer. Occasional intrusions, often as warm-core rings, from the Gulf Stream, which flows to the south of the LME, may bring waters of over 10°C onto the southern portion of the Shelf (Drinkwater *et al.* 1998).

Time series of water column temperature and salinities are best documented for Station 27, just outside St. John's harbour, in the inner arm of the Labrador current. This station has been sampled several times each season back to the 1940s, and measurements have been shown to be typical of the Labrador Current in this region (Petrie *et al.* 1992; Colbourne *et al.* 1994). Additional stations and oceanographic transects along the 47° latitude, northeastward from Cape Bonavista, and northeastward from Seal Island off southern Labrador, have been occupied nearly annually back to the 1940s as well. Additionally bottom temperatures are mapped during multispecies groundfish surveys. These surveys give more thorough spatial coverage of the water column in the area, with thorough coverage during fall in NAFO Divisions 2J3KL, spring in 3LNO, and winter in 3Ps, but all these time series only begin in the late 1970s.

Physical oceanography – Time trends

Compared to long-term averages, sea ice coverage was more extensive, arrived earlier, and stayed longer in the early 1970s, the mid-1980s, and the early 1990s (Drinkwater 1994 – Figure 3-4). Water temperatures generally were slightly warmer than average in the early 1950s, then near normal until about 1963 (Figure 3-5, top panel). Temperatures at all depths then increased sharply and remained high for five years, before dropping precipitously between 1968 and 1972 to what were, for the time, record cold conditions, particularly in the surface layer. Over the following 10 years, average temperatures then rose steadily, so by the early 1980s temperatures at all depths were again warmer than average. Temperatures again dropped sharply between 1982 and 1984, and remained at or near record lows until the mid-1990s. However, between 1994 and 1996 temperatures returned to normal or even warmer than average conditions, and have remained there for the second half of the 1990s.

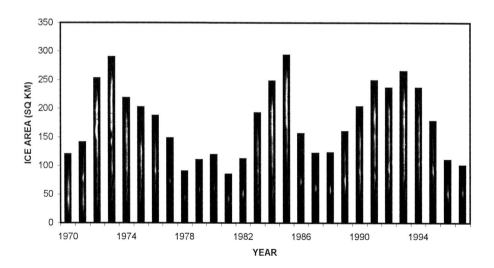

Figure 3-4. Extent of annual 10% or greater average ice cover on Newfoundland Shelf south of 55° N. Data from Drinkwater *et al.* (1998).

The annual volume (both areal coverage and depth) of the CIL shows a similar pattern, with a relatively small areal coverage from the mid-1950s to the mid-1960s, and occupying a large area in the late 1960s and early 1970s, and again in the late 1980s and early 1990s (Figure 3-6; Colbourne 1999). It has been possible to calculate the full volume of the CIL back to 1980. Again the widespread occurrence of sub-0° C water in the late 1980s and early 1990s is apparent. The CIL was also anomalously cold during its periods of greatest volume.

Strong negative salinity anomalies occurred in the late 1960s-early 1970s, mid-1980s, and early 1990s (Figure 3-5, lower panel). These were particularly marked in the surface waters but were of generally comparable magnitude in the remainder of the water column. Each coincided with a period of low temperatures, resulting in cold, fresh oceanographic conditions. There were periods of less strong positive anomalies from the mid-1940s to the mid-1950s, through most of the 1960s, and in the late 1980s.

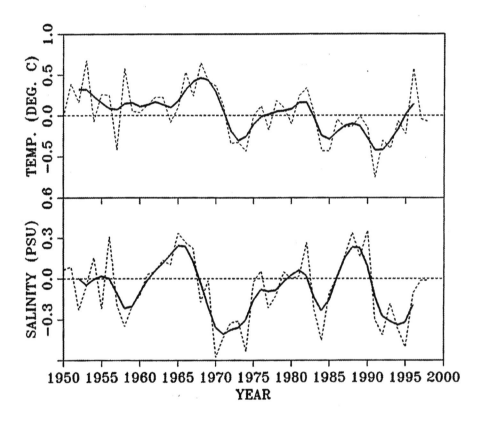

Figure 3-5. Time series of bottom temperature (top panel) and salinity (lower panel) anomalies from 1950 to present at Station 27, off St. John's, Newfoundland. Data from Colbourne (1999).

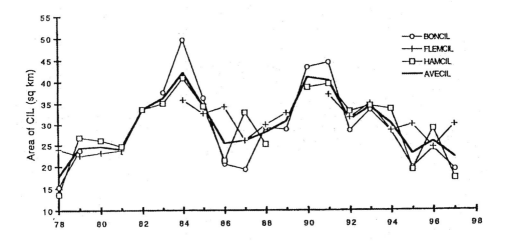

Figure 3-6. Annual extent of the Cold Intermediate Layer for three transects along the southern (plus signs), middle (circles), and northern (squares) portions of the LME, and their average. Data from Colbourne (1999), Drinkwater (1994), and Petrie *et al.* (1992).

The multi-annual trends in temperature, salinity, sea ice, and extent of the CIL have been linked causally to dominant meteorological features, in particular the Icelandic low and the Azores High. Physical oceanographers are still exploring the detailed mechanisms by which the oceanography is linked to the climate. Nonetheless, the major mechanisms are explained in, for example, Prinsenberg *et al.* (1997), Drinkwater (1994), and other contributions in this volume.

THE FIRST COLLAPSE – OVERFISHING IN THE 1960s AND 1970s

During the late 1950s and early 1960s, major fisheries developed on the offshore Banks of Newfoundland and Labrador for cod, haddock, American plaice, witch flounder and redfish. These fisheries did not all develop at once. More importantly for our purposes, the patterns of increase, peak, and subsequent decline did not coincide for different stocks of the same species over time, nor for the major target species in a given area (Figure 3-7). Nor were the declines centered on the brief cold period between 1970 and 1974.

Haddock catches peaked soonest, and subsequently showed among the most dramatic collapses. On St. Pierre Bank (NAFO 3Ps), an essentially new fishery in 1948 produced catches in excess of 27,000 t, 57,000 t, and 29,000 t respectively, between 1954 and 1956, but have exceeded 3,000 t only five times since 1957. On the Grand Bank (NAFO

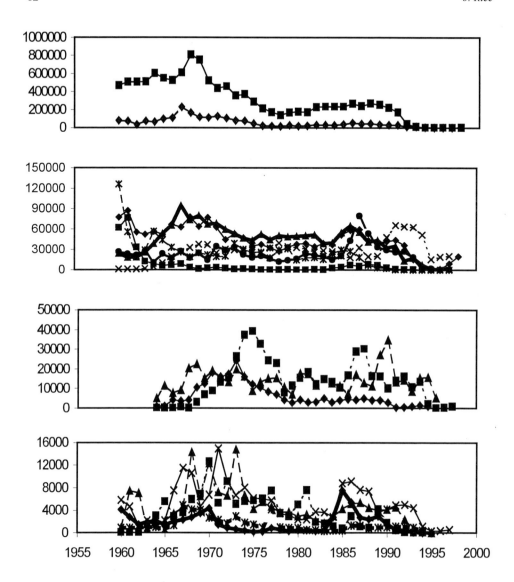

Figure 3-7. Time series of catches of major targetted groundfish stocks in Newfoundland - Labrador Shelf area. Top panel - 2J3KL cod (squares), 3NO cod (diamonds). Second panel - 3NO Haddock (squares), 3Ps cod (diamonds), 3LN redfish (circles), 2+3K redfish (x's), 3LNO American plaice (thick dashed line), Greenland halibut (whole area - thin dotted line). Third panel - 3LNO yellowtail (squares), 2J3KL witch (diamonds), 3O redfish (dashed line). Bottom panel - 2+3K American plaice (squares), 3Ps haddock (diamonds), 3Ps witch (stars), 3NO witch (thin solid line), 3Ps American plaice (dashed line). Panel groupings based on similar magnitudes of maximum catches (in tonnes) over time series. Data from most recent assessment research document of CSAS or NAFO.

3LNO) haddock catches went from negligible to over 8,000 t between 1943 and 1950, and surpassed 50,000 t by 1956. Catches dropped to under 30,000 t by 1959, and peaked again in 1961 at over 75,000 t with the recruitment of two strong year-classes from the mid-1950s. Such catches were not sustainable, and declined to below 10,000 t by 1964.

The offshore cod fishery showed the next great cycle of increase and collapse. From the historical patterns of catches described earlier, catches of all four cod stocks around Newfoundland showed two cycles of increases and subsequent declines. The first peak in the early 1950s was seen in all four stocks; from the more northerly stocks (2GH in 1953 at >53,000 t), through the northern (2J3KL in 1954 at >300,000 t) and southern Grand Banks (3NO in 1954 at >130,000 t), and in 1960 (>85,000 t) for the most southerly stock (3Ps). Catches by fleets from additional participants produced a second and usually higher peak between 1966 and 1968 for all four stocks, followed by continual declines through the mid-1970s (Figure 3-7, Table 3-1), despite persistently high although poorly quantified fishing effort (ICNAF 1952-1978).

Major directed fisheries for the four species of flatfish started later than offshore fisheries for cod and haddock, with relatively low catches of all stocks in 1960. The American plaice fishery on the Grand Bank (3LNO) expanded soonest, and peaked first at nearly 95,000 t in 1967. That stock was in continual decline when the fisheries for other stocks of American plaice, witch and yellowtail (*Limanda ferruginens*) expanded in the late 1960s, peaked between 1970 and 1975, and generally collapsed even more precipitously than American plaice in 3LNO later in the 1970s (Table 3-1). Greenland halibut (*Reinhardtius hippoglossoides*) catches also went from a minor amount to over 30,000 t during the 1960s, but varied between 25,000 and 35,000 t for nearly two decades thereafter, without any marked peaks or collapses (Figure 3-7, Table 3-1).

The final stocks receiving significant effort over this period of offshore fishing were redfish stocks. The northern stock (2+3K) was heavily exploited during the 1950s, with several years of catches in excess of 150,000 t. However, by 1961 catches fell to 55,000 t and for the rest of the period prior to extension of jurisdiction in 1977, it was rare for catches to exceeded 35,000 t or to fall below 25,000 t (Power 1995). Catches of the two stocks along the southern Grand Bank never showed the period of rapid expansion, fluctuating without major trends through the 1960s and 1970s (Power and Maddock Parsons 1998a; Figure 3-7).

Extension of jurisdiction in 1977 brought an end to the major international offshore fisheries on the Newfoundland Banks. However, the new jurisdictional regime found most of the gadoid and flatfish stocks at or near record low biomasses. Were these low biomasses solely the consequence of overexploitation during the preceding two decades or more, or had environmental factors also played an important role in the collapses? Lacking data from extensive biological sampling of catches during much of that period,

Table 3-1. Timings of first peak in reported catch after1958 and subsequent large declines until extension of jurisdiction (1977), for the major target stocks in Newfoundland-Labrador LME. Peak catches in Kt. Reported catches are vulnerable to several sources of inaccuracies, but ICNAF applied same standards to national reporting of catches over the period 1958 to 1977.

SPECIES	STOCK	YEAR OF PEAK	PEAK CATCH	YEAR CATCH WAS: 50% MAX	25% MAX	10% MAX
HADDOCK	3Ps*	1960	4.1	1962	1972	1974
	3NO	1961	76.4	1962	1964	1968
COD	3Ps	1961	86.8	1975	--	--**
	3NO	1967	226.8	1969	1975	1977
	2J3KL	1968	810.0	1971	1976	--
AM. PLAICE	3LNO	1967	94.4	1975	--	--
	2J3K	1970	12.7	1973	1979	--
	3Ps	1973***	14.8	1974	1978	--
WITCH	2J3KL	1970	24.4	1973	1976	--
	3NO	1971	15.0	1972	1977	--
	3Ps	1967	4.8	1971	1976	--
YELLOWTAIL	3LNO	1972	39.9	1976	--	--
GREENLAND. HALIBUT	All	1978	39.1	--	--	--
REDFISH	2+3K	1960	126.6	1961	1962	--
	3LN	1971	34.4	1975	--	--
	3O	1965	22.4	1976	--	--

* 3Ps haddock actually had a peak catch of 57.8 Kt in 1955, but consistent catch records for all   stocks were not being kept at that time.
** -- means catches did not fall to as low as corresponding % of maximum catch during the interval
*** 3Ps American plaice catches were 14.4 Kt in 1968, which almost matches the 1973 peak. Catches declined consistently after 1968, except for the one-year spike of catches in 1973.

and systematic scientific surveys of all but the most southern portion of the LME, it is not possible to provide a definitive answer to that question. However, several pieces of circumstantial evidence suggest that fishing was the only major factor in the collapses of commercial species.

Throughout the period of explosive growth and subsequent collapse of catches, the major oceanographic conditions generally were average or above average. Particularly through the 1960s when catches of many stocks peaked and began rapid declines, ocean temperatures were above normal, ice coverage was less than average, and the area occupied by the CIL was relatively small (Figures 3-4, 3-5, 3-6). The Great Salinity Anomaly (GSA) shows in several oceanographic signals between 1969 and 1972, such as record cold surface temperatures and strong negative salinity anomalies. Nonetheless, the catches of many flatfish stocks increased over that interval, and the declines in gadoid stocks were already well underway prior to the onset of the GSA. Mertz and Myers

(1994) have argued that the GSA does not show up as a coherent temperature signal through the full water column in the Newfoundland LME, and that there is no evidence the GSA affected productivity nor trophic coupling in the area. Although Anderson and Gregory (in press) do find exceptionally high cod larval and juvenile mortality during the GSA, the overall decline in stocks has not been attributed to the GSA phenomenon, whose two year time course in this region is much shorter than the period of declines in the individual stocks.

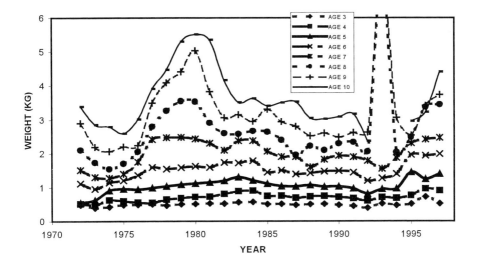

Figure 3-8. Weight (kg) at various ages for 2J3KL cod from 1971 to 1997, based on catch samples prior to 1977, and commercial and research survey catches thereafter. Figure shows great increase in weight at age and, by inference, growth rates in 1970s, and continued decline until 1994 or 1995. The spike for the oldest ages in 1993 is likely to be an artifact due to very small sample sizes of older cod in the 1990s. Data from Lilly *et al.* (1998a).

Growth rates or surrogates such as length or weight at age cannot be reconstructed for stocks during the interval in the 1960s, with the exception of cod in 2J3KL (Figure 3-8). However, with the initiation of comprehensive research surveys and associated sampling in the late 1970s, growth rates of cohorts produced in the late 1960s and 1970s can be inferred. For the cod stocks, where catches had peaked in the 1950s and 1960s, and declined throughout the 1970s, lengths at age at the beginning of the survey series in the late 1970s were larger than the average for the period from the 1950s to the present, and as large or larger than any subsequent lengths at age (Figure 3-9). Moreover, the effect was greatest for the oldest ages. Hence, throughout the period of stock decline in abundance, individual growth rates were among the highest observed. For flatfish, catches generally peaked somewhat later. Correspondingly, length at age of American

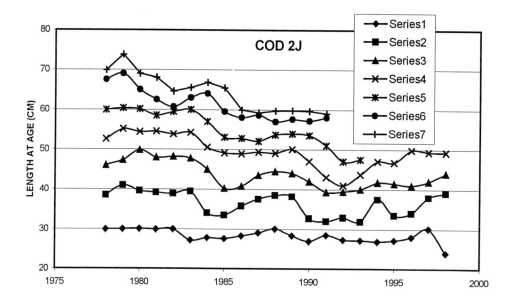

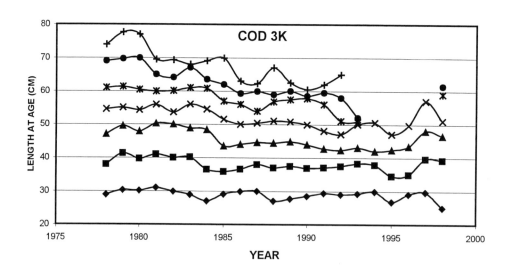

Figure 3-9 (this and next four pages). Length or weight at age from survey catches of several stocks from 1977 to the 1990s. This page: Cod from divisions 2J and 3K. Data from Lilly *et al.* (1998), Brattey *et al.* (1999), and Brodie *et al.* (1998a, b).

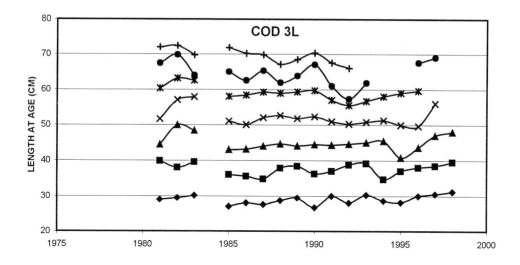

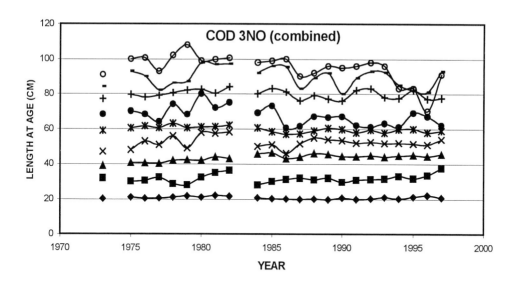

Figure 3-9 (contd.). Length or weight at age from survey catches of several stocks from 1977 to the 1990s. This page: Cod from divisions 3L and 3NO (combined). See legend in panel for cod division 2J. Data from Lilly *et al.* (1998), Brattey *et al.* (1999), and Brodie *et al.* (1998a, b).

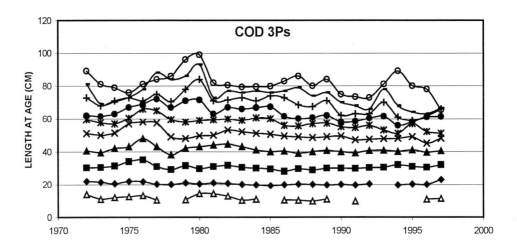

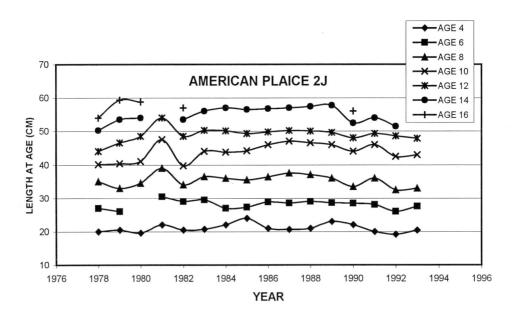

Figure 3-9 (contd.). Length or weight at age from survey catches of several stocks from 1977 to the 1990s. Top panel: Cod from division 3Ps. See legend in panel for cod division 2J. Bottom panel: American plaice from division 2J (females only - males show similar trends but are smaller at age). Data from Lilly *et al.* (1998), Brattey *et al.* (1999), and Brodie *et al.* (1998a, b).

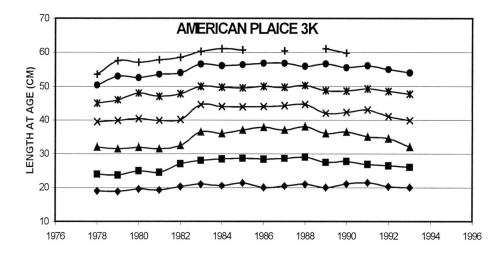

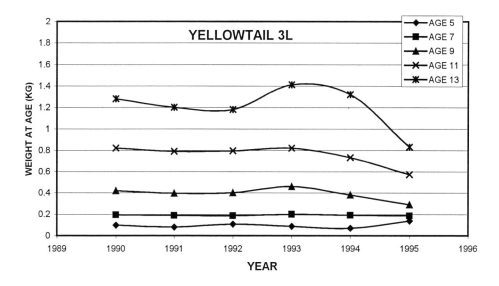

Figure 3-9 (contd.). Length or weight at age from survey catches of several stocks from 1977 to the 1990s. Top panel: American plaice from division 3K (females only - males show similar trends but are smaller at age). See legend in panel for American plaice division 2J. Bottom panel: Yellowtail in division 3L (time series begins in 1990). Data from Lilly *et al.* (1998), Brattey *et al.* (1999), and Brodie *et al.* (1998a, b).

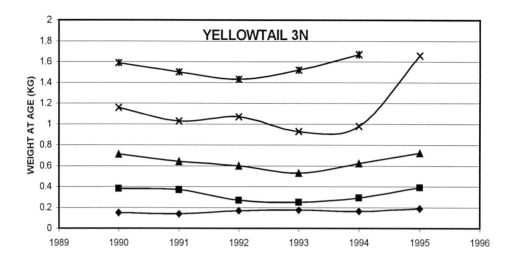

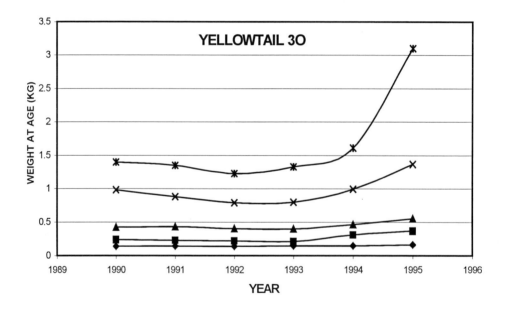

Figure 3-9 (contd.). Length or weight at age from survey catches of several stocks from 1977 to the 1990s. This page: Yellowtail in divisions 3N and 3O (time series begins in 1990). See legend in panel for Yellowtail division 3L. Data from Lilly *et al.* (1998), Brattey *et al.* (1999), and Brodie *et al.* (1998a, b).

plaice in 2J and 3K was average or below at the beginning of the research survey series, and increased sharply in the succeeding years.

These patterns of length at age are consistent with the hypotheses that density dependent factors affect growth rates for cod and flatfish. For cod, densities had been severely reduced by overfishing through the 1960s, so growth rates accelerated in the face of a food supply that was not being fished and hence was hypothesised to not be diminishing. The reduction in density of flatfish occurred later than that for cod, and the density dependent growth responses appear to have occurred somewhat later as well. Because of the very slow growth rates of redfish, and the difficulties in accurate aging, comparable consideration of changes over time in length at age are not possible for redfish stocks. Older haddock were caught in numbers too few for reliable estimates of population weights at age over much of the survey period.

Finally most stocks of cod and several stocks of flatfish produced relatively good year-classes through the 1970s. There is some evidence that year-classes between 1969 and 1972 were small, given the sizes of the spawning biomasses, consistent with the GSA having some direct effect on recruitment processes (Anderson & Gregory in press; Rice & Evans 1987). However recruits per spawner were high during the period of marked decline for the few stocks with population reconstructions that extend back to the 1960s.

None of these lines of evidence prove that overfishing was the only factor associated with the major declines in the gadoid, flatfish, and redfish stocks on the Newfoundland and Labrador Shelves. Systematic sampling of the lower trophic levels did not occur, although at least capelin stocks are thought to have remained generally healthy through the 1960s and 1970s (Jangaard 1974; Carscadden 1983). Absent systematic data other than catches of major groundfish stocks, attempts to model the first period of stock collapses in an integrated way would be purely speculative exercises, supporting whatever preconceptions were built into the models. However, all the circumstantial evidence supports the view that the ecosystem productivity did not change markedly over the period of the first collapses of these stocks. Rather, the collapses were brought on by overfishing, and their distribution in space and time reflects the entry of new fleets to the region, or the retargeting of fleets already active but experiencing declining catch rates as stocks became successively overfished.

Most importantly, once fishing pressure was reduced markedly by ICNAF efforts in the mid-1970s and excluding foreign fleets from most of the Grand Banks after 1977, stocks began to rebuild quickly. To varying degrees almost all stocks increased in biomass and numbers in the years following extension of jurisdiction. Both recruits produced per spawner and growth rates were high (Figures 3-8, 3-9). The stocks whose catches had declined the most, and for the longest period, showed gains in the years following extension of jurisdiction comparable to the gains of stocks which had declined much less (Figure 3-10). These stock increases were accompanied by increasing domestic fishing effort and gradual returns of recruits per spawner and somatic growth rates to values similar to historic rates (Figures 3-8, 3-9). This combination of biological circumstances, combined with increasing efficiency of all parts of the fishing fleet (Murawski *et al.* 1997; Sinclair and Murawski 1997) allowed fishing mortality to increase once more to rates well in excess of the rebuilding targets. This set the stage for the second collapse of groundfish on the Newfoundland and Labrador Shelves.

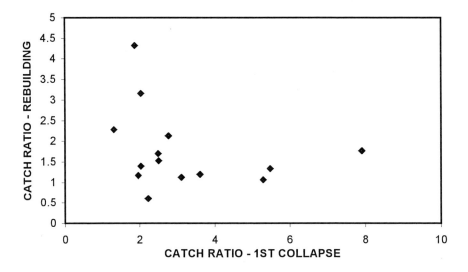

Figure 3-10. Degree of rebuilding (reflected by ratios of maximum catch in 1980s and early 1990s to mean catch of 1977-1979: y axis), vs. degree of stock decline prior to 1977 (reflected by ratio of maximum catch prior to 1977 to mean catch 1975 to 1977: x axis). Annual variation in commercial and catches and changes in target exploitation rates over time make these ratios only coarse estimators of rebuilding and degree of decline, but suggest greates rebuilding was in stocks that declined by less than a factor of three. Data from most recent annual assessment documents (see references).

THE SECOND COLLAPSE – OVERFISHING AND ECOSYSTEM CHANGES

Through the 1980s assessments of most groundfish species on the Newfoundland Shelves concluded all stocks were rebuilding (or in the case of haddock, stable). Naturally not every evaluation concluded all stocks increased annually, but the overall trajectories were up, and the messages were positive (Parsons 1993). Controversy did exist about stock trajectories, particular for cod, and the intensity of the controversy increased once many assessments were found to have retrospective patterns (cf. Keats *et al.* 1986; CAFSAC 1986) (A "retrospective pattern" refers to a situation where successive assessments indicate that the preceding assessment had overestimated the increase in biomass, and underestimated mortality rates; see Mohn 1999 for details.).

However, the conclusions of the January 1989 assessment of 2J3KL cod by DFO scientists were dramatically different from previous assessments (CAFSAC 1989). The rebuilding had not reached nearly the level estimated in earlier assessments. Moreover, there had been no growth at all, and possibly a decline, in biomass since 1986. In the next three years, assessments of essentially every groundfish stock in the LME concluded the stocks were in a state of decline, and in some cases, dramatic collapse. Successive assessments concluded stock statuses were progressively worsening, as previously good year-classes seemed to disappear without contributing significantly to reported catches or the mature biomass (Shelton and Lilly in press; Figure 3-11).

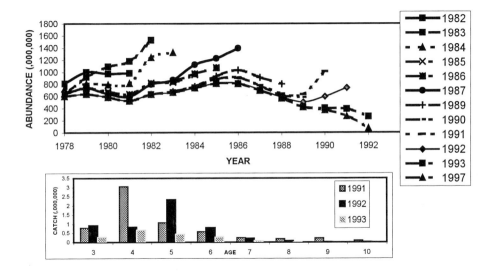

Figure 3-11. Difficulties in assessing status of 2J3KL cod, as reflected in (top panel) retrospective pattern (from Lilly *et al.* 1998) which shows effects of changing CPUE series in early 1980s, the high 1986 survey index value, and the disappearance of the 1986 and 1987 year-classes in the early 1990s; and (lower panel) the changing representation of cohorts from the late 1980s in successive assessments from 1991 to 1993. Data from Shelton and Lilly, in press.

Controversy over the cause(s) of the collapse of Northern cod reached intensities rare in the community of scientific scholars. All commentaries attributed some portion of the collapse to overly optimistic assessments and overfishing, but opinions differed on the role of other factors. Some attributed the excessive mortality the cod stocks were experiencing solely to fishing (Myers and Cadigan 1995; Myers *et al.* 1996, 1997; Hutchings 1996), whereas others implicated the harsh environmental conditions of the late 1980s and early 1990s as a factor possibly contributing directly or indirectly to the stock reductions (Parsons 1993; Taggert *et al.* 1994; Colbourne *et al.* 1997). Some analyses suggested that changes in prey availability may have occurred (Warren 1997b), and interest in the magnitude of predation by seals grew (Stenson *et al.* 1995, 1997). Likewise, the inaccuracies in the assessments and advice were attributed by some simply to inadequate tools and weak practices (Walters and Maguire 1996), by others to lack of vision and competence by the scientists (MacKenzie 1995), and by still others to intentional mismanagement of information – and stocks (Hutchings *et al.* 1997).

Summarizing the controversy and alternative hypotheses for the cause of the collapse of 2J3KL cod has already taken full volumes of journals (Campbell 1997), and will not be repeated here. Nor is it necessary to recapitulate the evidence documenting that fishing mortality was excessive for 2J3KL cod, and many other stocks, in the period from approximately 1988 to 1992. Rather, largely lost in the debate about the collapse of one cod stock is information about what happened to the other stocks and species in the LME, and what changed for cod, aside from its total biomass and harvest.

Since the improvements in data collection that coincided with extension of jurisdiction, scientists are in a better position to examine stock trajectories and their causes. Research survey series are available, so insights into stock trajectories can be separated from the many social and economic factors that influence commercial catches. On the other hand, the survey catches have great variance and occasionally surveys appear to catch anomalously many or few fish of most ages of a given species, compared to adjacent years, for reasons which are not fully understood (commonly referred to as a "year effect"). Hence, comparing individual high and low values in a time series will exaggerate the changes in stock status. However, rather than smooth data which contain both trend and variance, only large scale patterns and timings are needed for this synthesis. Such patterns are reflected reasonably in the contrasts between individual index values; say, the largest index value observed between 1982 and 1989 (a period when rebuilding under Canadian jurisdiction could reasonably have been expected to have occurred) and 1988 – 1995 (when the excessive mortality in 2J3KL cod occurred), even though absolute magnitudes are likely to be over-estimated. We will look for evidence of changes in biological factors other than simply harvest, to obtain indications whether or not more was changing than simply abundance of target species, and for more reasons than just excessive fishing mortality.

Survey Abundance Indices - The survey indices for all but one commercial groundfish stock with centers of distribution north of the southern Grand Bank (i.e. 3L and northward) showed at least a 90% decline from their peaks in the mid-1980s to their troughs in the mid-1990s (Table 3-2). The exception was the widespread Greenland halibut stock, whose total index declined by less than 50% from the peak in 1984 to the trough in 1992. However, for the part of the stock in 2J and 3K, the survey index

Table 3-2. Ratio of maximum RV survey index during 1980s to minimum survey index between 1990 and 1997, by stock. High inter-annual variance in survey indices means that these ratios likely over-estimate true declines in total stock size, but underscore the general magnitudes of change between the periods of rebuilding and collapse. Survey catches of haddock not consistently available in 1990s, so they are deleted from table.

SPECIES	STOCK	RATIO	YEAR OF MAXIMUM	YEAR OF MINIMUM
COD	2J3KL	114.9	1986	1994
	3NO	122.8	1987	1995
	3Ps	6.4	1985*	1997
AM. PLAICE	3LNO	24.3	1981	1995
	2J3K	8.7	1982	1994
	3Ps	11.5	1982	1993**
WITCH	2J3KL	51.7	1984	1994
	3NO	13.0	1988	1996
	3Ps	3.25	1986	1994
YELLOWTAIL	3LNO	8.35	1984	1993
GREENLAND HALIBUT	ALL	12.2	1983	1993
REDFISH	2+3K	366.5	1983	1994
	3LN	28.1	1984	1994
	3O***			

*   Survey index in 1991 is largest in series, 33% larger than 1985 value.
**   Survey index series last reported for 1993.
*** Survey index series only begins in 1991, ratio not calculated.

declined by more than 75% between those two years (2.7×10^5 in 1994 to 0.6×10^5 in 1992; Brodie *et al.* 1998b). For stocks whose distributions include 3NO, the cod and haddock declined as greatly as stocks to the north (Stansbury *et al.* 1998a; Bishop and Murphy 1992).

However, the time course for haddock was quite different, with no noticeable rebuilding until 1980 (instead of the immediate response seen for most other stocks after 1977). A single sharp peak in 1984 was followed by a rapid decline, with survey catches at their nadir by the late 1980s (Bishop and Murphy 1995; Bishop *et al.* 1987). For the flatfish, yellowtail declined by slightly less than 60% (Walsh *et al.* 1998b), witch declined by 80% (Bowering and Orr 1998), and for American plaice the decrease again was over 90% (Brodie 1992; Brodie *et al.* 1998a). Redfish survey indices declined over 95% in 3LN, but the survey series only begins in 1991 for 3O (Power and Maddock Parsons 1998a).

South of Newfoundland, in 3Ps, the decline in the cod stock was much less marked, with the lowest value (1992) only a third the highest (1988) (Stansbury *et al.* 1998b; Brattey *et al.* 1999). Moreover, the index showed particularly large interannual variation, maintaining high values until 1991. Plaice and haddock survey indices both declined by over 90% from the mid-1980s to the early 1990s, with haddock showing the same time course of increase and decrease in the 1980s that is seen for the 3NO stock (Morgan *et al.* 1995; Bishop and Murphy 1995; Bishop *et al.* 1987). The lowest survey index for witch occurred in 1994, and like that for cod, was no lower than a third the maximum index value, with the index again remaining high through 1992 (Bowering 1995).

Overall different things seem to be occurring in different parts of the LME. Of these major commercial stocks, more species appear to be in decline, and the declines are more marked in more northerly areas. In the 1980s and early 1990s much more fishing effort was concentrated in 3LN than in 2J, particularly for flatfish such as American plaice. Hence, if fishing were the sole cause of the collapse, this pattern would not be expected. Fortunately, the research surveys and the SPAs they permit to be calibrated allow us to examine several other data sets to see if they indicate similar patterns.

Size at age decreased throughout the second period of collapse for many but not all stocks for which data are available (Figure 3-9). For cod in 2J3KL, length at age dropped most strongly for 2J and for the oldest ages (with the most years to accumulate differences in growth rates among cohorts), less for 3K, and there was hardly a change for cod in 3L. For cod in 3NO there was minimal decline for the younger ages, but some decrease after about 1985 for cod ages 6 and above (Lilly *et al.* 1998a,b). For cod in 3Ps, again only cod older than 6 show any decline in length at age over the period, and younger cod show some increase during the time of decline (Brattey *et al.* 1999).

The decline in length at age is not restricted to cod, however. American plaice from 2J3K and 3NO both show some evidence of a decline in size during the period of decline in abundance, and again more markedly for the older fish in more northerly components of each stock. This pattern is very different than was observed in the first decline. In that case, growth appeared to increase as abundance decreased, so the greatest differences in size at age were at the beginning of the series in 1977/8, having been accumulated over the period of low abundance in the 1970s. The 1970s pattern is consistent with the occurrence of classic density dependent population responses; fewer fish sharing a fixed (or surplus) food supply individually will eat more and grow faster. In the second collapse abundance and growth rates were often dropping in unison. For yellowtail, length at age data are only available after 1990, and an upward trend from a minimum in 1990 is apparent.

Condition factor is not measured regularly for most of these stocks, but for 2J3KL cod, condition as well as size at age deteriorated during the mid to late 1980s (Figure 3-12), as reflected in both body and liver indices. Again this was most apparent for cod in 2J, and there was little trend in 3L. Condition factor of cod in 3Ps was stable (body) or increased (liver) through the 1980s, before dropping dramatically from 1990 to 1993.

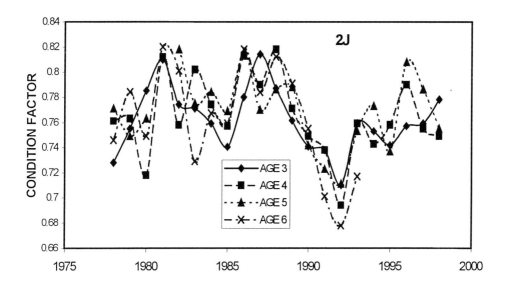

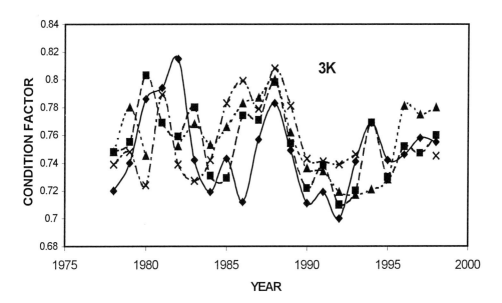

Figure 3-12 (this page and next). Condition index for cod in 2J and 3K, from cod taken in survey catches. Data from Lilly *et al.* (1998) and Brattey *et al.* (1999).

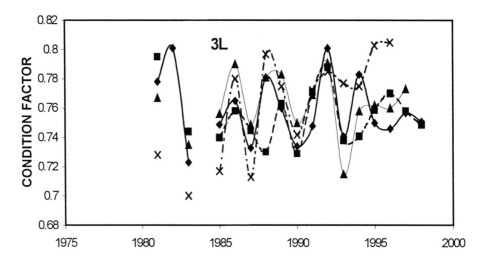

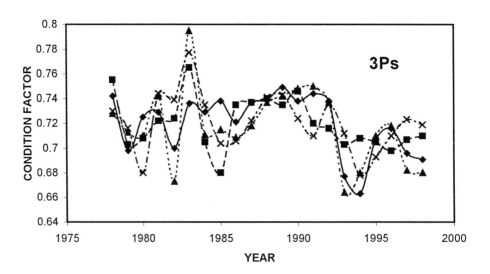

Figure 3-12 (cont'd.). Condition index for cod in 3L and 3Ps, from cod taken in survey catches. See legend in chart 2J, prior page. Data from Lilly *et al.* (1998) and Brattey *et al.* (1999).

Recruitment - Not only was total mortality excessive during the period of collapse, but recruiting year-classes were among the weakest ever observed for many of these stocks. Depressed spawning biomasses possibly had played a role in the poor recruitment success (Myers *et al.* 1994, 1995), but cohort survival at young ages was also poor for the few stocks where it has been estimated (Figure 3-13; Ings *et al.* 1997; Anderson and Gregory in press; Morgan and Walsh 1997). As with many other factors, however, stocks which showed smaller overall percentage declines from their peak in the mid-1980s to their later trough, also showed less depression of cohort sizes, suggesting both mortality and recruitment were less depressed than for stocks which declined more. Cod in 2J3KL and 3NO both show substantial declines in recruits per spawner, reaching record low values in 1991. Recruits per spawner are somewhat lower for 3Ps cod as well in the late 1980s and early 1990s, but values greater than 1 are scattered through the period, and the lowest values (1991, 1992) are not nearly as low as for the other cod stocks. Year-classes spawned in the second half of the 1980s were only a third less numerous than year-classes during the first half of the decade.

Haddock produced only one or two successful year-classes in the early 1980s; and although they were strong at ages 1-3 in research surveys, very few survived to contribute to the spawning biomass. Otherwise haddock year-classes were poor with no apparent trend.

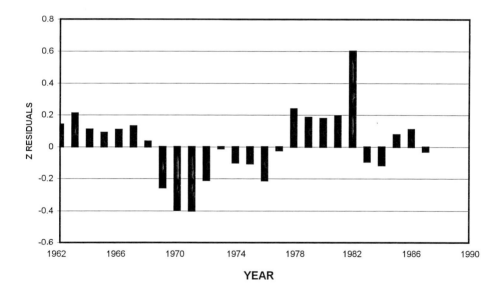

Figure 3-13. Estimates of relative annual juvenile mortality (expressed anomalies from mean) for 2J3KL cod between ages 1 and 3. From Anderson and Gregory, in press.

For American plaice, recruits per spawner were low through the 1980s and early 1990s for all stocks. In 3LNO and 3Ps, year-class strengths declined from the beginning of the time series, bottoming out at low values by the mid-1980s (Brodie *et al.* 1998a; Morgan *et al.* 1995). Yellowtail in 3LNO, whose low is only 60% less than its high, tended to produce average or above average cohorts during the period (Morgan and Walsh 1997). Recruitment to all the witch stocks is poorly estimated in recent assessments, but has been sporadic during the full time series (1977-present). Greenland halibut, which declined less than other flatfish stocks also showed strong increases in recruits per spawner and year-class strengths in the last half of the 1980s, and increases again in the 1990s.

For redfish, recruitment had been below average to poor for all stocks since the entry of one or more year-classes from the 1970s into the populations. However, the 1985 or 1986 year-class of redfish in 3LN was the best of the past two decades, and in 3O the 1984 and 1988 year-classes are markedly better than others produced since the 1970s.

Geographic Distribution – Maps of distributions of survey catches over years cannot be displayed concisely for all of these species. However, for almost every stock, assessments or overview evaluations have concluded that there were major alterations of distribution in the years leading up to and/or during the collapse.

Cod in 2J3KL – Atkinson *et al.* (1997), examined cod distribution in detail. They report "Our data demonstrate that the area cod inhabit from the northern Grand Bank to Labrador (NAFO divisions 2J+3KL) is positively correlated with stock abundance ... declines in occupied areas preceded declines in biomass by approximately two years. ... Our analyses confirm that southward and seaward distribution changes of the northern cod stock have occurred since about 1988, and that lesser southward movements occurred in the cold years of 1984 and 1985" (pg. 137).

Cod in 3NO – Survey mean numbers per tow were highly variable over the entire time series, making it difficult to search for patterns of distribution over the series (Stansbury *et al.* 1998a).

Cod in 3Ps – Again, interannual variability in survey indices has been high over the series, with occasional very large catches making identification of spatial patterns particularly difficult. Cod were reported to be scarce on St. Pierre Bank over most of the 1990s, with greater concentrations of cod in Placentia Bay and occasionally on the offshore slope of the Bank. Again, however, no regular transitions in distribution over time have been apparent (Stansbury 1998b; Brattey *et al.* 1999).

American Plaice in 2J3K – "During the later 1970s and early 1980s, American plaice were widely distributed in Divisions 2J and 3K ... By 1990 the level of abundance was negligible in the areas of previous high concentration. They were most abundant during this time in the deep area … From 1992 … there were no identifiable concentrations of American plaice in Divisions 2J or 3K" (Bowering *et al.* 1997, page 205).

American plaice in 3LNO – Analyses in Brodie *et al.* (1998a – Figure 3-8) illustrate that during the mid-1980s more than 80% of the stock biomass estimated from survey catches occurred north of 45°. It began to decline from 1988 onward and dropped precipitously to less than 40% between 1990 and 1992 and has remained at that level since. This represents a substantial contraction of range.

Plaice in 3Ps – Assessments simply note that in several years surveys did not give good coverage to the areas where American plaice tend to concentrate. However, through the 1990s, most of the biomass was found in deeper areas than usual (Morgan *et al.* 1995).

Yellowtail in 3LNO – Analyses in Walsh *et al.* (1998b) illustrate that until the mid-1980s approximately 30% of the stock biomass estimated from survey catches occurred north of 45°. It became highly variable but declined from 1984 onward and dropped precipitously to less than 5% after 1989. This represents a substantial contraction of range. It only began to recover in late 1995 and appears to have stabilized more recently around 25%.

Witch in 2J3KL – Bowering (1997) reports: "The data from the late 1970s and early 1980s … indicate that witch flounder were widely distributed throughout the shelf area in deep channels. … primarily in Div. 3K. By the mid-1980s, they were rapidly disappearing and by the early 1990s had virtually disappeared from the area entirely …". Results since 1994 "clearly indicate that witch flounder remain virtually absent from the shelf and channel areas and are only located along the deep continental slope area, especially in Div. 3L."

Witch in 3NO – Survey indices are highly variable interannually. It is conjectured that migration into deeper strata in some years causes anomalously low index values, and such movements have happened more frequently in the late 1980s (Bowering and Orr 1998).

Witch in 3Ps – Survey indices have been highly variable with little trend. However, "during the late 1970s and 1980s there was considerable biomass in depths less than 183 m whereas during the 1990s there was none" (Bowering 1995).

Greenland Halibut – Prior to about 1987, "Greenland halibut were relatively abundant in the deep channels … especially in Divisions 2J and 3K. They were also plentiful along the slope of the continental shelf. … after 1987 a decreasing trend in abundance was clearly apparent, detected first in Div. 2J. This was followed by a similar trend in 3K by 1990. By 1993 catches in Div 2J and 3K were extremely low." (Brodie *et al.* 1998b). There has been some re-expansion of range since 1995.

Redfish in 2+3K – Survey indices in 2J and 3K both declined steadily through the 1980s, but there was no apparent shift in range. However, the decline in 2J was nearly as marked as in 3K, despite the much lower effort for redfish in 2J, due to the presence of a parasite which makes infected fish undesirable to harvesters (Power 1995).

Redfish in 3LN and 3O – The survey abundance indices are variable, but with no apparent trends in distribution (Power and Maddock Parsons 1998a,b).

Non-targeted Groundfish - The groundfish surveys since 1977 allow the abundance of other groundfish not targeted by commercial fisheries to be tracked. Although limited resources have meant many of these data remain unanalyzed, some analyses have been undertaken. Time series of several species of no commercial value and low bycatch rates in commercial fisheries also showed declines of at least 80% in their survey indices, between 1985 and 1992, although the individual time series showed some minor variation.

Gomes *et al.* (1995), in a study of assemblage membership, noted that both broadhead wolffish (*Anarhichas denticulatus*) and thorny skate (*Raja radiata*) showed declines in abundance beginning in the mid-1980s, and in more northerly areas before more southerly ones. They also reported shifts in distribution of the four assemblages they had identified with multivariate methods, beginning in the mid-1980s. Their main and northern assemblage groups were displaced further offshore to the outer edge of the Grand Banks, with their coastal group more than doubling its area. Because of the substantial overlap in species composition of their four assemblages, it is difficult to evaluate unambiguously the sources of the changes in distribution of their assemblages. However, it is possibly crucial that Arctic cod (*Boreogadus saidi*) is a characteristic species only of their coastal assemblage. Prior to the change of survey gear in 1994, catchability of Arctic cod was very low, but a few individuals were regularly taken in annual fall surveys, particular in deeper strata of 2J. However, abundance of Arctic cod increased regularly through the 1980s, and more markedly from 1990 onward. Moreover, its distribution extended much further south than previously. By the early 1990s it was the most common species encounter over much of 2J and 3K, particularly in the hydroacoustic surveys.

Roundnose Grenadier (*Coryphaenoides rupestris*) and Roughhead Grenadier (*Macrourus* berglax) are deep-water species whose range is poorly sampled in the traditional research surveys. The two species supported a directed fishery of around 20,000 t annually prior to extension of jurisdiction. However, after 1977 catches fell to well below the TAC of 11,000 t with harvests largely as bycatch in the Greenland halibut fishery. Directed deepwater surveys since 1987 by Canada, Russia, and Japan found that the abundance in the northerly areas (2GHJ) decreased by more than 50% over that recent period, and increased somewhat in the areas of 3LM, where traditionally it was encountered only as incidental catch. The northern declines occurred with an exploitation rate (estimated as ratio of catch to survey "swept area" biomass estimate) less than 1% (Power and Maddock Parsons 1998b).

Other species such as silver hake (*Merluccius bilinearis*) and pollock (*Pollachius virens*), characteristic of more southerly regions, showed temporary pulses of abundance in the most southerly areas, particularly 3Ps, during the early or mid-1980s. These stocks are contiguous with stocks whose centers of distribution are on the Scotian Shelf, however, so presence in 3Ps and 3NO may be a consequence of immigration by an unknown but large extent. Short-finned squid (*Iex illecebrosus*), which also migrate into the southern portion of the LME from the south, were abundant during the mid-1980s as well. Beyond inferring that conditions in the early 1980s at least allowed more southerly species to be present in relatively high abundance, it is inappropriate to infer a great deal of information about population trends of these species from their relatively short periods of abundance.

Harvested Invertebrates and Capelin - As noted earlier, no regular monitoring program for invertebrates has been maintained in this area. Moreover, even studies based on remotely sensed data are of short duration and largely in the developmental stages (Prasad and Headrich 1993). However, significant fisheries have existed for lobster for over a century, and in the past two decades new fisheries have developed for shrimp (*Pandalus borealis*) and snow crab (*Chionoecetes opilio*). The lobster fishery is almost exclusively inshore, and dynamics appear to be associated with local environmental conditions (DFO 1998a). Nonetheless, there is widespread evidence of local increases in lobster throughout the late 1980s. Abundances peaked in the early 1990s, but remained above the average of recent decades through the 1990s.

Shrimp and crab populations both extend throughout the offshore, and there is significant crab abundance in the inshore. Time series of crab and shrimp abundance could be constructed in several ways from series of commercial and survey catches in the 1980s and 1990s. Such series would convey a false sense of quantification, because both the commercial fisheries and the survey coverages expanded several times through the 1990s (DFO 1997, 1998b; Parsons and Veitch 1998). These expansions in survey coverage kept uncovering new and high concentrations of shrimp and crab in areas progressively further from the original centres of the fisheries. It cannot be ruled out that the populations had always been there, and it was only when commercial and research effort was displaced with the collapses of traditional groundfish species that these populations were "discovered". That explanation is considered unlikely, however (Parsons and Veitch 1998), for at least two reasons. Catch rates of crab through the late 1980s and 1990s were increasing in traditional areas despite increasing harvests and a tripling of effort. Catch rates of shrimp increased threefold, while effort showed at best a modest decline, as a consequence of catching a stable quota with progressively less effort. Moreover, the abundances of these invertebrates in the 1990s are high and so widespread that had they always been there in equivalent numbers, fisheries for other species and research surveys for groundfish would have frequently taken excessive bycatches. The ratios of commercial catches to survey biomasses indicate very low exploitation rates for both stocks, despite three to five-fold increases in catches in the 1990s. These are again circumstantial pieces of evidence, but they strongly suggest crab and shrimp have both increased during the period of collapse of groundfish in this LME. Despite record high crab catches, "estimates indicate a stable biomass of legal sizes crabs in 3KLNO and an increase in 2J [and 3Ps]" (DFO 1998b).

Capelin is a key species at the second trophic level. Although it has been fished for nearly four decades, exploitation rates have been kept very low because of its important trophic role. Abundance has proven hard to quantify, with an occasional lack of accord between different hydroacoustic estimates and aerial surveys, and low catchability to the research trawl gear used before 1994 (Carscadden *et al.* 1994). Notwithstanding the difficulties, capelin showed a marked change in spatial distribution after 1989 (Lilly and Davis 1993; Frank *et al.* 1996), with much lower abundances in much of its traditional range in 2J and 3L. At the same time capelin schools were found relatively frequently in the Northeast Scotian Shelf, an area where they had previous been rare, and on the Flemish Cap, where they had almost certainly emigrated from elsewhere, probably 3LN (Frank *et al.* 1996). Starting in 1991 capelin spawning times have been later than the previous average dates, and size at age is also smaller than the historic average (Carscadden *et al.* 1997). Changes in distribution and biological traits are associated with cold environmental conditions but have persisted until at least 1998, despite the return to normal oceanographic conditions by 1996 (Carscadden *et al.* 1997; Carscadden and Nakashima 1997).

POST COLLAPSE PATTERNS

By 1994 fisheries for all gadoids and most flatfish and redfish in the Newfoundland Shelf LME were either closed or operating under greatly reduced quotas. The exception was Greenland halibut which, after experiencing record high catches through the early 1990s, has had a TAC of 27,000 t (slightly below average catches from the 1960s to the 1980s) since 1994 (Figure 3-7; Brodie *et al.* 1998b). The absence of commercial catch for most other groundfish stocks has meant that in recent years SPAs cannot be conducted for many of these stocks. Hence, for the species of historic commercial importance, the research survey indices and associated biological samples are nearly the sole source of information on recent stock status. Because of the change in research survey trawl gear to a much finer mesh in 1994, however, the exact comparability of data from the two parts of the time series is open to question. Although research vessel catch rates have been calibrated for major species, the more subtle challenges of ensuring that length at age series, for example, are corrected for size dependent changes in catchabilities have yet to be overcome.

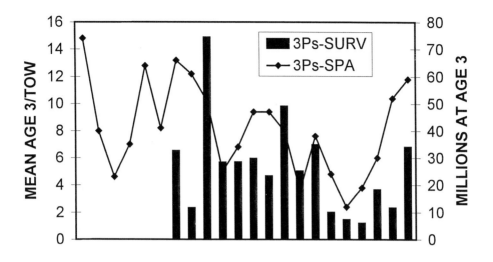

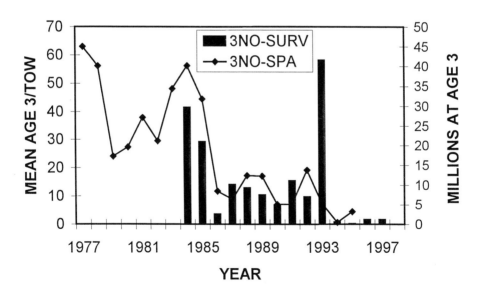

Figure 3-14. Strengths of year-classes produced by 3Ps cod (top panel) and 3NO cod (lower panel) as estimated by SPA numbers at age 3 (lines), and mean catch of age 3 per tow from surveys (bars).

Many of the traditional stocks in 3Ps appear to have returned to historic average levels. Cod, in particular, have rebuilt strongly, with above average year-classes from 1989 and 1990 protected from fishing for three years by a moratorium and contributing strongly to present spawning biomass (Figure 3-14). On the southern Grand Banks, American plaice, witch, and yellowtail in 3LNO are showing some improvement in recruitment since 1995. The improvements are much more apparent for yellowtail and, in the southernmost portions of 3NO, for all stocks (Brodie *et al.* 1998a; Bowering and Orr 1998; Walsh *et al.* 1998b). In contrast, haddock have not shown any signs of improved recruitment since the early 1980s in either 3Ps or the southern Grand Bank (Bishop and Murphy 1992, 1995).

Despite closures of directed fisheries and stringent, carefully enforced bycatch limits in invertebrate fisheries for over half a decade, none of the finfish stocks in 2J and 3K are showing noticeable signs of improvement. Cod in 2J3KL is showing up occasionally in dense aggregations in some inshore bays, but together these aggregations do not represent a significant biomass, when compared to the historic status of the resource. Offshore, cod are extremely rare in surveys (DFO 1999a), as are American plaice, witch and redfish (DFO 1999b).

Although the changes in survey gears has made it risky to compare time trends in size, growth rate, and other biological factors from before and after the mid-1990s, the moratoria have made it possible to estimate natural mortality of many of these stocks. For a given stock, cohort specific Z's (total instantaneous mortality) can be estimated by the rate of decrease of catches of successive ages fully recruited to the fishing gear. In some cases cohort effects can be removed by a multiplicative model and annual Z's for a stock can be estimated as well (Sinclair 1998). The excessively high Z's for many of these stocks during the period of collapse are well known. An important result is that following the moratoria Z's, where they have been estimated, have remained very high. With fishing mortality very nearly zero, the conclusion is that natural mortality in the 1990s has been much higher than had been assumed previously for cod and flatfish stocks. For the eight cod stocks in the Canadian Atlantic north of Halifax, assessments in the late 1990s have concluded that M for cod has been at least 0.4 since the mid-1980s (DFO Proceedings 1999) for six of them (Figure 3-15). The exceptions are cod in 3Ps, the stock showing the strongest rebuilding following its decline in the late 1980s, and 3NO, where year effects in the survey series make it impossible to estimate Z for cohorts. (The data produce positive estimates for Z in 1995, bracketed by estimates well below -0.5 for most ages in 1993, 1994, and 1996.) Mortality rates for cod at young ages (1-3) have also been higher for cod after the mid-1980s than before the mid-1980s (Anderson and Gregory in press).

Z's for American plaice and yellowtail in 3LNO have also been estimated, and have remained at record high values, often above 0.6, during the fishery closures. In the case of American plaice, mortality seems to have continued to increase after fishing mortality was reduced to negligible (Figure 3-15). Finally, marine survival of Atlantic salmon from coastal rivers of Newfoundland and Labrador, and multi-sea-winter salmon that migrate through this LME from mainland rivers, has been declining through the 1990s (ICES 1999). Despite the closure of all commercial salmon fisheries in Atlantic Canada, and the most stringent catch-and-release regulations in history, marine mortality of most stocks is at or near the highest rate ever recorded.

The finding that natural mortality remained more than double the historic estimated rate throughout the 1990s for cod, salmon, and at least some flatfish is important. The initial documentation of a natural mortality around 0.2 for cod on the Newfoundland Shelf was sound (Pinhorn 1976). The change in that value could not be documented until fishing mortality was essentially zero. However, once documented, the elevated level was found to apply back well into 1980s, and to other stocks at least in 2J3KLNO. Maintaining m at 0.2 in the assessments when in reality it had increased to at least 0.4 can also account for some of the inaccuracies and strong retrospective pattern in the assessments of the late 1980s, assessments whose over-optimistic forecasts contributed to quotas which, in retrospect, were too high.

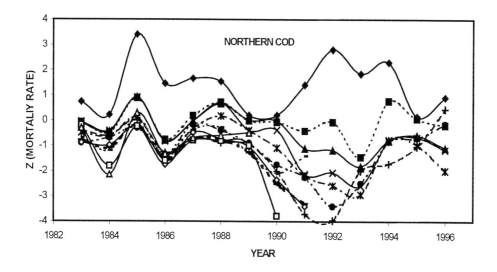

Figure 3-15 (this and next two pages). Estimates of total mortality based on survey catches at age, for cod (2J3KL, 3NO, and 3Ps), 3LNO American plaice, and Greenland halibut, showing increasing Z in northern stocks, and strong year effects in 3NO cod survey.

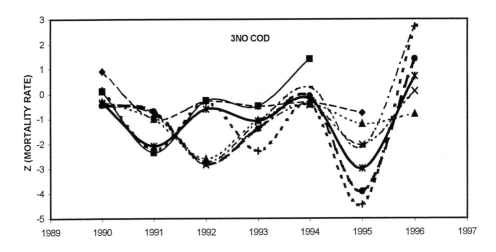

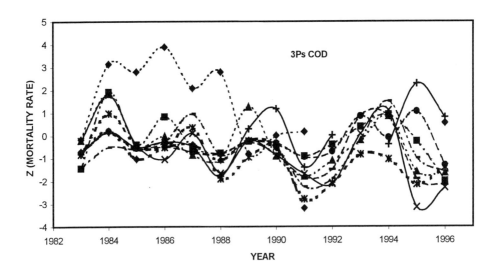

Figure 3-15 (cont'd.).   Estimates of total mortality based on survey catches at age, for cod (2J3KL, 3NO, and 3Ps), 3LNO American plaice, and Greenland halibut, showing increasing Z in northern stocks, and strong year effects in 3NO cod survey.

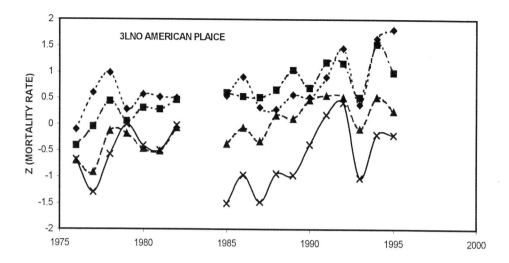

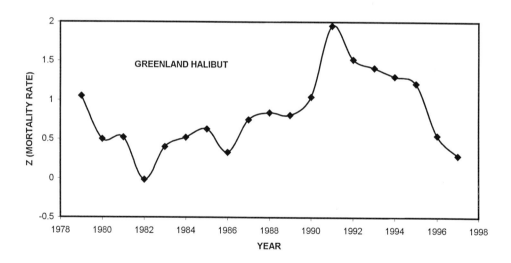

Figure 3-15 (cont'd.). Estimates of total mortality based on survey catches at age, for cod (2J3KL, 3NO, and 3Ps), 3LNO American plaice, and Greenland halibut, showing increasing Z in northern stocks, and strong year effects in 3NO cod survey.

CONTRASTING THE COLLAPSES

The collapses of cod, haddock, redfish, and major flatfish in 1960s were almost certainly due to overharvesting. All stocks showed biological responses that were consistent with classic density-dependent responses to reduced abundance in the face of continued availability of food and suitable environmental conditions. Inferring growth rates from length at age data, growth rates increased as abundance declined in all stocks. While the foreign fisheries were operating, year-classes appeared to get progressively weaker, with reduced spawning stock biomass playing a role (Rice and Evans 1987). However, aside from a period of high pre-recruit mortality during and immediately after the GSA, juvenile survivorship of cod was exceptionally high during the period of lowest spawning biomass (Anderson and Gregory, in press).

The important conclusion is that the capacity of the stock to produce high numbers of recruits per spawner was not compromised during this period of collapse. For the cod and flatfish, strong year-classes occurred in the mid-1970s in almost all stocks. Environmental conditions were favourable, with average temperatures and average to above average salinities. These year-classes grew rapidly to a size where they began to recruit to the fishable population. The low fishing effort in the years following extension of jurisdiction allowed them to be protected from the excessive fishing mortalities characteristic of the preceding two decades. Moreover, the major potential marine predators, harp seals (*Phoca groenlandica*) and cod (as predator or cannibal) had both been reduced to their lowest abundances in the historic record. These strong year-classes, growing quickly and surviving at a high rate, provided the basis for rapid rebuilding of the depleted spawning biomasses through the early 1980s.

It is much more problematic to account completely for the second collapse and its persistence through the 1990s. As argued most forcefully by Myers *et al.* (1996, 1997) and Hutchings and Myers (1994), overfishing unquestionably played a major role in the excessive mortality in the late 1980s and early 1990s. With sufficient argumentation, it is possible to explain the coincident drop in year-class strengths, size at age, abundance of non-target species, and changes in distribution solely as a consequence of excessive fishing. The explanations require invoking some first and second order effects of fishing, including extensive discarding and bycatch mortality, but they are not impossibly convoluted (Table 3-3).

It becomes much more difficult to account for the persistence of high mortality and low growth rates in at least the cod and flatfish after the moratorium without invoking factors in addition to fishing. The extremely harsh environmental conditions, with extensive and persistent sea ice, deep and broad distribution of the CIL, and extreme cold anomalies all suggest the physical environment contributed in some manner to the population collapses. This line of argument has been developed in various ways by Parsons (1993), Rose *et al.* (1994), Gomes *et al.* (1995), Kulka *et al.* (1995, 1996), Bowering *et al.*

Table 3-3. Qualitative evaluation of the ease (Low – to high ***) of accounting for various observed changes in biological properties of Newfoundland-Labrador Shelf LME, using alternative classes of processes.

	CAUSAL PROCESS		
Property of Stock/System	Fishing	Harsh Envt.	Trophodynamics
Collapsed abundances of target species	***	*	-
Poor somatic growth & condition	*	**	***
Poor year-classes	**	***	***
Declines in non-target species	*	***	**
Distributional changes of stocks	*	***	*
High mortality after closures	-	*	***

(1997), Anderson and Gregory (in press), and others. How much can be explained by the inclusion of environmental factors to the explanation of any changes in the LME? Changes in distribution of many species, abundance of non-target species, and biological factors such as size at age and condition factor can be explained at least as readily or more so than with excessive fishing as the only factor. The general gradient of severity of declines from 2J (severe and enduring for almost all groundfish) to 3Ps (lowest percentage drops, and generally rapid recoveries) is also consistent with a role for environmental stress, rather than fishing effort, which was most intense in 3L and 3N. However, even if one does invoke a harsh environment as a contributing factor to the second great collapse, it is important to stress that the productivity of the ecosystem did not fail totally. Boreal species such as Greenland halibut declined relatively little, and stocks such as Arctic cod, crab, and shrimp increased greatly in abundance and distribution.

Invoking harsh environmental conditions is not specifying mechanisms for the change however, nor is it directly finding and explaining the reasons for dead fish. The increased unaccounted for mortality in the second half of the 1980s could be unreported landings and bycatch mortality, and hence due to fishing. However, such mortality should have ended with the moratoria, yet Z stayed high. Moreover, observer coverage of offshore fishing was essentially 100% by late in the 1980s, so bycatch and discard mortalities in the offshore have been quantified and accounted for (Kulka 1996). Mortality caused by environmental stress could have persisted beyond the moratorium, but becomes unlikely after 1995.

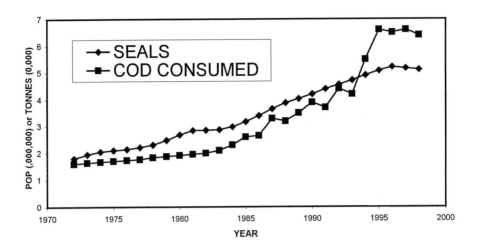

Figure 3-16. Trends since 1970s for harp seal abundance, and estimates of quantities of cod consumed by harp seals in this LME. Consumption estimates prior to 1985 assume average diet; thereafter, annual diet data are used. Data from Stenson *et al.* (1997).

Hence, attention is turning to seal predation as a major cause of the elevated mortality. With annual harvests well below replacement production, the harp seal population has reached the highest abundance in the time series (Figure 3-16). Moreover, the herd is staying in the Newfoundland Shelf LME much longer than historically (Stenson *et al.* 1995, 1997). Although consumption estimates are only available for cod, and numerous uncertainties remain, seal are estimated to be consuming at least 80% of the production of cod in 2J3KL in recent years, and represent a source of significant mortality to cod up to at least age 5 (DFO 1999a). Seals contributing to the high natural mortality is consistent with the observations that M did not increase in 3Ps, where the presence of seals has not increased greatly, and the beginnings of improved recruitment on the extreme southern parts of 3NO, where seals penetrate least.

Again, once trophic factors have been invoked in one aspect of the ecosystem changes, it can be asked how many other changes can be accounted for more directly and simply by trophodynamic processes. The species of invertebrates that have increased the most are favored prey of cod and other collapsed major predators. It can be argued that reducing groundfish stocks by overfishing and increased predation by seals led to a reduction in predation and subsequent population increases in large invertebrates. These new population densities may have, in turn, redistributed energy flows in ways that have made it difficult to return to earlier system configurations.

Trophodynamic models could undoubtedly be built which reflect many of the ecosystem changes observed in this LME. However, initial attempts have been difficult to balance unless a number of *ad hoc* parameter adjustments are made (DFO Proceedings 1999). Moreover, Gomes *et al.* (1992) showed for at least 3LNO that data do not allow differentiation among several quite different trophodynamic representations of the ecosystem – representations with quite different properties. Lacking quantitative information on almost all species below the highest trophic level makes trophodynamic modelling approaches to the dynamics of this LME possibly of interest intellectually, but of questionable practical value. With so many parts of the ecosystem biomasses and flows unconstrained by reliable data, it becomes an exercise in creativity to design configurations with a variety of properties, stable or unstable.

The same difficulty of explanatory flexibility applies to determining which explanation of the causes of the collapse and subsequent pattern of strong (3Ps), weak (southern 3NO), or nearly non-existent (2J and northern 3K) rebuilding is the correct one. By now, almost no one denies that overfishing, almost certainly associated with poor fishing practices, was a major cause of the collapse in the late 1980s, as it was the collapse in the 1960s. However, it becomes awkward and convoluted to account for all the things that changed as consequences of this single process. Adding some role for the particularly cold conditions in the late 1980s and early 1990s makes several observations more readily explained. However, the partitioning of causality of the collapses in abundance among the two candidate explanations becomes difficult, and has been characterized by strident argumentation on each side. Finally, explaining the continued high mortality with closed fishery and average to above-average environmental conditions becomes difficult unless trophodynamic factors are invoked. Again, once they are given some role, giving them primacy means one would have to reduce greatly the explanatory role of the two other processes.

It is attractive to provide some platitude about ecosystem complexity, and the reality of multiple causal factors in system dynamics. That strategy is unhelpful, however, in light of the importance of learning from the ecological, economic and social tragedies of the Newfoundland LME in the late 1980s and 1990s. We need to learn as much as we can from the events of that period (Hutchings and Myers 1994), to know what signals are key ones to react to in the fishery, in the physical environment, and in the predator - prey linkages. We should continue to hope that further analyses of existing data, expanded monitoring of this ecosystem (particularly now that research gears efficiently sample things smaller than 10-15 cm) and temperate, professional debate among scientists with differing viewpoints on the primacy of the various classes of processes, will clarify further both the causality and the reliable early signals, of the major changes in this LME.

I will conclude by offering a couple of personal speculations on concerns we should carry forward from the last two decades in this LME. By now everyone is aware of the notion of alternate stable states of ecosystems (May 1974; Pimm 1991). The stability of the current configuration is key to the near and medium term future of this LME. There are still no signs that boreal species like shrimp and crab have started to have weaker year-classes, and Arctic cod remains common at least in the northern half of the system. On the other hand, major components in 3Ps were not perturbed as far and returned to pre-perturbation conditions quickly. We did not need this unplanned experiment to know, as scientists and managers, that big perturbations are likely to have greater consequences that little ones. Nor did we need the experiment to know that ecosystems have a lot of inertia - it is hard to change quickly the direction of large trends in many ecosystem components (even ones which in theory we *should* have substantial control over, such as fishing fleet capacity). The important question is: When a big perturbation has occurred for whatever cause, are there ecosystem assembly rules (sensu Diamond 1975; Conner and Simberloff 1979; Strong at al.1984) with some "memory" of the previous state of the ecosystem at all? This is a different viewpoint than multiple stable equilibria. Invoking multiple stable equilibria intrinsically assumes a memory of the previous state exists. In that framework, the difficulty in returning to the ecosystem characteristic of "the good old days" is simply breaking out of the local configuration, to head back in the direction of some historic, preferred one. I think some serious exploration of assembly rules of marine ecosystems is overdue, to evaluate if there is any empirical support for the belief that some earlier configuration can be re-attained after large perturbations.

The other idea we need to focus on is what it truly means to be precautionary in uses of marine systems. The precautionary approach is another idea that is much in vogue right now (FAO 1995a, b). In the recent history of this LME, the assessments between 1988 and 1992 were particularly uncertain. Knowing now that M had increased - possibly as much as doubling - by that time for at least some of the key species, we can understand better why, and why the uncertainty was not due to any incompetence or malevolence of the assessment scientists alluded to earlier. With survivorship overestimated, indices could not be reconciled, a strong retrospective pattern was present, and catch limits were set too high. It is a fact that had M not changed, fishing to TACs would not have resulted in such severe overfishing, and stock collapses would have been slower or possibly even averted. But M apparently did change, and the uncertainty was a cause for review panels and deferred social and economic pain. Deferred was not avoided.

We need to remember that it is inescapable that when an ecosystem is undergoing some fundamental change, the uncertainty will always increase greatly. This will be very unsettling for managers and decision-makers, who will have become used to uncertainty decreasing as the system appeared to follow a constant trajectory. So just when the future is least well known, precisely because the ecosystem is undergoing a major and rapid change, science gives the managers advice with the greatest uncertainty. As scientists we need to advertise the message from this LME that the Precautionary Approach is essential in these situations. The uncertainty alone is cause for particularly conservative actions, until the new system trajectory is understood, and appropriate management strategies can be identified and implemented.

CONCLUSIONS

The large marine ecosystem of the Newfoundland – Labrador Shelf has shown major changes over time, but the changes have been greater in recent decades than in any other period in history.

The 1985-1993 collapses were different from fisheries collapses in 1960 and the early 1970s. There is strong evidence that overfishing played the greatest role in both collapses. However, the overfishing was aggravated by, and partly caused by, ecosystem dynamics (increased m) and biological changes that are hard to account for without invoking effects of the ocean environment.

The slowness of the system to return towards an historic "typical" configuration when stresses from both fishing and the physical environment are largely eliminated raises new and deeply troubling questions about managing ecosystem trajectories. Ecosystems could have much more inertia than we thought (and than we put in our models). Systems may not have "assembly rules" (sensu community ecologists) strong enough to drive them back to an "historic" configuration once they have been perturbed severely enough.

To learn from experience, and detect future changes in time to react wisely, it is vitally important to monitor many system components and use precaution when faced with making major perturbations to the system.

ACKNOWLEDGMENTS

I would like to thank the many scientists, biologists, and technicians in Newfoundland Region, DFO, for decades of work at sea and laboratories. I would particularly like to thank the many authors of assessment documents who allowed me to draw data, information, and figures from their works. I also thank the IUCN for funding support for travel to the LME Symposium in Bergen, and a reviewer who made a large number of very helpful suggestions for clarifying presentation. I would like to thank Ann Royko for exceptional patience and help in preparing the copy for printing.

REFERENCES

Anderson, J. and B. Gregory. In press. Factors Regulating Survival of Northern Cod (NAFO 2J3KL) During Their First Three Years of Life. ICES Mar. Sci. Symp. 202.

Anderson, J.T. and G.G. Gardner. 1986. Plankton communities and physical oceanography observed on the Southeast Shoal region, Grand Bank of Newfoundland. J. Plankton Res. 8: 1111-1135.

Atkinson, D.B., G.A. Rose, E.F. Murphy, and C.A. Bishop, 1997. Distribution changes and abundance of northern cod (*Gadus morhua*), 1981-. Can. J. Fish. Aquat. Sci 54, no. Suppl. 1: 132-138.

Bengtson, S-A. 1984. Breeding ecology and extinction of the great auk (*Pinguinus impennis*): Anecdotal evidence and conjectures. Auk: 101 1-12.

Bishop, C.A., J.W. Baird, and H.F. Hicks. 1987. Biomass and abundance of haddock in NAFO Divisions 3NO and Subdivison 3Ps from research vessel surveys along with estimates of commercial catch and effort. CAFSAC Res. Doc. 87/48. 22 p.

Bishop, C.A. and E.F. Murphy. 1992. A review of stock status, 3NO and 3Ps haddock. CAFSAC Research Document 92/113. 25 p.

Bishop, C.A. and E.F. Murphy. 1995. The status of 2GH cod, 3LNO haddock, 3Ps haddock, and 3Ps pollock. CAFSAC Research Document 95/33. 30 p.

Bowering, W.R. 1995. Witch Flounder in Subdivision 3Ps: A Stock Status Update. DFO Atl. Fish. Res. Doc 95/38.

Bowering, W.R. 1997. A Review of the Status of Witch Flounder in Divisions 2J, 3KL. NAFO, N2895, NAFO SCR Doc. 97/61.

Bowering, W.R., M.J. Morgan, and W.B Brodie. 1997. Changes in the population of American plaice *(Hippoglossoides platessoides)* off Labrador and northeastern Newfoundland: A collapsing stock with low exploitation. Fish. Res. 30:199-216.

Bowering, W.R. and D.C. Orr. 1998. An Evaluation of the Current Status and Historic Review of the Witch Flounder Resource in NAFO Divisions 3NO. NAFO, N3040. NAFO SCR Doc. 98/49.

Brattey, J., N.G. Cadigan, G.R. Lilly, E.F. Murphy, P.A. Shelton, and D.E. Stansbury. 1999. An assessment of the cod stock in NAFO Subdivision 3Ps. CSAS Res. Doc. 99/18.

Brodie, W.B. 1992. Analysis of American Plaice, Cod and Yellowtail Flounder Catches in Research Vessel Surveys on the Tail of the Grand Bank, 1971-89. J. Northw. Atl. Fish. Sci.14: 47-58.

Brodie, W.B., W.R. Bowering, D. Orr, D. Maddock Parsons, and M.J. Morgan. 1998a. An Assessment Update for American Plaice in NAFO Divisions 3LNO. NAFO, N3061, NAFO SCR Doc. 98/69.

Brodie, W.B., W.R. Bowering, D. Power, and D. Orr. 1998b. An Assessment of Greenland Halibut in NAFO Subarea 2 and Divisions 3KLMNO. NAFO, N3038, NAFO SCR Doc 98/74.

Buch, E., S.Aa. Horsted, and H. Hovgard. 1994. Fluctuations in the occurrence of cod in Greenland waters and their possible causes. ICES Mar. Sci. Symp. 198: 158-174.

Campbell, J.S. 1997. Selected proceedings of the symposium on the biology and ecology of northwest Atlantic cod. Can. J. Fish. Aquat. Sci. 5 4. Suppl. 1:1-2.

Canadian Atlantic Fisheries Scientific Advisory Committee (CAFSAC). 1986. Advisory Document 86/25: Advice on the Status and Management of the cod stock in NAFO Divisions 2J, 3K, and 3L. 41 p.

Canadian Atlantic Fisheries Scientific Advisory Committee (CAFSAC). 1989. Advisory Document 89/1: Advice for 1989 on the Management of Cod in Divisions 2J3KL. 14 p.

Carscadden, J.E. 1983. Population dynamics and factors affecting the abundance of capelin (*Mallotus villosus*) in the Northwest Atlantic. In: G.D. Sharpe and J. Csirke. Proceedings of the Expert Consultation to Examine Changes in the Abundance and Species Composition of Neritic Fish Resources. FAO Fish. Rept. 789-811.

Carscadden, J., B. Nakashima, and D.S. Miller. 1994. An Evaluation of Trends in Abundance of Capelin (*Mallotus villosus*) from Acoustics, Areal Surveys and Catch Rates in NAFO Division 3L, 1982-89. J. Northw. Atl. Fish. Sci. 17:45-57.

Carscadden, J. and B.S. Nakashima. 1997. Abundance and changes in distribution, biology, and behavior of capelin in response to cooler waters of the 1990s. In: Forage Fishes in Marine Ecosystems, Lowell Wakefield Fisheries Symposium Series, no. 14. Amer. Fish. Soc. Fairbanks AK. 457-468.

Carscadden, J., B.S. Nakashima, and K.T. Frank. 1997. Effects of fish length and temperature on the timing of peak spawning in capelin (*Mallotus villosus*) AU: T Can. J. Fish Aquat. Sci 54:781-787.

Casey, J.M. and R.A. Myers. 1998. Diel variation in trawl catchability: Is it clear as day and night? Can. J. Fish. Aquat. Sci. 55: 2329-2340.

Colbourne, E. 1999. Oceanographic conditions in NAFO Divisions 2J 3KLMNO during 1998 with comparisons to the long term (1961-1990) average. DFO Can. Stock. Ass. Sec. 99/48.

Colbourne, E., B. deYoung, S. Narayanan, and J. Helbig. 1997. Comparison of hydrography and circulation on the Newfoundland Shelf during 1990-1993 with the long-term mean. Can J. Fish. Aquat. Sci. 54 (Suppl. 1) 68-80.

Colbourne E., S. Narayanan, and S. Prisenberg. 1994. Climate Change and Environmental Conditions on the Northwest Atlantic, 1970-1993. ICES Mar. Sci. Symp. 198:311-322.

Conner, E.F. and D. Simberloff. 1979. The assembly of species communities: Chance or competition? Ecology 60: 1132-1140.

Davidson, F.J.M. and B. deYoung. 1995. Modelling advection of cod eggs and larvae on the Newfoundland Shelf. Fish. Oceanogr. 4:1, 33-51.

deYoung, B. and C.L. Tang. 1989. An analysis of Fleet Numerical Oceanographic Center winds on the Grand Banks Atmosphere-Ocean 27: 414-427.

DFO Proceedings. 1999. Proceedings of the cod zonal assessment process, Rimouski, Quebec, 1-12 March 1999. DFO Proceedings 99/5; 140 p.

DFO Science Stock Status Report. 1997. Northern shrimp off Newfoundland and Labrador. DFO SSR #C2-05.

DFO Science Stock Status Report. 1998a. Newfoundland Lobster. DFO SSR #C2-03/1998.

DFO Science Stock Status Report 1998b. Newfoundland and Labrador snow crab. DFO SSR #C2-01/1998.

DFO Science Stock Status Report 1999a. 2J3KL Cod. DFO SSR #A2-01.

DFO Science Stock Status Report 1999b. Overview of Newfoundland and Labrador Shelf Ecosystem. DFO SSR #A1-01.

Diamond, J.M. 1975. Assembly of species communities. In: M.L. Cody and J.M. Diamond (eds.). Ecology and Evolution of Communities. Harvard University Press, Cambridge MA. 342-444.

Drinkwater, K. 1994. Atmospheric and Oceanographic variability in the northwest Atlantic during the 1980s and early 1990s. J. Northw. Atl. Fish. Sci. 18: 77-97.

Drinkwater, K.F., E. Colbourne, and D. Gilbert. 1998. Overview of environmental conditions in the northwest Atlantic in 1996. Northw. Atl. Fish. Org. Sci. Counc. Stud. 31: 111-146.

FAO 1995a. Precautionary Approach to Fisheries. Part I: Guidelines on the precautionary approach to capture fisheries and species introductions. FAO Fisheries Technical Paper No. 350, Part I: Rome, FAO. 52 p.

FAO 1995b. Precautionary Approach to Fisheries. Part II: Scientific Papers. FAO Fisheries Technical Paper No. 350, Part 2: Rome, FAO. 210 p.

Forsey, R. and W.H. Lear. 1987. Historical Catches and Catch Rates of Atlantic Cod at Newfoundland During 1677-1833. Can. Data Rep. of Fish. and Aquat. Sci. 662. 52 p.

Frank, K.T., J.E. Carscadden, and J.E. Simon. 1996. Recent excursions of capelin (*Mallotus villosus*) to the Scotian Shelf and Flemish Cap during anomalous hydrographic conditions Can. J. Fish. Aquat. Sci. 53: 1473-1486.

Gomes, M.C., R.L. Haedrich, and J.C. Rice. 1992. Biogeography of Groundfish Assemblages on the Grand Bank. J. Northw. Atl. Fish. Sci. 14:19-27.

Gomes, M.C., R.L. Haedrich, and M.G. Villagarcia. 1995. Spatial and temporal changes in the groundfish assemblages on the north-east Newfoundland/Labrador Shelf, north-west Atlantic, 1978-1991. Fish. Oceanogr. 4:2, 85-101.

Gough, J. 1993. A historical sketch of fisheries management in Canada. In: L.S. Parsons and W.H. Lear. Perspectives on Canadian Marine Fisheries Management. Can. Bull. Fish. Aquat. Sci. no. 226. 5-53.

Head, G.C. 1976. Eighteenth century Newfoundland: a geographer's perspective. McClelland and Stewart, Ltd. 286 p.

Helbig, J., G. Mertz, and P. Pepin. 1992. Environmental influences on the recruitment of Newfoundland/Labrador cod. Fish. Oceanogr. 1: 39-56.

Hutchings, J.A. 1996. Spatial and temporal variation in the density of northern cod and a review of hypotheses for the stock's collapse. Can. J. Fish. Aquat. Sci. 53: 943-962.

Hutchings, J.A. and R.A. Myers. 1994. What can be learned from the collapse of a renewable resource? Atlantic cod, *Gadus morhua*, off Newfoundland and Labrador. Can. J. Fish. Aquat. Sci. 51: 2126-2146.

Hutchings, J.A., C. Walters, and R.L. Haedrich. 1997. Is scientific inquiry incompatible with government information control? Can. J. Fish. Aquat. Sci. 54:1198-1210.

ICES 1999. Extract No. 7, Report of the Advisory Committee on Fisheries Management; North Atlantic Salmon Stocks. International Commission for Exploration of the Seas. 44 p.

ICNAF 1952-1978. Statistical Bulletins (28 Annual Volumes 1952-1978). Intern. Comm. N. Atl. Fish. Dartmouth, Canada.

Ings, C.P., D.W. Schneider, and D.A. Methven. 1997. Detection of a recruitment signal in juvenile Atlantic cod (*Gadus morhua*) in coastal nursery areas. Can. J. Fish. Aquat. Sci. 54, Suppl. 1: 25-29.

Jangaard, P.M. 1974. The capelin (*Mallotus villosus*): biology, distribution, exploitation, and composition. Bull. Fish. Res. Board Can. 186: 70 p.

Keats, D.A., D.H. Steele, and J.M. Green. 1986. A report to the Newfoundland Inshore Fisheries Association on scientific problems in the northern cod controversy. Revision of the recent status of the northern cod stock (NAFO Div. 2J, 3K, and 3L). Memorial University of Newfoundland. St. John's, Newfoundland.

Kulka, D.W. 1996. Discarding of cod (*Gadus morhua*) in the northern cod and northern shrimp directed fisheries from 1980-1994. NAFO Sci. Counc. Rep. 12 p.

Kulka, D.W., A.T. Pinhorn, R.G. Halliday, D. Pitcher, and D. Stansbury. 1996. Accounting for changes in spatial distribution of groundfish when estimating abundance from commercial fishing data. Fish. Res. 28: 321-342.

Kulka, D.W., J.S. Wroblewski, and S. Narayanan. 1995. Interannual patterns in the winter distribution and recent changes and movements of Northern cod (*Gadus morhua*) on the Newfoundland-Labrador Shelf. ICES J. Mar. Sci. 52:889-902.

Lear, W.H. 1998. History of fisheries in the Northwest Atlantic: the 500 year perspective. J. Northw. Atl. Fish. Sci. 23: 41-73.

Lear, W.H. and L.S. Parsons. 1993. History and management of the fishery for Northern cod in NAFO Division 2J, 3K, and 3L. In: L.S. Parsons and W.H. Lear. Perspectives on Canadian Marine Fisheries Management. Can. Bull. Fish. Aquat. Sci. no. 226. 55-90.

Lilly, G.R. and D.J. Davis. 1993. Changes in the distribution of capelin in Divisions 2J, 3K, and 3L in the autumns of recent years, as inferred from bottom trawl catches and cod stomach examinations. NAFO SCR. Doc 93/54. No. 2237. 14 p.

Lilly, G.R., P.A. Shelton, J. Brattey, N. Cadigan, E.F. Murphy, D.E. Stansbury, M.B. Davis, and M.J. Morgan. 1998a. An assessment of the cod stock in NAFO divisions 2J+3KL. NAFO, N3037, SCR Doc. 98/46:67-94.

Lilly, G.R., P.A. Shelton, J. Brattey, N. Cadigan, E.F. Murphy, D.E. Stansbury, M.B. Davis, and M.J. Morgan. 1998b. An assessment of the cod stock in NAFO Divisions 2J+3KL. CSAS Res. Doc. 98/15.

MacKenzie, D. 1995. The cod that disappeared. New Scientist 147(1995): 24-29.

May, R.M. 1974. Stability and Complexity in Model Ecosystems. Princeton Univ. Press, Princeton, NJ. 265 p.

Mertz, G., J.A. Helbig, and E. Colbourne. 1994. Revisiting Newfoundland capelin (*Mallotus villosus*) recruitment: Is there a wind effect? J. Northw. Atl. Fish. Sci., 17: 13-22.

Mertz, G. and R.A. Myers. 1994. The ecological impact of the Great Salinity Anomaly in the northern North-west Atlantic. Fish. Oceanogr. 3: 1-14.

Mitchell, E. and R.R. Reeves. 1983. Catch history, abundance, and present status of Northwest Atlantic humpback whales: Workshop on historical whaling records. Sharon, Massachusetts, September 12-16, 1977, Int. Whaling Comm. Spec. Issue, No. 5: 153-212.

Mohn, R. 1999. The retrospective problem in sequential population analysis: an investigation using cod fishery and simulated data. ICES J. Mar. Sci. 56: 473-488.

Morgan, M.J., W.B. Brodie, W.R. Bowering, D. Maddock Parsons, and D.C. Orr. 1998. Results of data conversions for American plaice in Div. 3LNO from comparative fishing trials between the Engel otter trawl and the Campelen 1800 shrimp trawl. NAFO SCR. Doc. 98/70. Ser. No. 3062. 10 p.

Morgan, M.J., W.B. Brodie, and G.T. Evans. 1995. Assessment of the American plaice stock in NAFO Subdiv. 3Ps. DFO Atl. Fish. Res. Doc. 95/36.

Morgan, M.J. and S.J. Walsh. 1997. Observations on maturation, recruitment and spawning stock biomass in yellowtail flounder on the Grand Bank. NAFO Sci. Counc. Res. Doc. No 97/73. 7 p.

Munn, W.A. 1938. A Harbour Grace History. Chapter Seventeen – the Sixties. The Newfoundland Quarterly. 37:13-16.

Murawski, S.A., J.-J. Maguire, R.K. Mayo, and F.M. Serchuk. 1997. Groundfish stocks and the fishing industry. In: J. Boreman, B.S. Nakashima, J.A. Wilson, and R.L. Kendall (eds.). Northwest Atlantic Groundfish: Perspectives on a Fishery Collapse. Am. Fish. Soc. Washington, DC. 27-69.

Myers, R.A., N.J. Barrowman, J.A. Hutchings, and A.A. Rosenberg. 1995. Population dynamics of exploited fish stocks at low population levels. Science (Washington) 269 no. 5227: 1106-1108.

Myers, R.A. and N.G. Cadigan. 1995. Was an increase in natural mortality responsible for the collapse of Northern cod? Can. J. Fish. Aquat. Sci. 52: 1274-1285.

Myers, R.A., J.A. Hutchings, and N.G. Barrowman. 1996. Hypotheses for the decline of cod in the North Atlantic. Mar. Ecol. Prog. Ser. 138: 293-308.

Myers, R.A., J.A. Hutchings, and N.G. Barrowman. 1997. Why do fish stocks collapse? An example of cod in Atlantic Canada. Ecol. Applications 7: 91-106.

Myers, R.A., A.A. Rosenberg, P.M. Mace, N.J. Barrowman, and V.R. Restrepo. 1994. In search of thresholds for recruitment overfishing. ICES J. Mar. Sci. 51: 191-205.

Parsons, D.G. and P.J. Veitch. 1998. Status of Northern Shrimp (*Pandalus borealis*) Resources in Areas off Baffin Island, Labrador and Northeastern Newfoundland – Interim Review. CSAS Res. Doc. 98/72.

Parsons, L.S. 1993. Management of Marine Fisheries in Canada. Can. Bull. Fish. Aquat. Sci. 225: 763 p.

Pedersen, S.A. and J.C. Rice. 2001. Dynamics of fish larvae, zooplankton, and hydrographic characteristics in the West Greenland large marine ecosystem. This volume.

Petrie, B., S.A. Akenhead, J. Lazier, and J. Loder. 1988. The cold intermediate layer on the Labrador and Northeast Newfoundland shelves, 1978-86. NAFO Sci. Counc. Stud. no.12: 57-69.

Petrie, B., J.W. Loder, J. Lazier, and S.A. Akenhead. 1992. Temperature and salinity variability on the eastern Newfoundland Shelf: The residual field. Atmosphere-ocean 30: 120-139.

Pimm, S.L. 1991. The Balance of Nature? Univ. Chicago Press. 464 p.

Pinhorn, A.T. 1976. Estimates of Natural Mortality for the Cod Stocks Complex in ICNAF Divisions 2J, 3K and 3L. ICNAF 11:31-35.

Power, D. 1995. Status of redfish in Subarea 2 + Division 3K. DFO Atl. Fish. Res. Doc. 95/25.

Power, D. and D. Maddock Parsons. 1998a. The Status of the Redfish Resource in NAFO Divisions 3LN. NAFO, N3066 NAFO SCR Doc. 98/74.

Power, D. and D. Maddock Parsons. 1998b. An Assessment of Roundnose Grenadier (*Coryphaenoides rupestris*) in NAFO Subareas 2+3 and Catch Information on Roughhead Grenadier (*Macrourus berglax*). NAFO, N3049, NAFO SCR Doc. 98/57.

Prasad, K.S. and R.L. Haedrich, 1993. Satellite observations of phytoplankton variability on the Grand Banks of Newfoundland during a spring bloom. Int. J. Remote Sens. 14:241-252.

Prasad, K.S., J.T. Hollibaugh, D.C. Schneider, and R.L. Haedrich, 1992. A model for determining primary production on the Grand Banks. Cont. Shelf Res.12: 563-575.

Prinsenberg, S.J., I.K. Peterson, S. Narayanan, and J.U. Umoh, 1997. Interaction between atmosphere, ice cover, and ocean off Labrador and Newfoundland from 1962 to 1992 Can. J. Fish. Aquat. Sci. 54, Suppl. 1: 30-39.

Rice, J.C. 2000. Evaluating fishery impacts using metrics of community structure. In: C.E. Hollingsworth (ed.). Ecosystem Effects of Fishing. ICES.J. of Mar. Sci. 57:682-688.

Rice, J.C. and G.T. Evans. 1987. Tools for embracing uncertainty in the management of the cod fishery of NAFO Divisions 2J+3KL. ICES J. Mar. Sci. 45: 73-81.

Robinson, G.A., J.M Colebrooke, and G.A. Cooper. 1973. The continuous plankton recorder survey: plankton in the ICNAF area 1961 to 1971, with special reference to 1971. ICNAF Res. Doc. 73/78. Internat. Comm. N. Atl. Fish. Dartmouth, Canada.

Rose, G.A., B.A. Atkinson, J. Baird, C.A. Bishop, and D.W. Kulka. 1994. Changes in distribution of Atlantic cod and thermal variations in Newfoundland waters, 1980-1992. ICES Mar. Sci. Symp. 198: 542-552.

Shelton, P.A. and G.R. Lilly. In press. Interpreting the decline and collapse of the northern cod stock from survey and catch data. Can. J. Fish. Aquat. Sci. 45 p. + 30 tables.

Sinclair, A.F. 1998. Estimating trends in fishing mortality at age and length directly from research survey and commercial catch data. Can. J. Fish. Aquat. Sci. 55: 1248-1263.

Sinclair, A.F. and S.A. Murawski. 1997. Why have groundfish stocks declined. In: J. Boreman, B.S. Nakashima, J.A. Wilson and R.L. Kendall (eds.). Northwest Atlantic Groundfish: Perspectives on a Fishery Collapse. Am. Fish. Soc. Washington, DC. 71-93.

Stansbury, D.E. 1997. Conversion factors for cod from comparative fishing trials for the Engel 145 otter trawl and the Campelen shrimp trawl used on research vessels. NAFO SCR Doc. 97/73, Ser. No. 2907: 10 p.

Stansbury, D.E., P.A. Shelton, E.F. Murphy, G.R. Lilly, and J. Brattey. 1998a. Scientific Council Meeting – June 1998, An Assessment of the Cod Stock in NAFO Divisions 3NO. NAFO, N3057, NAFO SCR Doc. 98/65.

Stansbury, D.E., P.A. Shelton, J. Brattey, E.F. Murphy, G.R. Lilly, N.G. Cardigan, and M.J. Morgan. 1998b. An assessment of the cod stock in NAFO Subdivision 3Ps. CSAS Res. Doc. 98/19.

Stenson, G.B, M.O. Hammill, and J.W. Lawson. 1995. Predation of Atlantic cod, capelin, and Arctic cod by harp seals in Atlantic Canada. NAFO SCI. Counc. Res. Doc., 1 95/95, 17 p.

Stenson, G.B., M.O. Hammill, and J.W. Lawson. 1997. Predation by harp seals in Atlantic Canada: Preliminary consumption estimates for Arctic cod, capelin and Atlantic cod. J. Northw. Atl. Fish. Sci. 22:137-154.

Strong, D.R. Jr., D. Simberloff, L.G. Abele, and L.A. Thistle (eds.). 1984. Ecological Communities: Conceptual Issues and the Evidence. Princeton University Press. Princeton, NJ. 613 p.

Taggert, C.T. and 10 co-authors. 1994. Overview of cod stocks, biology, and environment in the Northwest Atlantic region of Newfoundland, with emphasis on Northern cod. ICES Mar. Sci. Symp. 198: 140-157.

Templeman, W. 1966. Marine Resources of Newfoundland. Bull. Fish. Res. Board of Can. #154. 171 p.

Therriault, J.-C., B. Petrie, P. Pepin, J. Gagnon, D. Gregory, J. Helbig, A. Herman, D. Lefaivre, M. Mitchell, B. Pelchat, J. Runge, and D. Sameoto. 1998. Proposal for a northwest zonal monitoring program; Can. Tech. Rep. Hydrogr. Ocean Sci. No. 194, 64 p.

Walsh, S.J., D. Orr, and W.B. Brodie. 1998a. Conversion factors for yellowtail flounder survey indices derived from comparative fishing trials between the Engel 145 otter trawl and the Campelen 1800 shrimp trawl. NAFO SCR Doc. 98/60: 10 p.

Walsh, S.J., W.B. Brodie, M. Veitch, D. Orr, C. McFadden, and D. Maddock Parsons. 1998b. An assessment of the Grand Bank yellowtail flounder stock in NAFO Divisions 3LNO. NAFO SCR Doc. 98/72.

Walters, C. and J.-J. Maguire. 1996. Lessons for stock assessment from the northern cod collapse. Reviews in Fish Biology and Fisheries 6: 125-137.

Warren, W. 1997a. Report on the comparative fishing trials between the Gadus Atlantica and the Teleost. NAFO Sci. Coun. Stud. 2:81-92.

Warren, W. 1997b. Changes in the within-survey spatio-temporal structure of the northern cod (*Gadus morhua*) population, 1985-1992. Can. J. Fish. Aquat. Sci.. 54, Suppl. 1: 139-148.

Large Marine Ecosystems of the North Atlantic
K. Sherman and H.R. Skjoldal (Editors)
© 2002 Published by Elsevier Science B.V.

4

Decadal Changes in the Scotian Shelf Large Marine Ecosystem

K.C.T. Zwanenburg, D. Bowen, A. Bundy, K. Drinkwater,
K. Frank, R. O'Boyle, D. Sameoto, and M. Sinclair

ABSTRACT

The eastern and western portions of the Scotian Shelf were treated separately based on differences in physical oceanographic characteristics and differences in fish and other faunal compositions. The central and western shelf are warmer, experienced a cold period during the 1960s, and then warmed to above average until 1998 when an intrusion of cold Labrador Shelf water caused significant cooling. The eastern shelf cooled from about 1983 to the early 1990s and bottom temperature has remained below average since then. This cold period is associated with increased abundance of cold water fish (capelin, turbot) and invertebrates (snow crab, shrimp) usually more prevalent in colder Gulf of St. Lawrence and Newfoundland waters. Colder temperatures are also implicated in reductions in growth rates of demersal fishes. Phytoplankton greenness and total copepod indices are higher and show peaks earlier in the year during the early 1990s than during the 1960s. An increase in the abundance of arctic copepod species also accompanied the cooling of the eastern shelf.

Trawlable demersal biomass has declined on both portions of the shelf but most rapidly on the eastern shelf since the early 1980s. However, total finfish biomass (corrected for relative catchabilities) shows a variable increase since the mid-1980s. Pelagic fish biomass has increased, and the ratio of demersal to pelagic biomass has decreased, especially in the east. Average size of demersal fishes decreased by 60 to 70% since 1970 in both systems, and both have shown significant changes in the integrated community size frequency (ICSF) with a reduction in numbers of large fish. Fishing effort, which increased rapidly following the 1977 establishment of Canada's 200-mile exclusive economic zone, was negatively correlated with community size structure. A reduction in effort in the early 1990s is associated with an apparent reversal in the trend of decreasing ICSF. The decrease in ICSF was more rapid in the colder eastern shelf and was considered to be related to differences in growth rates between the two systems. Grey seal abundance has increased exponentially since the 1960s, especially on the eastern shelf; however, the impact of this increase on levels of predation remains uncertain due to the paucity of comparative diet information prior to the mid-1980s.

Human exploitation of these systems parallels biomass trends with declines in demersal landings since the early 1980s, leading to a virtual fishery closure on the eastern shelf in the early 1990s and significant reductions in allowable removals on the southern shelf. Management of these systems has been under solely Canadian jurisdiction since 1977. The dominance of fishery development objectives over conservation objectives has resulted in documented over-exploitation of fish resources. Collateral effects of fishing have to date been implicitly discounted. Although the concept of large marine ecosystems is appealing, ecosystem level management must first define operational objectives and ecosystem level metrics by which attainment of management objectives can be evaluated. Monitoring programs to track the dynamics of these metrics are essential. Furthermore, an effective body for the integration of the many single function management agencies now in operation must be established. The lack of long-term data for many of the trophic levels within the system (primary/secondary production, benthic invertebrate composition and abundance) and the rudimentary understanding of the connections both between these levels and with higher trophic levels indicate the need for continued expenditures on monitoring and research.

INTRODUCTION

The concept of large marine ecosystems (LMEs) as developed by Sherman and others (see for example Sherman and Tang 1999; Alexander 1989) provides a spatial and temporal framework within which to develop effective monitoring and management approaches. At present, some 50 LMEs have been defined which encompass about 95% of annual global fisheries landings (Sherman and Duda 1999). The boundaries of the Scotian Shelf LME, as envisioned by (Sherman *et al.* 1996), are not sharply defined but correspond roughly to the Scotian Shelf proper and the Southern Gulf of St. Lawrence, and exclude the Bay of Fundy on a line running roughly from Yarmouth south to the Fundian Channel. The Scotian Shelf LME is bounded by the Newfoundland Shelf LME to the north and by the Northeast U.S. Continental Shelf LME to the south.

An analysis of demersal fish of the Scotian Shelf (Mahon 1997) indicates that there are assemblages unique to the eastern and western portions of the shelf as well as those that span the entire shelf. This makes the decision of whether to treat the shelf as a single system or two systems problematic. This analysis also shows that the demersal fish assemblage of the Bay of Fundy is more closely allied to that of the eastern shelf than the adjacent southwestern shelf. The southern Gulf of St. Lawrence marks a transition zone for pinnipeds. It is the southern limit for northern species like harp (*Phoca groenlandica*) and hooded seals (*Cystophora cristata*) that occupy the Gulf of St. Lawrence and Grand Banks in winter. Southern species like the harbour (*Phoca vitulina*) and grey seals (*Halichoerus grypus*) have the center of their distribution in the southern Gulf and further south. This is also true for cetaceans – belugas are present in the Gulf and further north but not in the south. In this paper, we do not consider the Southern Gulf of St. Lawrence as a part of the Scotian Shelf LME proper. Although the Bay of Fundy has been included, its association to the shelf proper is somewhat problematic. Because of the differences in demersal fish assemblages and the fact that there are significant hydrographic differences between the northeast shelf and the central to southwest shelf, we analyze them separately. This decision is supported by the findings of Mahon and Sandeman (1985) who showed transitions in benthic community composition at Halifax and Cape Cod.

Brown *et al.* (1996) examined trawl survey data for the continental shelf of the east coast of North America from Cape Hatteras (NC, USA) to Cape Chidley (Nfld, Canada) and showed that some assemblages transcend the boundaries of the Scotian Shelf LME. This indicates that in the case of fishes, the present boundaries are approximations at best. Temporal analysis of this same data set also shows that the location of the assemblage boundary between the Scotian Shelf and the adjoining Newfoundland Shelf LME, shifts in response to changing environmental conditions.

We present an overview of the information available on long-term (decadal or longer) changes in physical oceanographic environment, biological oceanography, finfish communities, and their intensity and patterns of exploitation, and the changes in marine mammal populations, which have occurred within the Scotian Shelf LME. We give an overview of an ecosystem level modeling exercise presently underway to explore data deficiencies and trophic interactions within these systems. We also discuss the longer-term history of exploitation and management of the shelf and how fisheries management objectives in particular have influenced this exploitation. Finally we discuss the requirements for, and progress toward, management of human activities in the context of ecosystem level effects.

PHYSICAL OCEANOGRAPHIC FEATURES OF THE SCOTIAN SHELF

Circulation

The continental shelf off the coast of Nova Scotia is characterized by complex topography consisting of numerous offshore shallow (< 100 m) banks and deep (> 200 m) mid-shelf basins (Figure 4-1). It is separated from the southern Newfoundland Shelf by the Laurentian Channel and borders the Gulf of Maine to the southwest. The shelf width varies from approximately 130 km in the southwest to 230 km in the northeast. The surface circulation is dominated by a general southwestward flow interrupted by clockwise movement around the banks and counterclockwise around the basins with the strengths varying seasonally (Figure 4-1; see also Han *et al.* 1997). Low salinity waters from the Gulf of St. Lawrence discharge onto the Scotian Shelf on the south side of the Cabot Strait. Part of this flow rounds Cape Breton to form the southwestward moving Nova Scotia Current on the inner half of the shelf. The remainder of the flow continues along the Laurentian Channel, turning at the shelf break. It continues southwestward, temporarily diverting inshore at the western end of Western Bank, joining the anticlockwise movement around Emerald Basin and then moves back to the shelf break

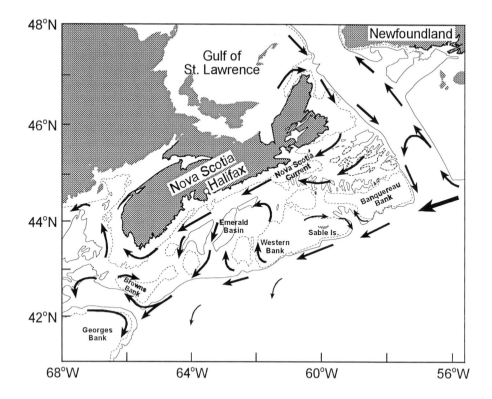

Figure 4-1. Scotian Shelf topography showing the general near surface circulation.

at a point offshore of LaHave Bank. This water eventually enters the Gulf of Maine through the Northeast Channel while the Nova Scotia current enters closer inshore.

Off the Scotian Shelf lies the slope water region with a relatively slow anticlockwise circulation pattern caused by interactions of the northeast-flowing Gulf Stream at its offshore edge and the southwest-flowing shelf currents. Large and variable meanders develop along the Gulf Stream and occasionally these meanders break off and separate from the Stream to form eddies. The eddies which develop on the north side of the Stream rotate clockwise, entrapping warm Sargasso Sea water from south of the Stream in their centers, giving rise to the term "warm-core" rings. They extend from the surface to depths in excess of 1000 m and they normally move slowly westward through the slope water region. Rings that "bump up" against the continental shelf cannot move into shallower waters because of their depth. They do, however, draw off large amounts of water from the shelf into the slope water region with compensating flows of slope water onto the shelf occurring, usually through the gullies and channels. These eddies play an important role in the heat and salt budgets of the shelf (Smith 1978) and may be a cause for decreased recruitment in several important commercial fish stocks through transport of eggs and larvae off the shelf (Myers and Drinkwater 1989).

Average Hydrographic Conditions

Hydrographic properties on the Scotian Shelf vary spatially due to the complex bottom topography, transport from the Gulf of St. Lawrence, and exchange with the adjacent, offshore slope waters. They tend to be characterized by moderate annual mean temperatures, relatively low upper-layer salinities and strong summer stratification. Mean annual surface temperatures on the Scotian Shelf range from 6°-10°C, with large seasonal changes (range of 16°C) that reflect seasonal variability in atmospheric heat fluxes (Petrie *et al.* 1996). Indeed, the amplitude of the annual cycle in sea surface temperature in the region from the Laurentian Channel to the Middle Atlantic Bight is the largest anywhere in the Atlantic Ocean (Weare 1977) and one of the largest in the world. Mean surface salinities range from 30-32 ppt. but also undergo a large seasonal variation (range of 0.5-2) due to the influence of the low salinity Gulf waters and the melting and freezing of ice in the northeastern shelf. Surface salinities (both their mean and the seasonal range) increase across the shelf and southwestward along the shelf, again reflecting the effects of low salinity Gulf waters (Petrie *et al.* 1996).

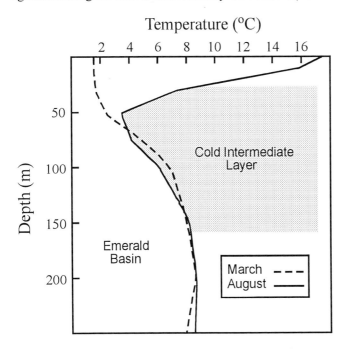

Figure 4-2. Vertical temperature structure in Emerald Basin showing the presence of the Cold Intermediate Layer in summer.

The vertical structure of the water column on the Scotian Shelf also undergoes a large seasonal variation. In winter, strong winds and cool air temperatures result in a large loss of heat from the ocean and the water is vertically mixed down to depths of 50-150 m. In spring and summer, solar heating, together with reductions in salinity due to ice melt and advection, make the surface waters less dense and lead to rapid stratification of the water column, trapping the cold, winter-mixed water below. In many shelf areas, relatively warm water of offshore origin moves in along the bottom beneath the winter-cooled waters. Although warmer, the offshore waters are more saline and hence more dense. Sandwiched between warmer layers, the winter-chilled waters are referred to as the cold intermediate layer (CIL; Figure 4-2). The absolute temperatures within the CIL typically vary from 2-5°C on the Scotian Shelf. On the northeastern Scotian Shelf, bottom topography prevents the warm offshore waters from penetrating inshore and hence waters typical of the CIL (temperatures less than 5°C) extend to the bottom (depths > 200 m). Where depths are shallower than 150 m, there is no warm bottom layer and in areas of strong tidal currents, such as off southwest Nova Scotia, the waters even in summer are vertically well mixed. On the shelf, bottom temperatures vary from 2°-4°C in the northeast to 8°-10°C in the deep basins in the central and southwest. There are little to no seasonal variations at these depths.

Time Trends

Year-to-year changes in water temperatures and salinities on the Scotian Shelf are also among the most variable in the North Atlantic Ocean. The temperature pattern in the deepest reaches of Emerald Basin (250 m) is representative of long-period trends in the deep waters throughout the central and western Scotian Shelf (Figure 4-3). Temperatures were near or above average in the 1950s and declined to approximately 2°C below average in the 1960s. The extended period with the lowest temperatures occurred during the mid-1960s and was related to the replacement of warm slope water by cold Labrador-type slope water along the shelf edge (Petrie and Drinkwater 1993). These slope waters subsequently penetrated onto the shelf through cross-shelf exchange, most likely forced by Gulf Stream eddies or local winds. In the late 1960s, temperatures rose rapidly by 2-3°C as warm slope water returned to the area of the shelf break. Emerald Basin waters generally remained warmer-than-average from the 1970s to 1997. Indeed, the highest sustained temperature anomalies in the approximate 50-year record were observed in the mid-1990s. In 1998, cold Labrador Slope water again appeared with a subsequent lowering of temperatures in Emerald Basin by over 3°C (Drinkwater *et al.* 1998).

In the cold intermediate layer (CIL) waters of the eastern shelf, as represented by the 100 m record on Misaine Bank (Figure 4-4), the amplitude of the long-period temperature trend is smaller (order 1°C) than for the deep waters of Emerald Basin. As previously stated, these CIL type waters extend to the bottom on the northeastern Scotian Shelf. Temperature anomalies during the 1960s in this region were not as low as in the central and southwestern portions of the shelf. From the late-1960s to the mid-1970s, temperatures in the northeast oscillated near or above average. They rose above normal around 1980 but

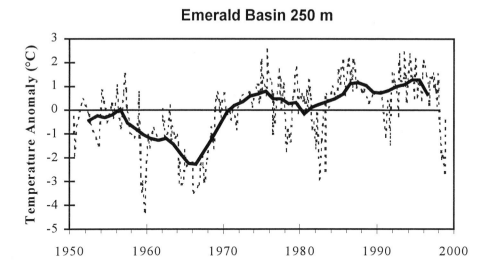

Figure 4-3. Monthly (dashed) and 5-year running means (solid line) of the temperature anomalies at 250 m in Emerald Basin on the central Scotian Shelf.

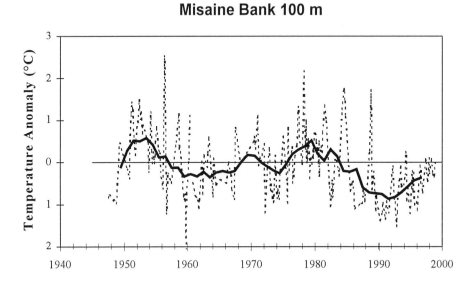

Figure 4-4. Monthly (dashed) and 5-year running means (solid line) of the temperature anomalies at 100 m on Misaine Bank in the eastern Scotian Shelf.

by the mid-1980s, temperatures fell sharply. Below approximately 50 m, temperatures have generally remained colder-than-normal and the coolest in the approximately 50-year record occurred in the early 1990s. In recent years, the waters have been warming and are now approaching normal. The long-term temperature trends over the eastern inshore areas (e.g. Sydney Bight) and offshore banks (e.g. Banquereau) are similar to those in the Misaine area. Cold conditions since the mid-1980s have also been observed in the inshore waters as far south as Lurcher Shoals. The cause of this large-scale cooling is thought to be related to downstream advection, as both the Gulf of St. Lawrence and southern Newfoundland waters experienced similar cold waters during this period. *In situ* cooling on the Scotian Shelf during the winter may also have contributed, however.

BIOLOGICAL OCEANOGRAPHY OF THE SCOTIAN SHELF

There are relatively few data available which describe long-term changes in phytoplankton or zooplankton abundance/production for the Scotian Shelf. Information from the Scotian Shelf Ichthyoplankton Program (SSIP, O'Boyle *et al.* 1984) has been used to compile seasonal distribution maps for *Calanus* spp. (Sameoto and Herman 1990), and distribution gradients of various zooplankton species within the central area of the Scotian Shelf were described for the summer months (Tremblay and Roff 1983). Although these data were spatially and seasonally synoptic in that most of the Scotian Shelf was sampled at various seasons over a number of years, they give no insight into longer-term changes.

The longest time series describing zooplankton and phytoplankton species distribution and abundance on the Canadian eastern continental shelf comes from the Continuous Plankton Recorder survey (CPR) collected for the Allister Hardy Foundation of Plymouth, England. These data were collected from 1959 to 1975 and again from 1991 to the present. Data are collected by commercial and weather ships, which tow a standard sampling device along generally fixed shipping routes at an average depth of 6.7 m (Hays and Warner 1993) and at speeds of up to 20 knots. The routes sampled differed somewhat between the earlier and present time periods. During 1961 to 1975 the route crossed the Gulf of Maine, whereas, during 1991 to the present the route was moved to the south so that it crossed Georges Bank and missed most of the Gulf of Maine. The data for the earlier period were analyzed by Myers *et al.* (1994). We compared CPR data from the two periods with an emphasis on detecting temporal and spatial differences in phytoplankton, and zooplankton abundance on the eastern and western portions of the Scotian Shelf. The eastern and western portions of the shelf were analyzed separately in recognition of the different bottom temperature regimes of the two areas (see Physical Oceanography above, also Sameoto *et al.* 1997). Data from the St. Pierre Bank region, the Gulf of Maine and Georges Bank were also analyzed.

The phytoplankton color index was significantly higher (in all seasons and on both portions of the shelf) in the more recent period, indicating higher concentrations of phytoplankton recently (Figure 4-5). This observation is consistent with observations from the North Sea and Eastern central Atlantic (Reid *et al.* 1998) that indicate that the color index in these areas increased between 1985 and 1995. The reasons for this increase in the phytoplankton color index remains a subject of debate but is likely related to changes in North Atlantic scale weather patterns (Taylor, this volume).

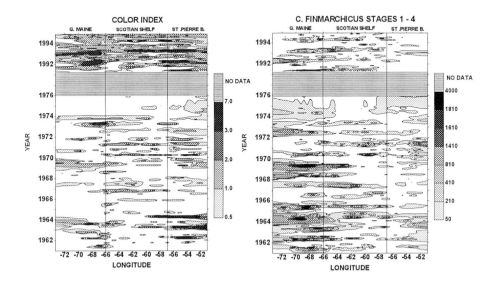

Figure 4-5. Contours of the phytoplankton color index and C. finmarchicus stages 1-4. Data were values for longitude (x-axis) and consecutive month (y-axis).

Seasonal patterns in the color index, and total abundance of copepods also differ between the two time periods (Figure 4-6). Values in all of these indices are higher in the first three to four months of the year in 1990 - 1994 than in 1961 - 1969, indicating a possible shift in the seasonal cycles on both halves of the shelf. It is postulated that the shift in the zooplankton cycle is in direct response to the change in the phytoplankton peak. Such a significant change in the timing of zooplankton abundance may have significant implications for the survival of fish larvae unless the timing of their production shifted coincidentally. The peak abundance of *Calanus* occurred about a month earlier and was appreciably lower in the 1991 – 1994 period than in the 1961 – 1969 period. The krill index was also lower during the 1991 – 1994 period.

We also noted that the concentrations of arctic zooplankton species (*Calanus glacialis*, and *Calanus hyperboreus*) which originate in the Gulf of St. Lawrence, increased significantly between the 1961 - 1976 and the period 1991 - 1994 (Figure 4-7). This influx of colder water species is consistent with the cooling trend observed on the eastern shelf in this period (see Physical Oceanography above).

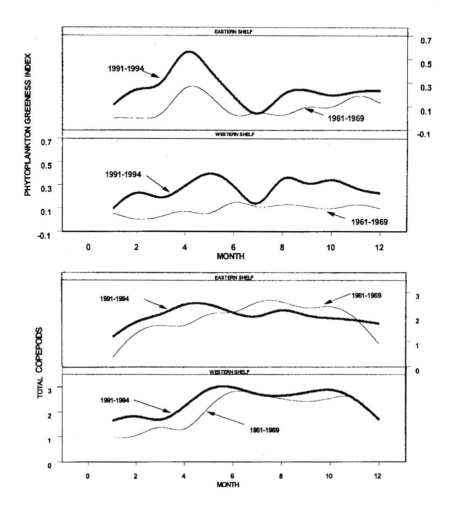

Figure 4-6 (a). Seasonal phytoplankton color and copepod indices for the Eastern and Western Scotian Shelf for two time periods, 1961-1969 and 1991-1994.

The long-term data on primary and secondary production are not extensive for the Scotian Shelf LME. Although additional data exist, these give essentially point estimates of production for a number of areas of the shelf. Where estimates of production have been available, linking these to the production of fish or other trophic levels has been problematic. In the section on ecosystem modeling below, we outline how we intend to use these point estimates of production in mass balance and other models to explore potential changes in overall productivity of the systems over time.

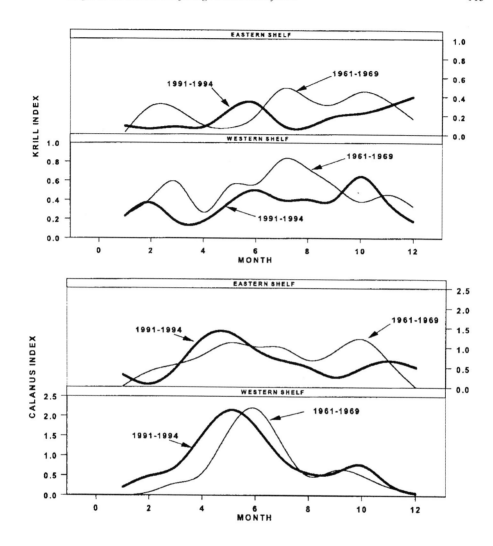

Figure 4-6 (b). Seasonal krill and *Calanus* indices for the Eastern and Western Scotian Shelf for two time periods, 1961-1969 and 1991-1994.

RECENT HISTORY OF THE FISHERIES OF THE SCOTIAN SHELF (1970 - 1997)

Total landings of demersal fish (mainly gadids) on the eastern shelf declined from a maximum of 450,000 metric tons (t) in 1973 to less than 15,000 t in 1997 (Figure 4-8).

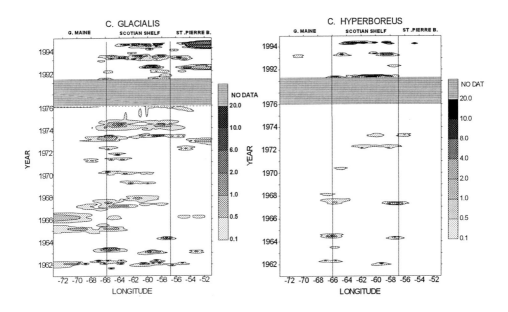

Figure 4-7. Contours for *Calanus glacialis* and *Calanus hyperboreus*. Data were values for longitude (x-axis) and consecutive month (y-axis).

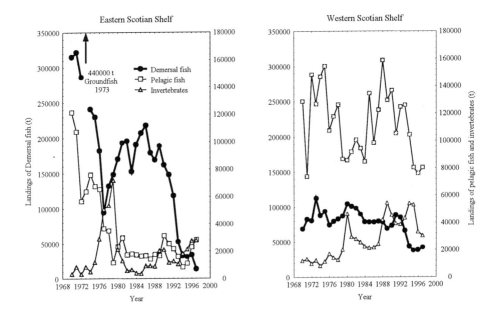

Figure 4-8. Recent history of landings from the eastern and western Scotian Shelf.

A moratorium on fishing, especially for cod, was imposed in 1993 and remains in effect. Longline fisheries for Atlantic halibut (*Hippoglossus hippoglossus*) and white hake (*Urophycis tenuis*), and a large trawler fishery for silver hake (*Merlucius bilinearis*) along the shelf edge, are presently operating on the eastern shelf. Landings of pelagic species (mainly Atlantic herring, *Clupea harengus* and Atlantic mackerel, *Scomber scombrus)* declined from about 120,000 t in 1970 to about 30,000 t in 1997. Landings of herring and mackerel averaged about 20,000 t during the 1980s, indicating a recent increase in landings. Landings of invertebrates (mainly northern short-fin squid, *Illex illecebrosus*) increased rapidly to a maximum of about 75,000 t in 1979. Invertebrate landings then declined substantially to less than 4,000 t in 1985 and have increased slowly to 30,000 t in 1997. This increase is mainly due to increased landings of snowcrab (*Chionoecetes opilio*) and northern shrimp (*Pandalus borealis*). It is noteworthy that both of these invertebrate species prefer cold water and that their increased landings coincide with the cooling of the eastern shelf noted above.

For the western shelf, small pelagic species, mainly Atlantic herring and Atlantic mackerel, dominate the landings which have fluctuated between about 75,000 and 155,000 t from 1970 to 1997. Landings in 1997 reached about 80,000 t. Demersal fish landings varied between 70,000 and 110,000 t from 1970 to 1992 and then declined to approximately 40,000 from 1992 to 1996. The fisheries for the commercially exploited gadids of primary interest, Atlantic cod (*Gadus morhua*), haddock (*Melanogrammus aeglefinus*), and pollock (*Pollachius virens*) remain active on the western shelf although total allowable catches (TACs) are low relative to the documented history of these fisheries. Invertebrate landings showed a variable increase from about 10,000 t in 1970 to over 50,000 t in 1994 and a subsequent decline to about 30,000 t in 1997.

With the decline in the more traditional commercial species since the middle of the 1980s there has been a tendency, in both areas, towards increased landings of formerly less utilized species. These include monkfish (*Lophius americanus*), cusk (*Brosme brosme*), white hake (*Urophycis tenuis*), and several species of skate (*Raja* spp.).

FISH AND INVERTEBRATE COMMUNITIES

There has been a growing interest in determining the effects of fishing on marine ecosystems (e.g. Dayton *et al.* 1995; Bundy 1997; Pauly *et al.* 1998; Bundy *et al.* 2000; Bundy 2001; Pauly *et al.* 2001). Recent works have demonstrated changes in species composition or size structure of demersal fish communities in response to exploitation (Sainsbury *et al.* 1997; Haedrich and Barnes 1997; Bianchi *et al.* 2000; Zwanenburg 2000).

We have divided the Scotian Shelf into eastern and western systems based on the work of Mahon and Sandeman (1985), Mahon and Smith (1989), Mahon (1997), and Mahon *et al.* (1998). We have used the boundary between NAFO Unit areas 4W and 4X (approximated by a north south line at about 63° 30' N) as the shared boundary of these two systems and the Fundian Channel and Laurentian Channel as the southwestern and northeastern boundaries of the two systems respectively (Figure 4-9).

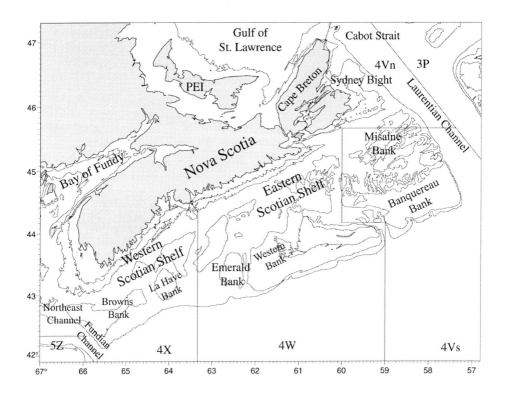

Figure 4-9. Map of the Scotian Shelf showing major submarine features.

The recent commercial (and perhaps biological) collapse of cod (*Gadus morhua*) on the eastern shelf (Fanning *et al.* 1996), and the significant declines in abundance for commercially exploited species on both portions of the shelf (Zwanenburg 2000), indicate that the limits of exploitation have been reached.

Systematic collection of fisheries statistics for the Scotian Shelf goes back to the late 1800s but systematic fishery independent trawl surveys of the shelf date to only 1970. The results of trawl surveys of the Scotian Shelf from 1970 to the present (Halliday and Koeller 1980), are used to describe temporal patterns of species composition and diversity, changes in population numbers and biomass, and trajectories of mean body size for all species. We also examine the changes in integrated community size structure and the relationship between these ecosystem metrics and exploitation. Zwanenburg (2000) gives a detailed description of these analyses.

Changes in Biomass

Trawlable demersal biomass in both systems is near the lowest observed in 30 years (Figure 4-10). On the eastern shelf, biomass declined by 80% from the early 1980s. Biomass in the western system has declined somewhat less precipitously due to increased biomass of spiny dogfish (*Squalus acanthias*). This species, which has a population that occupies both sides of the Canada and U.S. boundary, is heavily exploited in U.S. waters but virtually unexploited in Canadian waters. Declines in target biomass (defined as the biomass of those species actually landed and sold) are evident in both systems.

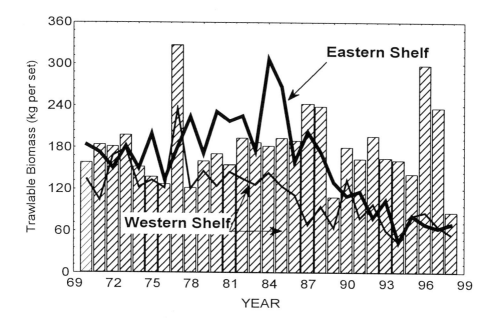

Figure 4-10. Trends in trawlable biomass for demersal fish species. The heavy line represents the trajectory for all demersal species on the eastern shelf, while the bars show this trend for the western shelf. The lighter line shows the biomass trajectory for the western shelf excluding dogfish (*Squalus acanthias*).

On the eastern shelf, Atlantic cod accounted for a long-term average of about 28% of the total biomass. At present, cod makes up about 11% (Figure 4-11). On the western shelf, spiny dogfish represents an increasing proportion of trawlable demersal fish biomass. Presently dogfish make up nearly 50 % of trawlable biomass.

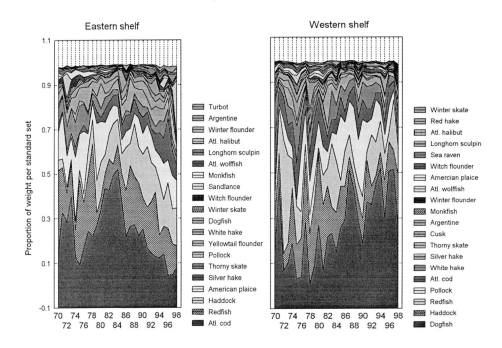

Figure 4-11. Relative species composition (by weight) of demersal finfish biomass for the eastern and western Scotian Shelf for the period 1970 – 1998.

The recent increase in the prevalence of sandlance (*Ammodytes dubius*) on the eastern shelf (Figure 4-12) is notable, especially relative to the abundance of cod, one of its main predators. In this case, the removal of cod biomass is correlated with an increase in sandlance biomass which then becomes available to other predators (see Marine Mammals below).

The foregoing discussion refers only to trawlable biomass of demersal species. A more complete description of changes in fish biomass would include both pelagic and demersal species and would convert trawlable biomass to estimates of true biomass. Since not all species are equally available or catchable by the survey gear, species and size-specific catchability correction factors must be applied to convert trawlable to true biomass. Not applying catchability factors makes estimating changes in relative biomass between species, and the construction of integrated biomass indices, problematic. Catchabilities vary by species and within species by body size, and have not been widely estimated. We derived a series of species-specific catchability correction factors from the work of Edwards (1968). We extended these estimates to other species by considering similarities in morphology and known behavioral patterns (e.g. burrowing behavior, or inhabiting untrawlable areas). In the absence of more

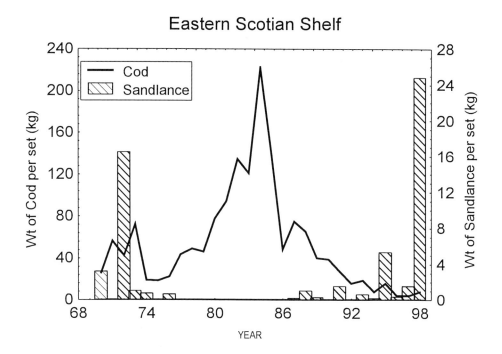

Figure 4-12. Trajectories of cod (*Gadus morhua*) and sandlance (*Ammodytes dubius*) biomass on the eastern shelf.

rigorous work using comparative catch rates of species between a number of gears fished simultaneously, these estimates will necessarily remain subjective. Another avenue of research is to use catchability correction factors derived from sequential population analysis to condition estimates for species of like morphology and behavior. This work is being planned under the ecosystem modeling umbrella (see below).

Applying our crude set of catchability coefficients, we see that the trend in total fish biomass shows a different trajectory than trawlable demersal fish biomass with a gradual increase until at least 1995 (Figure 4-13). Given the decline in demersal fish biomass (Figure 4-10), this indicates that the ratio of demersal to pelagic biomass has shown a steady decline on the eastern shelf since the early 1980s. Although one might argue that these two figures are not strictly comparable, since one contains demersal biomass only, the contribution of uncorrected pelagic biomass to Figure 4-10 would not be detectable at the scale measurement shown. On the western shelf, this decline is not apparent until 1990 (Figure 4-14). The most rapid shifts in this ratio have occurred in both systems since 1990.

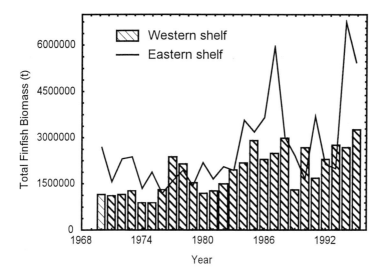

Figure 4-13. Estimates of total finfish (demersal and pelagic) biomass for the two portions of the Scotian Shelf. Estimated biomass for each species was adjusted by a catchability factor based on the work of Edwards (1968).

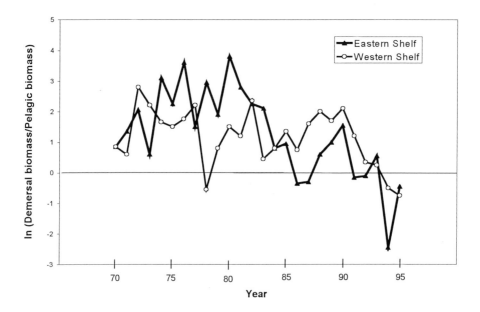

Figure 4-14. Trends in the ratio (ln) of demersal and pelagic biomass for the two portions of the Scotian Shelf.

Changes in Average Weight

The average size of a demersal fish (top 60 species caught by the trawl survey, combined) in these two systems declined significantly between 1970 and 1995. For the eastern shelf (Figure 4-15), this decline was more or less continuous between 1970 and 1995, over which time the average weight decreased by 66%. On the western shelf, average size appeared to be stable or even increasing between 1970 and 1983; however between 1983 and 1995, average size declined by 70%. With the exception of two years, the average weight of a fish on the western shelf has been consistently higher than on the eastern shelf.

Declines in aggregate mean size on both the eastern and western shelf are coincident with increasing fishing effort for the period 1977 to 1995. From 1995 to 1998, the trend is toward stable or increasing mean weights in both systems (Figure 4-15). This trend in increasing mean weights is coincident with the imposition of a moratorium on fishing for cod on the eastern shelf, and a significant reduction in total allowable catches on the western shelf.

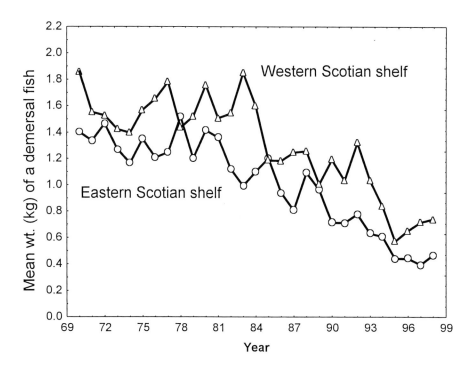

Figure 4-15. Trends in average weight of demersal fish (all species) since 1970 for the eastern and western portions of the shelf.

Changes in Integrated Community Size Frequency (ICSF)

The integrated community size frequency (ICSF) was estimated as the density (numbers per hectare) of all fish by size class (without regard to species). A comparison (Figure 4-16) of ICSFs shows that the eastern assemblage is composed of larger numbers of small fish (< 40 cm) than the western, while the latter houses larger numbers of larger fish (> 40cm). This may be a reflection of warmer bottom temperature regimes and higher growth rates in the western system. The slope of the descending limb of the ICSF is an integrated measure of growth and mortality for the component fish species. The slope was calculated for fish between 35 cm and 85 cm. Using only these length groups avoids the complications of incomplete catchability by the survey gear at smaller and larger size classes. These sizes represent the linear portion of the descending limb of the ICSF (Figure 4-16). We estimated the value of this slope for each portion of the shelf for each year of survey data.

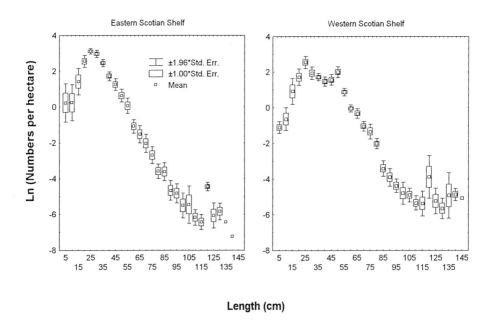

Figure 4-16. Integrated community size frequency (ICSF) for the eastern and western Scotian Shelf. Each point represents the average number of demersal fish per hectare caught at length over the 29 survey years. Note the overall difference in size composition for the two portions of the shelf. The eastern shelf contains larger numbers of smaller fish while the western shelf has larger numbers of larger fish.

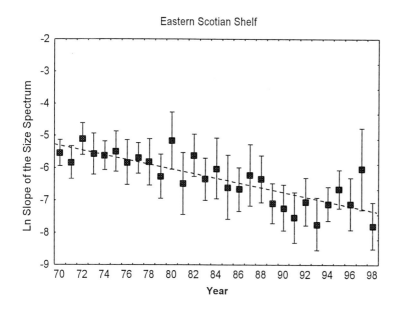

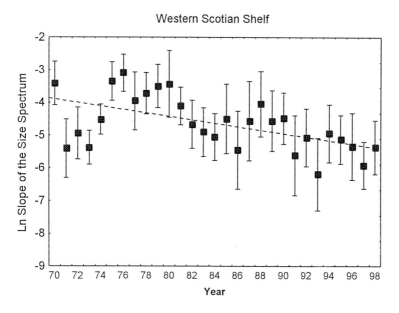

Figure 4-17. Trends in the slopes of the descending limb of the annual integrated community size frequency (ICSF) for the eastern and western Scotian Shelf. The slope was calculated for the linear portion of the descending limb for fish between 35 and 85 cm. The increased steepness of the slope of the ICSF in each system over the study period implies a long-term reduction in numbers of larger fish in both systems.

The slope of the ICSF shows long-term declines (increased steepness) for both portions of the Scotian Shelf (Figure 4-17). These increases in steepness of the ICSF reflect long-term reductions in the number of larger fish in both systems. This is consistent with the observed decreases in average weight of fish caught by the surveys. The slope of the ICSF is consistently steeper for the eastern Scotian Shelf, which may be a reflection of the overall lower temperature and resulting lower growth rates in this area (Zwanenburg 2000).

Reduced numbers of larger fish may have significant impacts on long-term recruitment success. Longhurst (1999) theorises that north-temperate demersal species rely on longevity and large maximum size of spawning females to bridge frequent gaps in recruitment success. These large females ensure a large supply of eggs each year. In years when conditions are not conducive to survival there are sufficient eggs to maintain some reproductive success for the population. In years of high survival rates, these large numbers of eggs provide the basis for the large year-classes that characterize the recruitment dynamics of this and similar species. The observed reduction in number of large fish and current small average size of fish is not conducive to such a strategy. In addition to the absence of large highly fecund cod, Trippel *et al.* (1997) have shown that the egg viability is lower for smaller, first time spawning cod than for older multi-year spawners. The reduced overall size therefore not only produces fewer eggs but less viable eggs. This combination of effects may seriously impair the population's ability to sustain itself or a fishery.

Changes in fish and invertebrate distribution

Changes in the oceanic environment have had measurable effects on several fish and shellfish species. Here we focus upon some responses to the cooling since the mid-1980s in the eastern shelf. Because each fish species or stock tends to prefer a specific temperature range (Scott 1982), long-term changes in temperature can lead to expansion or contraction of its distribution range. These are generally most evident near the northern or southern boundaries with warming (cooling) resulting in a northward (southward) shift. Associated with the cooling in the later half of the 1980s, capelin (Figure 4-18), a cold water species, began to appear in the north-eastern Scotian Shelf annual summer groundfish surveys (Frank *et al.* 1996). Through the intervening years, the abundance of capelin gradually increased and is presently near its peak value on the eastern shelf. These capelin are believed to have come from either the Gulf of St. Lawrence or eastern Newfoundland waters. Initially only adults were caught but over the years juveniles were found, indicating that capelin were successfully spawning on the shelf. The range expansion southward by capelin during cool periods is not unique. Frank *et al.* (1996) documented the arrival of capelin in the Bay of Fundy following the cooling in the 1960s and their disappearance when temperatures rose in the 1970s. Reported capelin catches in the Bay of Fundy prior to the 1960s all corresponded with periods of colder-than-normal water.

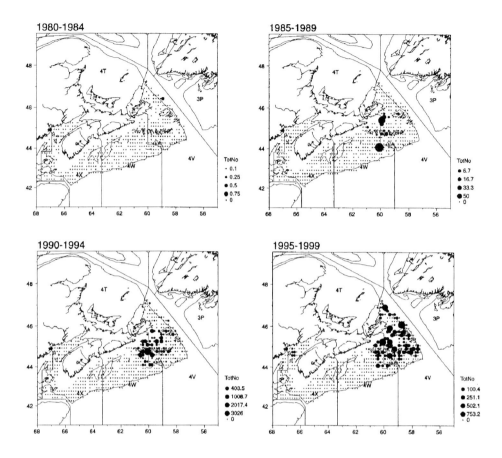

Figure 4-18. Catch per tow in numbers of capelin (*Mallotus villosus*) from summer groundfish surveys averaged over 5-year blocks. Note the increase in numbers from the warm period (1980-1984) to the peak of the cold period (1990-1994).

The distribution of snow crab (*Chionoecetes opilio*) also expanded during the recent cold period (Figure 4-19). It too is a cold water species preferring temperatures of -1° to 3°C. Tremblay (1997) documented the range extension of this species on the north-eastern Scotian Shelf and attributed it to colder temperatures. He also noted that reduced predation from groundfish may have increased the survival and growth of juvenile and adolescent snow crabs. Increases in the catch rates of shrimp (*Pandalus borealis*), another cold water invertebrate, have been observed coincident with the cooling of the eastern shelf (Figure 4-20, DFO 1998).

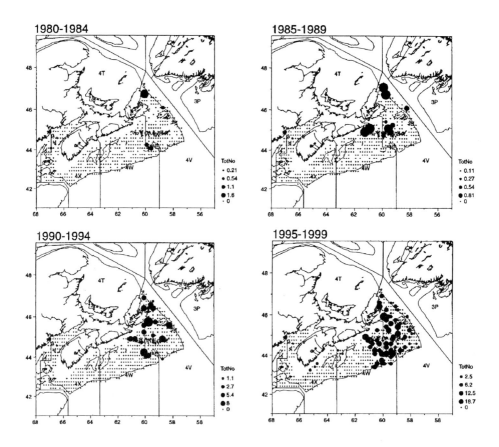

Figure 4-19. Catch per tow in numbers of snow crab (*Chionoecetes opilio*) from summer groundfish surveys averaged over 5-year blocks. Note the increase in numbers from the warm period (1980-1984) to the peak of the cold period (1990-1994).

Growth and Temperature

Growth is also temperature dependent. Recent studies have shown that mean bottom temperature is responsible for 90% of the observed (10-fold) difference in growth rates between different cod stocks in the North Atlantic (Brander 1995). Warmer temperatures lead to faster growth rates. There is a 5-fold difference in the lengths-at-age of cod at 4 years old in the NW Atlantic, the largest from Georges Bank (5Z) and the smallest being the northern cod (2J3KL) off Labrador and Newfoundland. Temperature accounts not only for the spatial differences in growth rates but also temporal patterns. Thus, the low bottom temperatures during the last decade have been shown to be partly responsible for the observed decrease in size-at-age of cod on the eastern Scotian Shelf (Campana *et al.* 1995). Not only cod have been affected; the size-

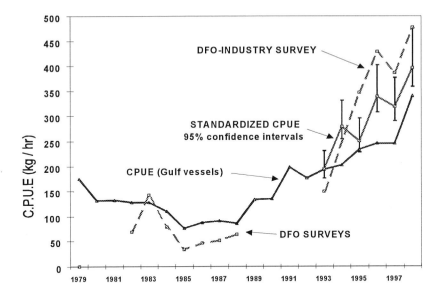

Figure 4-20. Trends in commercial and survey catch rates for northern shrimp (*Pandalus borealis*) on the eastern Scotian Shelf.

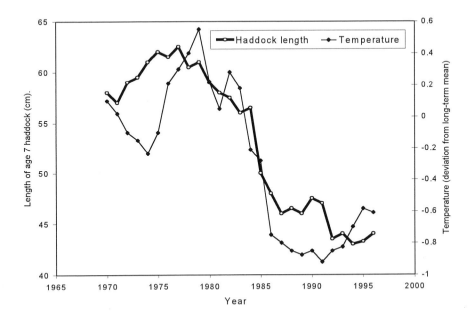

Figure 4-21. Mean length of age 7 eastern Scotian Shelf haddock and 5-year running mean bottom temperature for the eastern Scotian Shelf.

at-age of haddock has also decreased over the same period and is well modelled by changes in temperature (Figure 4-21).

Fishing Effort and ICSF

Cumulative fishing effort was negatively correlated with the slope of the ICSF on both the eastern and western Scotian Shelf (Figure 4-22). Using cumulative effort recognises that effort may have effects beyond the year in which it was applied, for example, by removing a significant proportion of a given year class, thereby creating a minimum in the growth and reproductive potential of the population over the longer term. In this instance, it was assumed that the effects of fishing in any one year would have a decreasing impact over the following 4 years. The reduction in fishing effort during the early 1990s is correlated with decreases in the steepness of the slopes of the ICSF in both portions of the Scotian Shelf.

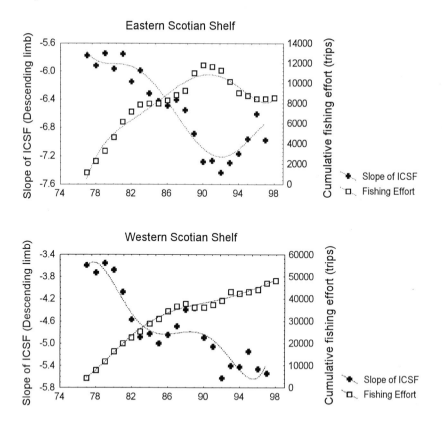

Figure 4-22. Relationship between cumulative fishing effort and slope of the descending limb of the integrated community size frequency (ICSF) for the two Scotian Shelf systems.

ENVIRONMENT VS. EXPLOITATION

In the previous two sections we have described the effects of fishing and changes in environmental conditions on size structure, growth, and distribution of a number of fish and invertebrate species of the Scotian Shelf as if each acted independent of the other. In fact they work in concert. The effects attributed to exploitation are really the effects of exploitation operating against the backdrop of changing distribution patterns and changing growth rates, which are in turn responding to changes in environmental conditions. The effects attributed to environmental changes have been modified by the underlying impacts of exploitation. These effects do not occur in isolation.

Since changing environmental conditions are beyond human control (save perhaps for the levels of greenhouse gas emission), the effects of these forcing factors on not only the exploited species but also on the exploited ecosystem as a whole need to be used to predicate patterns and levels of exploitation. At present neither environmental effects on individual populations nor the collateral effects of fishing on non-target organisms or the physical environment are taken into account in the formulation and implementation of harvest plans.

There are significant difficulties standing in the way of this objective. The relative effects of changes in environmental conditions to fishing (or other anthropogenic impacts) are difficult to quantify. In the previous sections we showed that decreased size at age of a single species (haddock) was highly correlated with decreased ambient temperature on the eastern Scotian Shelf. We also demonstrated that the change in the average size of a suite of exploited species is inversely related to fishing effort. This decrease in size occurred both on the eastern shelf where temperatures decreased in the latter 1980s through the early 1990s, and on the western shelf where temperatures remained relatively stable over this same period. It is noteworthy that the overall rate of decline in integrated community size frequency was more rapid on the eastern shelf than the western shelf despite much higher levels of effort on the latter. This may point to the mitigating impact of the warmer temperatures and higher growth rates of the western shelf. The relative impacts of exploitation and environmental conditions may be illuminated by a more detailed analysis of these differences in response.

A second problem relates to the establishment of ecosystem level metrics by which to judge both the impacts of exploitation and environmental changes. From 1977 to the present the only metric used related to single species (fishable biomass, recruitment, fishing mortality) and of these fishing mortality was paramount. The estimation of the rate of exploitation of target species as evidenced by the fishing mortality rate formed the basis of most management decisions. More recently, consideration has been given to the level of spawning stock biomass (SSB), in that levels of SSB below a predetermined threshold have resulted in the cessation of fishing until such time as the spawner biomass rebuilds. There are, however, no ecosystem level metrics or targets. If management of fisheries is to move beyond the isolationist single species model and take into account the collateral impacts of fishing and changing environmental conditions, it is essential that such metrics and targets be established.

MARINE MAMMALS

Two groups of marine mammals, the Order Cetacea (whales and dolphins), and the Suborder Pinnipedia (seals) both inhabit the Scotian Shelf and southern Gulf of St. Lawrence, including 23 species of cetaceans and 4 species of seals. In the case of cetaceans, the number of species is likely underestimated (Kenney *et al.* 1997). Data on the seasonal abundance of cetaceans are most complete for the southern portions of the region, but even here there are few data on long-term trends (Kenney *et al.* 1997). For most species, these data are based on surveys conducted between 1979 and 1982 during the CETAP program (CETAP 1982). Whitehead *et al.* (1998) provide a brief description of cetaceans on the eastern Scotian Shelf. In the Scotian Shelf and Southern Gulf of St. Lawrence long-term trends in the abundance of marine mammals are available for only several species. The best data are for the grey seal (*Halichoerus grypus*); however, there is also a series of estimates of harp seal (*Phoca groenlandica*).

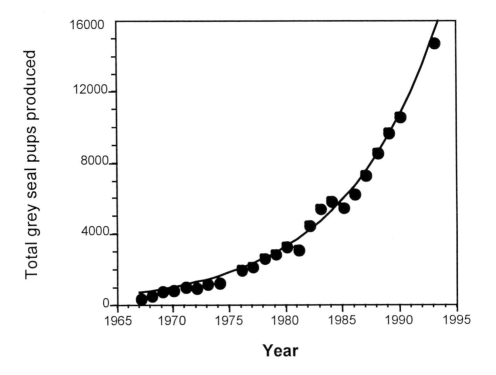

Figure 4-23. Counts of grey seal (*Halichoerus grypus*) pups born on Sable Island since the late 1960's.

The grey seal is a medium body-size member of the Family Phocidae (true seals) found in temperate continental shelves in both the northwest and northeast Atlantic. It is a size-dimorphic species with males weighing about 50% more than females. Although wide-ranging, the species is non-migratory, but rather shows seasonal changes in distribution. The changes are undoubtedly related to seasonal changes in distribution of prey (Bowen *et al.* 1993), but also reflect the aggregation of adults at a handful of land/ice breeding colonies throughout the region (Stobo *et al.* 1990).

An index of grey seal population trends, based on estimated pup production on Sable Island and the southern Gulf of St. Lawrence, covers the period 1962 to 1997 and from the early 1950s to 1997, respectively (Mansfield and Beck 1977; Zwanenburg and Bowen 1990; Mohn and Bowen 1996; Hammill *et al.* 1998; Bowen *et al.* submitted). Over the past three or four decades, the numbers of grey seals on the Scotian Shelf and Southern Gulf of St. Lawrence have increased dramatically. The greatest increase is associated with the Sable Island colony near the edge of the continental shelf in the central Scotian Shelf (Figure 4-23). Population numbers of the Sable Island colony have increased exponentially over the past four decades (Bowen *et al.* submitted). This exponential increase is not unexpected as grey seals in both areas were recovering from low numbers brought about by hunting. Short-term environmental influences are suggested in the Sable Island trends.

Associated with changes in the abundance of grey seals have been changes in the predation mortality exerted by grey seals on commercially harvested and other finfish. The nature of these changes is somewhat uncertain. However, as there is little information on grey seal diet prior to the mid 1980s and the functional form of grey seal predation to changes in prey abundance is unknown, so predicting diet in the absence of diet information is problematic. Estimates of the composition of grey seal diets near Sable Island indicate that sandlance was less common in the diet in the middle 1980s than during the 1990s (Mohn and Bowen 1996). Capelin was absent from estimates of the grey seal diet near Sable Island until 1994, but rapidly became a significant prey item in the diet of grey seals in 1995 and 1996, apparently partially replacing sandlance. Capelin virtually disappeared from the grey seal diet in 1997 and sandlance increased to account for about 80% of the diet (Bowen and Harrison, unpublished data).

Recent data from coastal sites along eastern Nova Scotia also point to changes in prey abundance as the cause of interannual differences in the diet of harbour seals (Bowen and Harrison 1996). Capelin was not identified in the diet from 1988 to 1990, but accounted for about 9% of the diet by wet weight in 1992. This increase in biomass was associated with a marked increase in the local abundance of capelin from about 1 to 9 capelin per tow in 1985 to 1990 to over 90 capelin per tow in 1992 (Frank *et al.* 1996).

Mohn and Bowen (1996) modeled the impact of grey seal predation mortality on Atlantic cod (*Gadus morhua*) on the Eastern Scotian Shelf off Nova Scotia over the period 1967 to 1994. They used two models to describe the response of grey seal predation to changes in cod abundance under two assumptions about the level and pattern of age-specific natural mortality of cod. Under either model (constant ration or proportional ration), grey seal predation mortality was only 10-20% of the estimated mortality caused by the fishery (Figure 4-24). The models generally predicted that future recruitment of age 1 cod to the cod population would be significantly reduced due to grey seal predation, but these predictions were quite sensitive to the assumption that seal predation mortality was additive. It is not known if this assumption is true. Mohn and Bowen (1996) concluded that uncertainty in assessing the impact of grey seal predation on Atlantic cod resulted from: the lack of information on the natural mortality rate of young cod, the nature of the interactions (i.e., additive vs. compensatory) among the components of natural mortality on young cod, and the functional response of grey seals to changes in prey abundance.

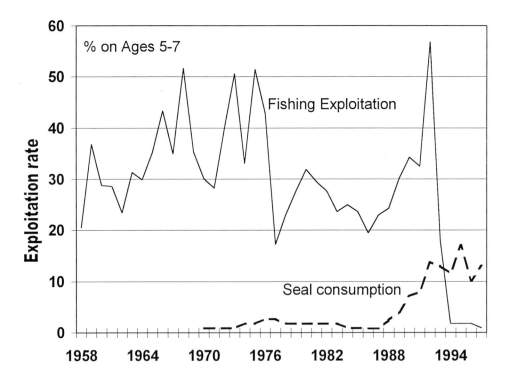

Figure 4-24. Effects of grey seal predation on exploitation of cod (*Gadus morhua*) on the eastern Scotian Shelf.

ECOSYSTEM MODELING

Over the past three to four decades we have observed changes, at times dramatic, in the abundance and distribution of many marine species on the Scotian Shelf. Some demersal fish populations have declined precipitously and others have collapsed completely. This is particularly true of the commercially exploited species. We have shown that these changes have been most dramatic on the Eastern Scotian Shelf. At the same time, the abundance of other species, both invertebrates and small pelagic fishes, has increased (see Section 3 on Fish Communities above). These changes suggest that there have been changes in the trophic structure of these ecosystems. Disentangling the fishery and environmental effects leading to ecosystem change has, so far, been inconclusive. Multidisciplinary and ecological approaches are being undertaken to better understand the changes that have occurred in the Scotian Shelf ecosystems. Specifically, this project will construct models of the structure and function of the Scotian Shelf ecosystems to determine how the physical and biological components of these ecosystems have changed over time and space. These models will then be used in conjunction with other analyses to support fisheries management and a pilot integrated planning initiative on the Eastern Scotian Shelf.

The foregoing synthesis of existing knowledge about the current state and recent changes in Scotian Shelf ecosystems provides the initial data for the comparative analysis of contrasting time periods within and between ecosystems. We have shown that at these spatial and temporal scales there have been changes in the physical environment, the biological environment, and in fishing effort. The project will use several modeling approaches to answer questions about how these ecosystems are structured, their dynamics, how these may have changed over time, and the implication of these changes. The project will consist of a number of components falling under the headings of biomass estimation, estimation of diet composition, and multi-species modeling.

Biomass estimates are fundamental inputs to ecosystem models. For most commercial fish, there exist formal estimates of population size. However, for the many other fish species, such estimates do not exist. Research vessel (RV) surveys provide estimates of trawlable biomass for all species caught by the survey. However, fish vary in their catchability to the RV survey gear, and the trawlable biomass is some unknown proportion of their actual biomass. We have used some initial approximations of catchability to estimate the overall biomass trajectories presented above. In order to obtain more realistic estimates of the biomass of these species, refined estimates of their catchability are required. We will use three strategies to estimate biomass over time, relative estimates, direct estimates and q-adjusted (i.e., correct for catchability) survey estimates.

Biomass estimates for pelagic species such as herring, squid, and mackerel that are not well sampled by the trawl survey, have extensive distributions, and migrate through the systems, are even less reliable. Estimates for the important small forage species (e.g. sand lance, and more recently capelin) also suffer from low catchabilities and high variances. We have noted that little or no information on biomass of primary or secondary production is available. Synoptic information on the abundance and distribution of most benthic macro-invertebrates is also scarce. Given the lack of information for these species, we will use the formal estimates for well studied species like cod, haddock, silver hake, and grey seals, to scale or limit the multi-species models, while working to improve our estimates of the poorly sampled components.

The other major area of poor information is the diets of species on the Scotian Shelf. Very little diet work has been conducted on the Shelf. The two main exceptions are the work by Waldron (1982, 1988) on silver hake in the 1980s, Bowen et al. (1993), and Bowen and Harrison (1994) on grey seals since the 1980s. There is, however, sufficient diet information for most species from adjacent ecosystems (Grand Banks, Gulf of St. Lawrence, Georges Bank) with which to develop a diet that may be representative of the Scotian Shelf. Extensive diet studies have also been initiated for the Shelf.

Mass-balance biomass models such as ECOPATH are being used globally as an efficient and useful method to systematise ecosystems and learn about their properties (Christensen and Pauly 1993; Christensen 1995; Pauly and Christensen 1996; Bundy 1997; Walters et al. 1997; Bundy et al. 2000). Christensen and Pauly (1993) describe three main benefits: (i) the researcher is required to review and standardise all available data on a given system; (ii) states and rates which are mutually incompatible are identified; (iii) species specialists and modellers are placed in a collaborative situation. A fourth benefit is that this approach can highlight the unknowns in systems, even those that are considered relatively well studied. Essentially, mass balance models are a simple approach to represent the complexity of an ecosystem that describes mass-balance trophic interactions; furthermore, they serve as diagnostic tools.

ECOPATH models are being developed for the eastern and western Scotian Shelf. These systems are modeled separately because they have different temperature regimes, bathymetry and have responded differently to perturbation (see Physical Oceanography and Effects of Fishing above). Each ecosystem will be modeled for three time periods: the late 1970s when some fish stocks declined, the early 1980s when the fish stocks had recovered and were buoyant, and the 1990s when the groundfish stocks collapsed. The models will encompass the decadal changes that have occurred in these ecosystems and will be used in a comparative way to understand them in terms of productivity and biomass, and changes in energy pathways. The models encompass the whole biotic system from primary productivity to top predators such as marine mammals.

These models will be useful diagnostic tools for three reasons. First, they will point very clearly to the parts of the ecosystem for which we have poor or no information. Second, they will allow evaluations of sensitivity in relation to different parameters, and thus provide a basis for prioritization of research efforts. Third, the mass balance constraint enables us to see how well the different groups and their parameters fit together. It may be that some rates are simply incompatible. This will force us to either re-examine interactions within the model or to re-examine various parameters. Regardless, we end up with a greater understanding of the ecosystems and focus our understanding in our knowledge gaps.

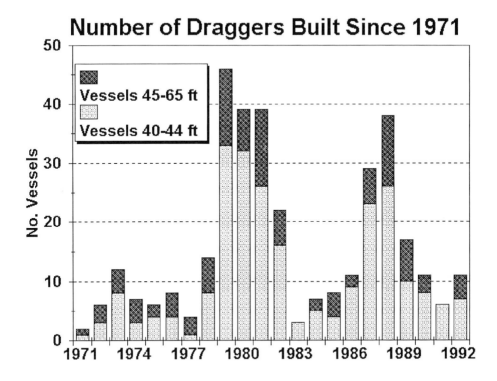

Figure 4-25. Post-1977 development of fishing effort on the Scotian Shelf.

GOVERNANCE OF THE SCOTIAN SHELF

The Scotian Shelf ecosystem has been exploited for some thousands of years. However, the most significant intensification of this exploitation started with the arrival of European interests in the 15th century. Their arrival saw the commercialization of the fishery (especially cod) which Great Britain, France and Spain used as a commodity in trade (Innis 1954). Since then, commercialization has continued and accelerated with the development of increasingly efficient extractive technologies and distribution of the extracted materials to ever-expanding markets. The expansion of these technologies and consumptive markets, especially the development of distant water international fleets in the post World War II period, increased the level of exploitation of the Scotian Shelf fisheries to the point of collapse in the mid 1970s.

The fisheries of the Scotian Shelf in the post-1945 era were regulated under the auspices of the International Commission for the Northwest Atlantic Fisheries (ICNAF) consisting of the industrialized fishing nations of the world operating in this area. The effectiveness of this organization to regulate was, however, limited by the voluntary nature of compliance to its rules. This lack of regulatory control and the limited development of the Canadian domestic fleet prompted the establishment of the Canada's 200-mile exclusive economic zone in 1977. Regulation of the Scotian Shelf fisheries was thus brought under Canadian jurisdiction.

Management of the post-1977 fisheries had at least two objectives, the development of the Canadian domestic fishing fleet and conservation. The conservation objectives were to rebuild the depleted stocks and to prevent subsequent "growth" and "recruitment" overfishing. Growth overfishing results when annual fishing mortality exceeds the maximum growth potential of the exploited population. Recruitment overfishing results when exploitation reduces or impairs the reproductive potential of the exploited resource. The latter is considered the most serious form of over-exploitation. Rogers (1995) has pointed out the essentially contradictory nature of these dual objectives and the influence that this has had on the exploitation of the Scotian Shelf over the past 25 years. The post-1977 expansion of the Scotian Shelf fishing fleet is testament to the fulfillment of the development objective (Figure 4-25), while the over-exploitation of fish resources leading to a number of fishery closures in the early 1990s attests to the failure of achieving the conservation objective.

Following the closure of a number of cod fisheries on the eastern shelf in 1993 two workshops were held to evaluate why fisheries conservation objectives for the Scotian Shelf had not been met (Angel *et al.* 1994), and to recommend improvements (Burke *et al.* 1996). Angel *et al.* (1994) concluded that conservation objectives for the period 1977 – 1993 had not been met for a number of reasons. These are as follows:

1. Conflicts between conservation and economic performance / employment objectives resulted in trade-offs favouring the latter. This included the setting of total allowable catch (TAC) levels higher than the scientific advice. Landings also exceeded TACs by allowing fisheries to continue on a "by-catch" basis when the quota for that species in a stock area had been taken, but while quota remained for other species.

2. Scientific advice frequently overestimated stock size, resulting in excessive TACs that were not limiting fishing mortality.

3. The use of single-species TACs in multi-species fisheries led to illegal behaviours such as mis-reporting (including non-reporting and reporting to adjacent stock areas), high-grading (selecting only the most highly values species or size classes), and dumping (usually of low value species / size classes). These practices compromised catch estimates and the accuracy of resource predictions.

4. The failure to protect the reproductive capacity of individual spawning components, and a lack of consideration for minimum spawning stock requirements for the population as a whole, which led to recruitment overfishing.

5. The assumption that economic constraints would restrain fishing effort prior to recruitment overfishing (i.e. that the cost of fishing at low stock abundance would become prohibitive before biological resilience was compromised), which was not borne out.

6. The refinement of harvesting technology and the expansion of domestic fishing effort were not effectively controlled and were likely encouraged by access to low interest loans, income subsidies (in the form of seasonal unemployment insurance), and optimistic forecasts of fish stock abundance (Canada 1981). In essence there was a shift in the open access equilibrium point, as a result of increased technological efficiency and government subsidization, to the point where these fisheries remained profitable, at high effort levels, to the point of stock collapse (Sinclair *et al.* 1997).

Rogers (1995) maintains that the market economic paradigm under which current fisheries management operates will always favor development and exploitation over conservation, and the many short-term trade-offs in favour of economic performance over conservation support this contention, at least from 1977 – 1993. The inability to meet conservation objectives undermined the sustainability of these commercial fisheries.

A second workshop (Burke *et al.* 1996) formulated a total of 32 recommendations meant to address the issues identified by Angel *et al.* (1994). These recommendations called for improvements to many aspects of the fisheries management structure, the most significant of which are outlined below.

1. That decision-making and management planning decisions be delegated to those most concerned with the management plans under consideration, with a clear division of responsibility between users and government.

2. That single species quota management continues as the core regulatory measure with the proviso that it be augmented with the use of closed areas, effort monitoring and gear restrictions. Effort monitoring would allow management to use the historical proportionality between fishing effort (days or hours fished) and fishing mortality, to estimate the number of days each fleet would require to catch their portion of the TAC. If the estimated total fishing effort were expended prior to the quota being taken, this would indicate either significant misreporting, or discarding, or overestimation of resource abundance.

3. That minimum spawning stock biomass levels be estimated for each management unit (stock) and that these limits, coupled with the maintenance of seasonal spawning and juvenile rearing area closures, would prevent recruitment overfishing and protect individual spawning components within a management unit.

4. That existing shares of resources between industry sectors be perpetuated to provide incentives for the industry to meet conservation objectives. This includes an evaluation

of whether enterprise allocations and individual transferable quota schemes (which impart quasi ownership rights to fishers) have reduced overall capacity or resulted in more conservation oriented fishing practices.

5. That catch and effort monitoring, including improved estimation of at sea discarding, be improved to support the combined quota / effort regulation scheme outlined in 2 above.

6. That sanctions and penalties be increased for fishers who violate regulations including permanent suspension from the fishery for repeat or habitual offenders.

It is not clear whether all of these recommendations will eventually be acted upon or, if implemented, that they will improve the ability of fisheries management to attain stated fisheries conservation objectives. There have been, however, a number of positive changes. Planning for the development of fishing plans and resource sharing has been partially devolved to the level of community management boards. Fisheries do not continue on a "by-catch" basis. Once the quota for any one species has been reached in an area, the area is closed for all species. Dockside monitoring, where an independent company identifies and weighs all landings, is becoming mandatory for all industry sectors and is improving the quality of catch statistics. Scientific advice more explicitly recognizes an historical trend of overestimating stock size and corrects for this when providing advice on TACs.

The effects of granting quasi ownership rights, especially in the form of individual transferable quotas (ITQs), remain equivocal. They appear to have resulted in some reduction in capacity within those fleet sectors where they have been implemented, but it is not clear whether this is due to ITQs or low resource levels (Creed 1996). There is some indication that the frequency of cheating (violation of fishing regulations) has been reduced but many feel that this is the result of the imposition of higher penalties (Creed 1996). Those in favour of ITQs feel that they allow fishers to rationalize their individual operations based on weather and market conditions rather than being forced to catch fish in a competitive atmosphere in bad weather and when markets are low. Opponents of ITQs point out that they have caused social unrest in coastal communities through concentration of quota in the hands of a small number of fishers.

The ambiguity in management objectives (development and exploitation vs. conservation) contributed to the documented overexploitation of many commercially harvested species on the Scotian Shelf. The wider ecological costs of this exploitation, as a result of by-catches of non-commercial species and the effects of these fisheries on the biophysical habitat, are not well understood. The analysis of fisheries management abstracted above, including the recommended changes, includes no conservation objectives for anything but the targeted commercial species. There are no explicit ecosystem conservation objectives. Present fishery regulations continue to discount the value of all but the targeted commercial species. Recent changes to fishing regulations state that all groundfish species caught must be landed (so that they can be accounted for and could be factored into the overall ecological costs of fishing). However, special dispensation is granted for any number of species (such as sculpins, skates, or dogfish, invertebrate species, and small pelagic species like herring or mackerel) considered nuisances by virtue of high by-catch rates and low economic value. These can then be discarded at sea and are therefore unaccounted for in terms of ecological costs.

There is now a stated desire to better understand the ecosystem wide effects of fishing and their impacts on the viability of exploited ecosystems as a whole. This is evidenced by the recent international symposium on "The Ecosystem Effects of Fishing" held under the auspices of the Scientific Committee for Ocean Research (SCOR) and the International Commission for Exploration of the Sea (ICES). Canada was a key participant in this symposium. The results of this work (Hollingworth 2000) will provide valuable guidance to managers of human activities in marine environments. Recent proposals by the fishing industry to harvest krill on the Scotian Shelf were denied because of concerns about the impacts of such a fishery on the viability of higher trophic levels. This is further testament to an explicit recognition of ecosystem level effects of harvesting. We are also developing programs to improve monitoring of exploited ecosystems and, by comparative studies, to learn more about their functioning and sensitivities (see section on Ecosystem Modeling). However, recent work by Jackson *et al.* (2001) reviews paleoecological, archaeological, and historical evidence that shows that time lags of decades or even centuries occurred before the effects of fishing were manifested in changes in ecological communities. These authors also point out that the great majority of currently exploited systems are in significantly altered states, far removed from their original unexploited states. The fact that these systems have likely already been significantly modified through fishing and other human activities must be explicitly recognized, particularly in establishing restoration or maintenance objectives.

Ecosystems cannot be managed other than to manage the nature and intensity of human activities within them. Human impacts must be constrained within the functional, historical, and evolutionary limits of the ecosystem (Pickett *et al.* 1992). Management at the scale of ecosystems requires that monitoring and assessment consider time scales that range from seasonal to decadal and beyond. It also requires that objectives be established which operate over intergenerational time scales and consider spatial scales ranging from local to near global. Management of anthropogenic effects at these scales requires at least three broad pre-requisites: 1) objectives, 2) information, and 3) integration.

Objectives

Global objectives of ecosystem level management have been articulated in documents resulting from the 1992 United Nations Conference on Environment and Development (Agenda 21, UNCED), and the 1994 United Nations Convention on the Law of the Sea (UNCLOS) more recently supplemented (for fisheries) by the United Nations Fisheries

Agreement (UNFA). The conservation objectives of fisheries management on the Scotian Shelf remain largely cast in a single species framework with no explicit objectives within fisheries management plans for the remainder of the ecosystem. At present there is the stated desire to change the system of fisheries management from an emphasis on single species to a more holistic or ecosystem level approach. Canada recently held a workshop to define ecosystem-based management objectives (Jamieson et al. 2001) and a process whereby these would be added to management plans. However, the metrics to monitor progress toward achieving them have yet to be defined, although initial steps have been taken for the Scotian Shelf (O'Boyle 2000). It is encouraging, however, that there is significant interest and discussion taking place within the marine community (both scientific and user), aimed specifically at articulating such goals, metrics, and tactics.

Even though ecosystem management regimes can manage only the anthropogenic impacts on them, they must monitor and assess the impacts of both human and environmental driving forces. Monitoring and assessing these impacts on the processes, process rates, and states of the ecosystem must occur through tracking of biological and environmental variables considered to be metrics of overall ecosystem health, even though this concept remains only loosely defined if at all. Such monitoring must also integrate the tracking of global environmental and anthropogenic impacts because these externalities may affect the system as much or more than the local or internal processes (e.g. El Niño and levels of greenhouse gas emissions). Changes to research vessel survey protocols, which broaden the collection of ecosystem monitoring data to include abundance and distribution of phytoplankton, zooplankton, as well as an increased suite of physical oceanographic parameters, have recently been implemented

In formulating goals for the management of human activities within ecosystems, we must recognize that ecosystems are essentially large-scale dynamic processes. Indeed the idea of maintaining or restoring ecosystems to "pristine" or "pre-disturbed" conditions is less appealing than the concept of "homeorhesis" (O'Neil et al. 1986) which implies a maintenance or return to normal dynamics rather than maintenance or return to some artificial undisturbed "state". The thrust of this concept is that the actual levels of abundance of various trophic components at any given point in time or space should not be an objective of management. Since these systems are essentially the expression of dynamic processes, it is their ability to return to or move through historically observed states, which existed prior to excessive exploitation or disturbance, that should be the goal of management of human activities in ecosystems (see also Pitcher and Pauly 1998; Pitcher et al. 1999).

Information

The information required to monitor anthropogenic and environmental effects at the level of ecosystems has not been clearly articulated. A recent initiative by Canada and the United States, entitled the East Coast of North America Strategic Assessment

Project (Brown *et al.* 1996; Mahon *et al.* 1998), attempted to gather together all data on the physical and biological components of the east coast of North America in support of several strategic environmental assessment initiatives. One of the findings of this initiative was the recognition that many pieces of essential information have not been collected. Ricketts *et al.* (in prep) point out that the successes and failures of this exercise provide valuable lessons for the kind of inter-agency and international cooperation on data and information management essential to strategic decision making for ecosystem level management.

Systematic surveys conducted in support of fisheries management between the 1960s and the present are the richest and most consistent source of information available. Canada is in the process of augmenting the scope of these long-standing surveys to include increased monitoring of physical and biological oceanographic information. Canada's Oceans Act calls for a scientific perspective on conservation objectives. Fisheries, and other activities within marine ecosystems, now need to be evaluated in relation to properties of the ecosystem as well as the dynamics of the target species.

Integration

The third pre-requisite to management at ecosystem scales is the establishment of an effective integrating body. In this sense, integration means the effective coordination and management of human activities, within established limits, and consideration of spatiotemporal scales ranging from local and immediate to nearly global and intergenerational. Management at the scale of the Scotian Shelf ecosystems presently involves a host of single function agencies from local community management boards, provincial government departments, and an array of federal government departments. Management of the terrestrial coastal zone (housing, tourism, and industry) will need to be integrated with what is now fisheries management, and these in turn will need to be integrated with management of non-renewable resource exploitation. An initial response to this requirement is the ongoing development of Canada's Ocean Strategy and the implementation of the Oceans Act in 1997, including the establishment of a federal Oceans Directorate. While the Oceans Act contains statements committing Canada to the development of a strategy for ocean management, the development of this strategy remains an ongoing and arduous process. Although this process is far from complete and will remain dynamic for the foreseeable future, the issues to be considered include: conservation of ocean biodiversity; marine environmental quality; integrated planning and management; oceans industries and related opportunities; and public and community awareness, understanding, and participation.

CONCLUSIONS

We have presented a description of some of the physical and biological attributes of the Scotian Shelf Large Marine Ecosystem and how these have changed over the past three decades. We conclude that increasing fishing effort over the period 1977 - 1993 took place against a background of changing environmental conditions and that both exploitation (fishing in particular) and large-scale changes in environmental conditions have significantly affected fish and invertebrate communities. Changes in spatial distribution patterns and size structure were apparent responses. It is likely that fishing pressure removed the larger individuals of many species of fish on both portions of the shelf, therefore contributing to changes in community size structure and average fish size. The cooling of the eastern shelf further reduced the size of fish living there and allowed for the increased abundance of a number of cold water fish and invertebrates. The reductions in size on the western shelf were not associated with decreases in bottom temperature and are therefore more directly the result of fishing. For the Scotian Shelf as a whole, these reductions in average body size and the numbers of larger specimens may have significant impacts on the ability of many species to rebuild or maintain their populations. These results indicate that the management of human activities such as fishing must be evaluated against a backdrop of changing environmental conditions. Both fishing and changing environmental conditions have had impacts on these communities. Expected yields must be predicated on observed or predicted environmental conditions. Our present inability to quantitatively apportion these observed changes to either changes in environmental conditions or increased fishing effort highlights the need to improve our understanding of the functioning of these ecosystems as a whole.

These descriptions of the Scotian Shelf Large Marine Ecosystem also point out that our information base is far from complete. In particular, long-term data on primary and secondary production, distribution and abundance of benthic invertebrates and small pelagic fishes, and the distribution and concentrations of pollutants are sparse or non-existent. Furthermore, our understanding of the linkages between primary and secondary production and the production of fishes (both demersal and pelagic) and benthic invertebrates is unsatisfactory. Work is in progress to address this shortcoming. It is likely that this will be a long-term (decadal) process, especially given the paucity of even simple diet information linking predators to prey. Information on the more complex energy transfer rates between primary / secondary production and fish production, required to make predictions about the ecosystem level effects of exploitation or changes in environmental conditions, is even less available.

The primacy of the fishery development and exploitation objective over the conservation objective of fisheries management has led to documented overfishing of most commercially targeted species. The recently stated objective of moving from the current single species management to a more ecosystem based management scheme, consistent with UNCED Agenda 21, demands the development of new management objectives. Such objectives must reflect the desires of human ecosystem users but respect both the historical and evolutionary limits of the ecosystem. In addition, this requires the development and monitoring of ecosystem level metrics or performance measures to be used as indicators of how well these objectives are being met. The development of these objectives and performance measures is an ongoing process.

REFERENCES

Alexander, L.M. 1989. Large Marine Ecosystems as global management units. In: K. Sherman and L.M. Alexander (eds.). Biomass Yields and Geography of Large Marine Ecosystems. AAAS Selected Symposium 111. Boulder, CO: Westview Press, 339-344.

Angel, J.R., D.L. Burke, R.N. O'Boyle, F.G. Peacock, M. Sinclair, and K.C.T. Zwanenburg. 1994. Report of the Workshop on Scotia-Fundy Groundfish Management from 1977 to 1993. Can. Tech. Rep. Fish. Aquat. Sci.: 1979 vi+175 p.

Bianchi, G., H. Gislason, K. Graham, L. Hill, K. Koranteng, S. Manichkand-Heileman, I. Paya, K. Sainsbury, F. Sanchez, X. Jin, and K.C.T. Zwanenburg. 2000. Impacts of fishing on the size composition and diversity of demersal fish communities. ICES J. Mar. Sci. 57: 558-571.

Bowen, W.D. and G.D. Harrison. 1994. Offshore diet of Grey seals (*Halichoerus grypus*) near Sable Island, Canada. Mar. Ecol. Prog. Ser., 112: 1-11.

Bowen, W.D. and G.D. Harrison. 1996. Comparison of harbour seal diets in two habitats in Atlantic Canada. Can. J. Zool. 74: 125-135.

Bowen, W.D., J.W. Lawson, and B. Beck. 1993. Seasonal and geographic variation in the species composition and size of prey consumed by Grey seals (*Halichoerus grypus*) on the Scotian Shelf. Can. J. Fish. Aquat. Sci. 50: 1768-1778.

Bowen, W.D., R. Mohn, and J. Macmillan. Submitted. Grey seal pup production on Sable Island: sustained exponential growth of a large mammal population. Can. J. Fish. Aquat. Sci.

Brander, K.M. 1995. The effects of temperature on growth of Atlantic cod (*Gadus morhua* L.). ICES J. Mar. Sci. 52: 1-10.

Brown, S.K., R. Mahon, K.C.T. Zwanenburg, K.R. Buja, L.W. Claflin, R.N. O'Boyle, B. Atkinson, M. Sinclair, G. Howell, and M.E Monaco. 1996. East Coast of North America groundfish: Initial explorations of biogeography and species assemblages. Silver Spring, MD: National Oceanic and Atmospheric Administration, and Dartmouth, NS: Department of Fisheries and Oceans. 111 p.

Bundy, A. 1997. Assessment and management of multispecies, multigear fisheries: A case study from San Miguel Bay, the Philippines. PhD. Thesis, University of British Columbia. 369 p.

Bundy, A. 2001. Fishing on ecosystems: the interplay of fishing and predation in Newfoundland-Labrador. Can. J. of Fish. Aquat. Sci. 58: 1153-1167.

Bundy, A., G. Lily, and P.A. Shelton. 2000. A mass balance model of the Newfoundland Labrador Shelf. Can. Tech. Rep. Fish. Aquat. Sci. No. 2310.

Burke, D.L., R.N. O'Boyle, P. Partington, and M. Sinclair. 1996. Report of the second workshop on Scotia-Fundy groundfish management. Can. Tech. Rep. Fish. Aquat. Sci. 2100: vii+247 p.

Campana, S.E., R.K. Mohn, S.J. Smith, and G. Chouinard. 1995. Spatial visualization of a temperature-based growth model for Atlantic cod (*Gadus morhua*) off the eastern coast of Canada. Can. J. Fish. Aquat. Sci. 52: 2445-2456.

Canada. 1981. Resource prospects for Canada's Atlantic fisheries 1981-1987. Communications Branch, Department of Fisheries and Oceans Canada. ISBN 0-662-11533-3. 60 p. + 30 figures.

CETAP. 1982. A characterization of Marine Mammals and Turtles in the Mid- and North Atlantic Areas of the US outer Continental Shelf. Cetacean and Turtle Assessment Program, Univ. Rhode Island. US Bureau of Land Manage. Contract No. AA551-CTB-48. Washington, DC.

Christensen, V. 1995. A model of trophic interactions in the North Sea in 1981, the Year of the Stomach. Dana 11: 1-28.

Christensen, V. and D. Pauly (eds.). (1993) Trophic models of aquatic ecosystems. ICLARM Conf. Proc. 26. 390 p.

Creed, C. 1996. Social responses to ITQ's: Cheating and stewardship in the Canadian Scotia-Fundy inshore mobile gear sector. In: D.L. Burke, R.N. O'Boyle, P. Partington, and M. Sinclair (eds.). Report of the second workshop on Scotia-Fundy groundfish management. Canadian Technical Report of Fisheries and Aquatic Sciences No. 2100. 72-82.

Dayton, P.K., S.F. Thrush, M.T. Agardy, and R.J. Hofman. 1995. Environmental effects of fishing. Aquat. Cons. Mar. Freshw. Ecosys. 5: 205-232.

DFO. 1998. Northern Shrimp on the Eastern Scotian Shelf. DFO Science Stock Status Report C3-15 (1998).

Drinkwater, K.F., D.B. Mountain, and A. Herman. 1998. Recent changes in the hydrography of the Scotian Shelf and Gulf of Maine - a return to conditions of the 1960s? NAFO SCR Doc. 98/37, 16 p.

Edwards, R.L. 1968. Fishing resources of the North Atlantic area. The future of the fishing industry of the United States. Univ. of Washington Publ. in Fisheries, N.S. 4: 52-60.

Fanning, L.F., R.K. Mohn, and W.J. MacEachern. 1996. Assessment of 4VsW Cod in 1995 with consideration of ecological indicators of stock status. DFO Atl. Fish. Res. Doc. 96/27.

Frank, K.T., J. Simon, and J.E. Carscadden. 1996. Recent excursions of capelin (*Mallotus villosus*) to the Scotian Shelf and Flemish Cap during anomalous hydrographic conditions. Can. J. Fish. Aquat. Sci. 53: 1473-1486.

Haedrich, R.L., and S.M. Barnes. 1997. Changes over time of the size structure in an exploited shelf fish community. Fish. Res. 31: 229-239.

Halliday, R.G. and P.A. Koeller. 1980. A history of Canadian groundfish trawling surveys and data usage in ICNAF Divisions 4TVWX. In: W.G. Doubleday and D. Rivard (eds.). Bottom Trawl Surveys. Can. Spec. Pub. Fish. Aquat. Sci. 58: 27-41.

Hammill, M.O., G.B. Stenson, R.A. Myers, and W.T. Stobo. 1998. Pup production and population trends of the Grey seal (*Halichoerus grypus*) in the Gulf of St. Lawrence. Can. J. Fish. Aquat. Sci. 55: 423-430.

Han, G., C.G. Hannah, J.W. Loder, and P.C. Smith. 1997. Seasonal variation of the three-dimensional mean circulation over the Scotian Shelf. J. Geophys. Res. 102: 1011-1025.

Hays, G.C. and A.J. Warner. 1993. Consistency of towing speed and sampling depth for the Continuous Plankton Recorder. J. Mar. Biol. Assoc. U.K. 73: 976 - 980.

Hollingworth, C.E. 2000. Ecosystem effects of fishing. Proceedings of an ICES/SCOR Symposium held in Montpellier, France, 16-19 March 1999. ICES J. Mar. Sci. 57, no. 3.

Innis, H.A. 1954. The Cod Fishery; the history of an international economy. University of Toronto Press. 522 p.

Jackson, J.B.C. and 18 co-authors. 2001. Historical overfishing and the recent collapse of coastal ecosystems. Science 293: 629-638.

Jamieson, G., R. O'Boyle, J. Arbour, D. Cobb, S. Courtenay, R. Gregory, C. Levings, J. Munro, I. Perry, and H. Vandermeulen. 2001. Proceedings of the national workshop on objectives and indicators for ecosystem-based management, Sydney, British Columbia. 27 February - 2 March 2001. Canadian Science Advisory Secretariat Proceedings Series 2001/09.

Kenney, R.D., G.P. Scott, T.J. Thompson, and H.E. Winn. 1997. Estimates of prey consumption and trophic impacts of cetaceans in the USA northeast continental shelf ecosystem. J. Northw. Atl. Fish. Sci. 22: 155-171.

Longhurst, A. 1999. Does the benthic paradox tell us something about surplus production models? Fish. Res. 867: 1-7.

Mahon, R. 1997. Demersal fish assemblages from the Scotian Shelf and Bay of Fundy, based on trawl survey data. Can. Man. Rep. Fish. Aquat. Sci. 2426. 38 p.

Mahon, R., S.K. Brown, K.C.T. Zwanenburg, D.B. Atkinson, K.R. Buja, L. Claflin, G.D. Howell, M.E. Monaco, R.N. O'Boyle, and M. Sinclair. 1998. Assemblages and biogeography of demersal fishes of the East Coast of North America. Can. J. Fish. Aquat Sci. 55: 1704-1738.

Mahon, R. and E.J. Sandeman. 1985. Fish distributional patterns on the continental shelf and slope from Cape Hatteras to the Hudson Strait - a trawl's eye view. In: R. Mahon (ed.). Towards inclusion of fishery interactions in management advice. Can. Tech. Rep. Fish. Aquat. Sci. 1347: 137-152.

Mahon, R. and R.W. Smith. 1989. Demersal fish assemblages on the Scotian Shelf, northwest Atlantic: spatial distribution and persistence. Can. J. Fish. Aquat. Sci. 46(Suppl. 1): 134-152.

Mansfield, A.W. and B. Beck. 1977. The grey seal in eastern Canada. Fish. Mar. Ser. Tech. Rep. 704: 1-81.

Mohn, R. and W.D. Bowen. 1996. Grey seal predation on the eastern Scotian Shelf: modeling the impact on Atlantic cod. Can. J. Fish. Aquat. Sci. 53: 2722-2738.

Myers, R.A., N. J. Barrowman, G, Mertz, J. Gamble, and H.G. Hunt. 1994. Analysis of continuous plankton recorder data in the northwest Atlantic 1959 - 1992. Can. Tech. Rep. Fish. Aquat. Sci. 1966.

Myers, R.A. and K.F. Drinkwater. 1989. The influence of Gulf Stream warm core rings on recruitment of fish in the northwest Atlantic. J. Mar. Res. 47: 635-656.

O'Boyle, R.N. 2000. Proceedings of a workshop on the ecosystem considerations for the eastern Scotian Shelf integrated management (ESSIM) area. Canadian Science Advisory Secretariat Proceedings Series 2000/14.

O'Boyle, R.N., M. Sinclair, R.J. Conover, K.H. Mann, and A.C. Kohler. 1984. Temporal and spatial distribution of ichthyoplankton communities of the Scotian Shelf in relation to biological, hydrological and physiographic features. Rapp. P.-v. Reun. Cons. perm. int. Explor. Mer. 183: 27-40.

O'Neil, R.V., D.L. DeAngelis, J.B. Wade, and T.F.H. Allen. 1986. A hierarchical concept of ecosystems. Princeton University Press. New Jersey, USA.

Pauly, D. and V. Christensen. 1996. Mass Balance Models of North-eastern Pacific Ecosystems. Fisheries Center Research Reports Vol. 4., No. 1. UBC, Canada. 131 p.

Pauly, D., V. Christensen, J. Dalsgaard, R. Froese, and J. Torres. 1998. Fishing down marine food webs. Science 279: 860-863.

Pauly, D., M.L. Palomares, R. Froese, P. Sa-a, M. Vakily, D. Preikshot, and S. Wallace. 2001. Fishing down Canadian aquatic food webs. Can. J. Fish. Aquat. Sci. 58: 51-62.

Petrie, B. and K. Drinkwater. 1993. Temperature and salinity variability on the Scotian Shelf and in the Gulf of Maine 1945-1990. J. Geophys. Res. 98: 20079-20089.

Petrie, B., K. Drinkwater, D. Gregory, R. Pettipas, and A. Sandström. 1996. Temperature and salinity atlas for the Scotian Shelf and the Gulf of Maine. Can. Tech. Rep. Hydrogr. Ocean Sci. 171. 398 p.

Pickett, S.T.A., V.T. Parker, and P.L. Fiedler. 1992. The new paradigm in ecology: implications for conservation biology above the species level. In: P.L. Fiedler and S.K. Jain (eds.). Conservation biology: the theory and practice of nature conservation, preservation, and management. Chapman and Hall. New York, New York, USA. 65-88.

Pitcher, T.J. and D. Pauly. 1998. Rebuilding ecosystems, not sustainability, as the proper goal of fishery management. In: T.J. Pitcher, P.J.B. Hart, and D. Pauly (eds.). Reinventing fisheries management, Chapman and Hall, London. 311-329.

Pitcher, T.J., D. Pauly, N. Haggan, and D. Preikshot. 1999. Back to the future: A method of employing ecosystems modeling to maximize the sustainable benefits from fisheries. Alaska Sea Grant Symposium 1998.

Reid, P.C., M. Edwards, H.G. Hunt, and A.J. Warner. 1998. Phytoplankton changes in the North Atlantic. Nature 391: 546.

Ricketts, P.J., S.K. Brown, R.N. O'Boyle, K.C.T. Zwanenburg, D. Basta, M. Monaco, G. Howell, J. Arbour. D. Leonard, and M.J. Butler. In prep. The East Coast of North America Strategic Assessment Project (ECNASAP): A case study of multi-disciplinary and multi-agency co-operation in support of coastal zone management.

Rogers, R.A. 1995. The Oceans are Emptying - Fish Wars and Sustainability. Black Rose Books, Montreal, Quebec, Canada. 176 p.

Sainsbury, K.J., R.A. Campbell, R. Lindholm, and A.W. Whitelaw. 1997. Experimental management of an Australian multi-species fishery: examining the possibility of trawl induced habitat modifications. In: E.K. Pikitch, D.D. Huppert, and M.P. Sissenwine (eds.). Global trends: fisheries management. American Fisheries Society, Bethesda Md. 107-112.

Sameoto, D.D. and A.W. Herman. 1990. Life cycle and distribution of *Calanus finmarchicus* in deep basins on the Nova Scotia shelf and seasonal changes in *Calanus spp.* Mar. Ecol. Prog. Ser. 66: 225-237.

Sameoto, D.D., M.K. Kennedy, and N. Cochrane. 1997. Zooplankton changes along the Halifax and Louisbourg transects in 1996. Dept. of Fish. and Oceans Can. Stock Assess. Secretariat Res. Doc. 97/82. 14 p.

Scott, J.S. 1982. Depth, temperature and salinity preferences of common fishes of the Scotian Shelf. J. Northw. Atl. Fish. Sci. 3: 29-39.

Sherman, K. and A.M. Duda. 1999. An ecosystem approach to the assessment and management of coastal waters. Mar. Ecol. Prog. Ser. 190: 271-287.

Sherman, K, N.A. Jaworski, and T.J. Smayda (eds.). 1996. The Northeast Shelf Ecosystem: Assessment, Sustainability, and Management. Blackwell Science, Inc. Cambridge, MA. 564 p.

Sherman, K. and Q. Tang (eds.). 1999. Large Marine Ecosystems of the Pacific Rim: assessment, sustainability, and management. Blackwell Science, Inc. Winnipeg. 465 pages.

Sinclair, M., R. O'Boyle, D.L. Burke, and G. Peacock. 1997. Why do some fisheries survive and others collapse? In: D.A. Hancock, D.C. Smith, A. Grant, and J.P. Breumen (eds.). Developing and sustaining world fisheries resources: The state of science and management. Second world fisheries congress, Brisbane, 1996. (CSIRO publishing, Melbourne). 23-25.

Sinclair, M.M., Chairman. 1998. Proceedings of the Marine Fisheries subcommittee; Regional Advisory Process, Maritimes Region. 26 - 30 October 1998, Dartmouth, Nova Scotia. Can. Stock Assessment Proc. Ser.: 99/2.

Smith, P.C. 1978. Low-frequency fluxes of momentum, heat, salt, and nutrients at the edge of the Scotian Shelf. J. Geophys. Res. 83: 4079-4096.

Stobo, W.T., B. Beck, and J.K. Horne. 1990. Seasonal movements of Grey seals (*Halichoerus grypus*) in the Northwest Atlantic. In: W.D. Bowen (ed.). Population biology of sealworm (*Pseudoterranova decipiens*) in relation to its intermediate and seal hosts. 199-213.

Taylor, A.H. In press. North Atlantic climatic signals and the plankton of the European continental shelf. This volume.

Tremblay, M.J. 1997. Snow crab (*Chionoecetes opilio*) distribution limits and abundance trends on the Scotian Shelf. J. Northw. Atl. Fish. Sci. 21: 7-22.

Tremblay, M.J. and J.C. Roff. 1983. Community gradients in the Scotian Shelf zooplankton. Can. J. Fish Aquat. Sci. 40: 598-611.

Trippel, E.A., O.S. Kjesbu, and P. Solemdal. 1997. Effects of adult age and size structure on reproductive output in marine fishes. In: R.C. Chambers and E.A. Trippel (eds.). Early life history and recruitment in fish populations. Chapman and Hall, New York. 31-62.

Waldron, D.E. 1982. Diet of the silver hake (*Merlucius bilinearis*) on the Scotian Shelf. J. Northw. Atl. Fish. Sci., 14: 87-101.

Waldron, D.E. 1988. Trophic biology of the silver hake (*Merlucius bilinearis*) population of the Scotian Shelf. Ph.D. thesis, Dalhousie University.

Walters, C., V. Christensen, and D. Pauly. 1997. Structuring dynamic models of exploited ecosystems from trophic mass-balance assessments. Rev. Fish Biol. Fish. 7: 139-172.

Weare, B.C. 1977. Empirical orthogonal analysis of Atlantic Ocean surface temperatures. Quart. J. R. Met. Soc. 103: 467-478.

Whitehead, H., W.D. Bowen, S.K. Hooker, and S. Gowans. 1998. Marine mammals. In: W.G. Harrison and D. G. Fenton (eds.). The Gully: A scientific review of its Environment and Ecosystem. Ottawa, Canada. Dep. Fish. Oceans. 186-221.

Zwanenburg, K.C.T. 2000. The effects of fishing on demersal fish communities of the Scotian Shelf. ICES Journal of Marine Science, 57: 503-509.

Zwanenburg, K.C.T. and W.D. Bowen. 1990. Population trends of the grey seal (*Halichoerus grypus*) in Eastern Canada. In W.D. Bowen (ed.). Population biology of sealworm (*Pseudoterranova decipiens*) in relation to its intermediate and seal hosts. Can. Bull. Fish. Aquat. Sci. 222.

Large Marine Ecosystems of the North Atlantic
K. Sherman and H.R. Skjoldal (Editors)

5

Dynamics of Fish Larvae, Zooplankton, and Hydrographical Characteristics in the West Greenland Large Marine Ecosystem 1950-1984

Søren A. Pedersen and Jake C. Rice

ABSTRACT

A relatively long series of plankton and hydrographic samples for the West Greenland ecosystem allows us to explore links between climate, physical oceanography, and abundance of major zooplankton and ichthyoplankton species. These linkages, manifest in food chains, fish stocks dynamics, and regime shifts in large marine ecosystems are important for fisheries management. Patterns of change in larval fish abundances and hydrographic characteristics were analyzed with data from three transects off southwest Greenland sampled between 1925 and 1984. Factor analysis and orthogonal varimax rotation on 8 physical oceanographic variables highlighted gradients in salinity, temperature, and temperature range (stratification). The period in the 1950s tended to be warmer than the 1970s, and a period of low salinity was present in the late 1960s and early 1970s. Non-parametric density estimation methods identified associations between the hydrographic characteristics and the distributions of fish and shrimp larvae: Atlantic cod (*Gadus morhua*), Redfish (*Sebastes marinus and S. mentella*), Greenland halibut (*Reinhardtius hippoglossoides*), Long rough dab (*Hippoglossoides platessoides*), Wolffish (*Anarhichas* sp.), Sandeels (*Ammodytes* sp.) and northern shrimp (*Pandalus borealis*). The fish larvae were generally distributed non-randomly to at least one of the hydrographic characteristics and to the indices of plankton and copepod abundance. The fish and shrimp larvae showed generally stronger associations with the indices of plankton and copepod abundance than with the hydrographic characteristics. The patterns suggest a suite of fish species whose larvae are weakly adapted to different oceanographic conditions and different ecological niches. For all but sandeels, June and July abundances are likely to be higher when water masses are relatively saline and well stratified, and when plankton, especially copepods, are abundant.

These relationships were then carried further to determine synchronous temporal patterns among oceanographic characteristics, abundance indices of zooplankton, fish larvae, and year-class strengths. There was a decreasing trend in the plankton abundance indices from the late 1950s and early 1960s to the 1970s as well as a weak significant positive association to sea temperature. This trend culminated in the year 1969 and 1970 the period of the "Great Salinity Anomaly". The variability in time series of zooplankton, fish and shrimp larvae could not be related to an impact of the GSA. There were weak significant associations between Atlantic cod year-class strength (given number at age 3) and both temperature and cod larvae abundance. The patterns explain reasonably well the decline in cod, redfish and long rough dab as consequences of changes in the relative influences of the East Greenland Polar Current and the Irminger Current, but do not explain the major increases in shrimp and Greenland halibut in the 1980s and 1990s. At least the former may be a consequence of reduced predation mortality. The associations are also generally consistent with a survey series of adult groundfish started in the 1980s in East and West Greenland waters.

INTRODUCTION

Changes in the West Greenland Fisheries

Annual landings of commercial fish species from West Greenland waters show major changes over the past century (Anon. 1995, 1998a,b). A rich Atlantic cod (*Gadus morhua*) fishery started in the 1920s after a general warming of the Arctic (Jensen 1939; Buch *et al.* 1994). Historically, the cod has been the most important fish species in Greenland waters, with annual catches peaking at levels between 400,000 and 500,000 tons in the 1960s. During the late 1960s, the annual catches of cod declined drastically as did the catches of other commercially important fish species - redfish (*Sebastes marinus and S. mentella*), Atlantic halibut (*Hippoglossus hippoglossus*) and wolffish (Atlantic wolffish, *Anarhichas lupus*, and spotted wolffish, *A. minor*) - mainly taken as by-catch in the fishery for cod. After 1969, catches of Atlantic cod and redfish fluctuated around a much lower mean than prior to the 1960s (Figure 5-1). Except for a temporary improvement of the cod fishery during 1988-1990 the catches of cod, redfish, Atlantic halibut and wolffish showed marked decreasing trends from 1980 onwards (Anon. 1998a; Anon. 1998b). The decline in amount caught does not tell the entire story, however, as the fisheries for cod also moved much further south in the 1980s, and the sizes of fish at age dropped greatly as well (Hovgård and Buch 1990). In the same periods, however, the catches of two other commercially important species, Greenland halibut (*Reinhardtius hippoglossoides*) and northern shrimp (*Pandalus borealis*) increased (Figure 5-1).

The shrimp fishery in West Greenland waters began in 1935 as a local fishery in a fjord south of the settlement Sisimiut (Holsteinsborg). After 1950 the inshore fishery expanded rapidly, with total annual catches between 7,000 and 8,000 t. from 1975 to 1987 and somewhat higher and more variable thereafter. The offshore northern shrimp fishery in the Davis Strait began about 1970. In the 1990s the annual landings were between 60,000 and 80,000 tons (NAFO Subarea 0+1), including between 2,000 and 7,000 tons taken on the Canadian side of the midline (NAFO Subarea 0A). Offshore and inshore landings from

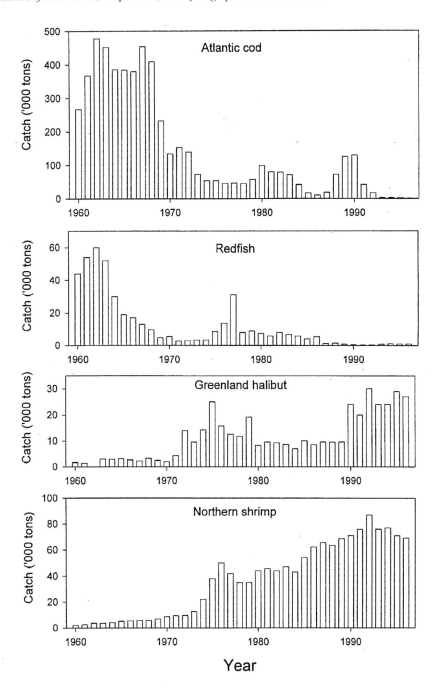

Figure 5-1. Nominal catches for the most important commercial species in West Greenland waters, 1960-1996.

West Greenland waters peaked in 1992 at about 87,000 and 21,000 t, respectively (Anon. 1998a). After 1992, only Greenland vessels were fishing shrimp in West Greenland waters.

An unknown additional amount of unreported fish and shrimp catches have been discarded at sea. The West Greenland Sea has often been observed as "red" due to large amounts of dead redfish floating in sea surface, especially in the 1950s and 1960s when several fleets did not land redfish (S. Aa. Horsted, former Director of Greenland Fisheries Research Institute, pers. comm.).

The explanatory hypotheses

The decline in the catches and abundances indices of Atlantic cod and other mainly boreal fish species on one hand and the increasing annual catches of northern shrimp and Greenland halibut on the other, raise several questions. To what extent are these changes due to changes in the oceanographic environment? Do the changes in catch composition reflect simply retargetting of commercial fleets after the initial target species were fished out? Does the apparent increase in shrimp and Greenland halibut reflect release from predation after the cod and redfish were fished down, or could shrimp and Greenland halibut be favored by the same environmental conditions which were unsuitable for cod and redfish? None of the above questions are new with this study; diverse hypotheses have been put forward to account for some or all of these changes:

Oceanographic Hypotheses: Several researchers proposed that decreased water temperatures reduced egg/larvae survival of several species directly or through changes in larval food production (e.g. Hermann *et al.* 1965; Bainbridge and McKay 1968). In addition variation in larval and juvenile drift has been shown to cause recruitment variability in West Greenland waters directly, and indirectly through effects on predator - prey distributions (Bainbridge and Corlett 1968; Bainbridge and McKay 1968; Smidt 1969; Buch *et al.* 1994; Pedersen 1994a). In terms of mechanisms linking oceanographic factors to recruitment in West Greenland, sea temperature, drift of larvae by surface currents, and stability of the water masses (oceanographic fronts) have been proposed (Meyer 1968; Horsted *et al.* 1978; Buch *et al.* 1994; Stein and Lloret 1995). Variability in these factors is related in turn to inflow of water from other parts of the North Atlantic (Lee 1968; Pavshtiks 1968,1972; Buch 1990) and possibly in more recent times, advective coupling (Stein and Lloret 1995).

Predation-release Hypothesis: In Greenland waters several fish species, e.g. Atlantic cod, Atlantic halibut, Greenland halibut, redfish, long rough dab (*Hippoglossoides platessoides*), and starry ray *(Raja radiata)* have been identified as important predators on shrimp (Jensen 1925; Horsted and Smidt 1965; Smidt 1969; Tiedtke 1988; Pedersen and Riget 1993; Grünwald and Köster 1994; Pedersen 1994a,b and 1995). The present abundance of shrimp in West Greenland waters may partly be a result of a lower abundance of cod and redfish, and reduced predation mortality (Horsted 1989; Pedersen 1994c).

<u>"Bycatch" Hypothesis</u>: The distribution of northern shrimp coincides with important nursery areas for several fish species. Large numbers of fish, mainly redfish, Greenland halibut, and polar cod (*Boreogadus saida*), but also cod, starry ray, long rough dab, and others were caught and discarded in the West Greenland shrimp fishery (Pedersen and Kanneworff 1995). Most of the fish in the bycatch are spawned up-current from the shrimp grounds and are transported to the nursery areas by prevailing currents. Although little quantitative information on the by-catches and the discards of fishes in the West Greenland shrimp fishery has been available, the annual effort of the Canadian and Greenland shrimp fishery in Davis Strait of close to 200,000 hours of trawling in the 1990s may be reducing the strengths of the recruiting year-classes of the demersal fish community (Pedersen and Kanneworff 1995).

Data are not available to allow definitive evaluation of the contribution of each of these processes to the variation in the West Greenland Sea ecosystem. Moreover, it is rarely easy to partition causality among alternative co-occurring factors. An assessment of the possible intensity of predation on fish larvae is particularly difficult because little is known of the relative importance of the various predators of young fish (Bainbridge and Corlett 1968). Moreover, strong associations between predators and their prey may not appear as high correlation coefficients of survey catches, if predators effectively grazed their prey down to low abundance.

Nonetheless, for the West Greenland ecosystem, a relatively long series of plankton and hydrographic samples allows us to explore links between climate, physical oceanography, and abundance of major zooplankton and ichthyoplankton species. These linkages, manifest in food chains, fish stocks dynamics, and regime shifts in large marine ecosystems are important for fisheries management (Mann 1993; Francis and Hare 1994; Steele 1995, Cushing 1995a,b). Furthermore, the long time series of samples of zooplankton and ichthyoplankton may allow temporal and spatial patterns of variation to be quantified (Colebrook 1982, 1985; Aebischer *et al.* 1990). When such patterns can be associated with hydrographic data and abundance of potential predators or prey it may be possible to relate the patterns to underlying processes, and begin to untangle the roles of multiple processes in stock and system dynamics.

In this paper, we analyze data on the abundance of several zooplankton taxa, six species of fish larvae: Atlantic cod, redfish, Greenland halibut, long rough dab, wolffish, and sandeels (*Ammodytes* sp.) and larval northern shrimp. Sampling was conducted as far back as 1925, in conjunction with collection of hydrographic data. Sampling intensity varied greatly over the period. However, most stations were allocated consistently along three hydrographic transects.

We approach the analyses in several steps. First we identify major patterns of variation in temperature and salinity profiles, in both space and time. We then use these patterns to characterize the hydrographic properties of sample locations where fish larvae and zooplankton of various taxa were either more common or more rare than expected

(Smith *et al.*, 1991; Perry and Smith, 1994). We then examine the data as a multi-decadal time series, to link population trends over time to the patterns present in the analyses of all samples. Both the effects of changes in spawning biomass over the full time series and the spatial distribution of spawners are addressed in the treatment of the time series. Hence, the paper is able to report on: (1) the relationships found between the abundances of fish larvae, zooplankton and the physical characteristics of the water masses, and (2) the extent to which changes over time in the oceanographic characteristics can account for changes in the zooplankton, fish larvae abundance, and year-class strength of the major fish stocks of the region. These are the first-order changes in the West Greenland Large Marine Ecosystem, on which additional effects of predation and bycatch mortalities must be superimposed.

MATERIALS AND METHODS

Survey design, oceanographic and plankton samples

Annual oceanographic surveys have been made in the Labrador Sea and Davis Strait off Southwest Greenland in 1950-94 by the Greenland (Danish) Fisheries Research Institute and in 1995-present by Greenland Institute of Natural Resources (Buch 1990, 2000). Combined with physical and chemical observations, zooplankton were sampled. Sampling and analysis of samples were made by technicians under the leadership of Mr. V.K. Hansen during 1952-62 and Dr. E.L.B. Smidt during 1963-85. Analysis ceased in 1985 as Smidt retired.

In 1963, the surveys were part of extended international surveys in Northwest Atlantic waters from Canada to Iceland. These surveys, named NORWESTLANT, were organised through the International Commission for the Northwest Atlantic Fisheries (ICNAF), with participation of research vessels from Canada, Denmark, Federal Republic of Germany, France, Iceland, Norway, UK, and USSR. - Results were published in 1968 (ICNAF 1968), and a Summary Report was given by Smidt (MS 1971). Smidt (1979) presented results of zooplankton investigations in inshore and coastal waters throughout the years 1950-66, mainly in the Nuuk (Godthåb) district. The present paper analyses data from hydrography and zooplankton sampling offshore.

Off West Greenland, hydrography and zooplankton samples were collected, generally at fixed positions, along transects from the end of June well into July. However, in some years additional samples were taken at positions off the transects, e.g. during the NORWESTLANT survey in 1963. Also the number of samples and distance from the coast varied greatly among years. Over all years sampling was performed consistently along 3 transects: *Store Hellefiske Bank* (1), *Sukkertop Bank* (2), and *Fyllas Bank* (3), nearly perpendicular to the Greenland coast, at approximately 66°45' N, 65°06' N, and

63°50' N (Figure 5-2). Analyses reported here were restricted to the area including the three transects consistently sampled over all years (Figure 5-3).

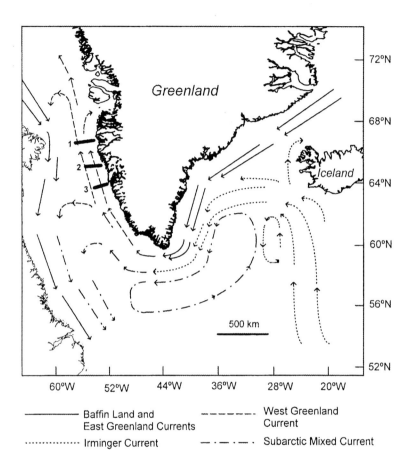

Baffin Land and
East Greenland Currents

West Greenland
Current

Irminger Current

Subarctic Mixed Current

Figure 5-2. Sea currents and locations of the three West Greenland transects where hydrography and zooplankton data were collected, 1950-1984. The currents around Greenland were simplified after Hachey *et al.* (1954).

Sea temperatures and salinities were measured using standard techniques at standard depths. The water-sampling device was a modified Nansen reversing water bottle equipped with protected reversing thermometers (accuracy to within 0.01°C.) (see e.g. Sverdrup *et al.* 1946). The sea temperatures and salinities at each plankton sampling station were calculated as simple means of temperature and salinity measurements in the range 10-50 m (normally the standard depths 10, 25, and 50 m or 10, 20, 30, and 50 m).

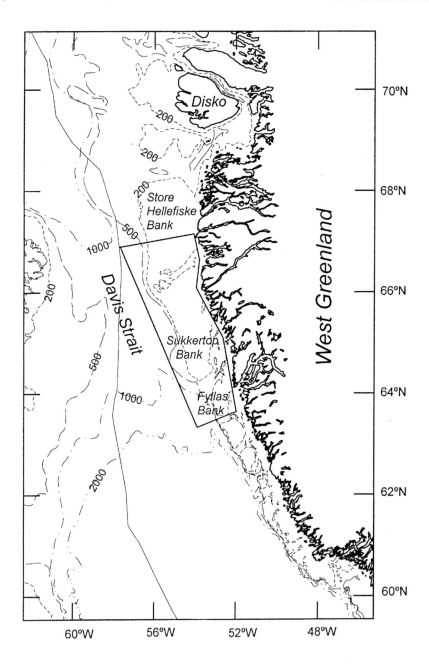

Figure 5-3. Map of major physiographic features and the study area (heavy line) referred to in the paper. The study area is approximately 54 000 km².

The zooplankton sampling gear was a 2 m (diameter) stramin ring-net with 1 mm mesh. Tows were made for 30 minutes at about 2 knots. Prior to 1963, and in 1964 and 1966, hauls were horizontal, stratified with two nets on the same wire, in three 10 minutes deployments. Wire lengths were 200, 150, and 125 m for the deeper net, and 100, 50, and 25 m for the shallower net. In 1963 all tows were made obliquely with a single net with a maximum wire length at deployment of 225 m (which corresponds to a depth of approximately 50 m). In 1964 and 1966, stratified hauls with two nets were again made, but in all the following years, 1968-1984, all tows were made obliquely with a single net as in 1963. No sampling was done in 1951, 1965, and 1967. To make all zooplankton samples comparable on the scale of a single haul, all samples were standardized to a 30 minutes tow. For years when two nets were used, all samples at a station were pooled and scaled to the standard time interval. When the zooplankton sampling was with stratified nets, some extra hauling time was generally allocated. Based on records from the research vessel logs and data sheets, a conversion factor of 0.75 was used for the stratified hauls before 1963 and 0.85 for hauls in the years 1964 and 1966, to scale these catches to 30 minutes. These conversion factors are rough, but improve the consistency of the longer data set.

Calibration experiments performed in 1984 on oblique tows, with a flowmeter attached to the net opening, estimated that a 30 minute tow at 2 knots filtered approximately 6125 m³ of seawater. This is used as the sampling unit for all counts and volumes of zooplankton and fish larvae in these analyses. Zooplankton abundance indices in volume or number per m² sea surface were found multiplying sampling units with $8.16*10^{-3}$ (50 m/6125 m³).

From each sample all "large" organisms (generally post-larval fish) were picked out. Because the gear is unlikely to retain salps and schyphomedusae effectively those organisms were removed from the catch without quantification. The displacement volume of all remaining plankton, scaled to a 30 minute tow, was recorded. Subsamples were taken for counts of the zooplankton taxa comprising the catch. The subsample size in ml depended on the size of the taxa and was adjusted so that all or a minimum of about 400 animals was counted per sample. In some years not all zooplankton taxa in the samples were sorted and counted. However all species of fish larvae (and eggs) in all samples were quantified. When processing of zooplankton samples was incomplete, only the data on fish larvae was analysed. The differences in quantification of samples did not show any major trends over time, so they are likely to represent added variance, but not a source of bias over years.

More consistent and detailed treatment of sampling, subsampling, and counting might have allowed estimation of within haul and among haul components of variation in abundance. However, more elaborate analyses could not be applied in a consistent way retrospectively across the data series. The additional noise due to subsampling differences among samples, which may be present in the data, would make patterns harder to detect. Therefore, we considered weak tests preferable to possibly introducing substantial bias through inconsistent analyses of different portions of the data set.

Analyses: Oceanographic measures

Our goal is to examine the patterns of occurrence of selected fish larvae and zooplankton over time and space and how those patterns relate to hydrographic features. To avoid inflated statistical error rates and to increase the efficiency of our estimators (Ludwig and Walters 1985), it was necessary to reduce the number of possible environmental covariates to a modest number (see the discussion in Smith *et al.* 1991). We used factor analysis to summarise the physical oceanographic measures over space and time because the statistical properties of the results are well known (Belsley *et al.* 1980). From each hydrographic profile we took the mean, maximum, minimum and the difference between maximum and minimum temperatures in the range 10-50 m, and the same four attributes for salinity. The differences in methods for sampling plankton before and after the mid 1960s made it impossible to associate plankton samples with specific depths. Therefore, we preferred these general characteristics of the hydrographic profiles to temperatures or salinities at particular depths.

Because of the large differences in sampling intensity among years, we weighted the mean, min, and max temperature/salinity of each profile by the number of samples taken in a year. We used a varimax rotation to obtain best interpretability of individual cases (sampling stations) (Anon., 1985, Factor procedure) and multi-way analysis of variance (ANOVA) to test for effects of year, transect, depth to bottom, month and interactions (Anon., 1985, GLM procedure). Use of factor scores reduces the potential for Type I errors in subsequent analyses and accommodates some of the measurement error due to depth-aggregated plankton samples. However, the use of factor scores means that the scales shown on Figure 5-4a-c are on relative gradients (anomalies) rather than absolute scales.

Analyses: Fish-environmental relationships

There are no *a priori* hypotheses for the exact functional form of the relationship between abundance of plankton taxa and oceanographic features. Nor is it attractive to specify a particular sampling distribution for the sample variance, particularly given the methodologies applied. Therefore we used methods which made very few assumptions about error structures or forms of functional relationships (Rice 1993).

For each of the important orthogonal factors from analysis of the hydrographic data, samples from all stations and years were sorted by increasing score, and the cumulative frequency distributions (cfds) of cases were constructed along each factor. For each

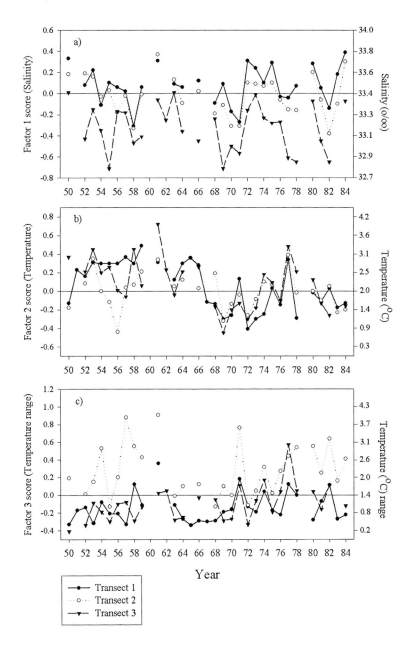

Figure 5-4. Mean factor 1, 2 and 3 scores corresponding to mean salinity (a), temperature (b), and temperature range (c) by transect and year. Only samples with bottom depths less than 1000 m are included.

species of fish larvae and zooplankton taxon, the cfds of log (base 10) transformed numbers (+1) also were accumulated across sampling sites as ordered on each of the factors. The cfds of samples and of log transformed abundances of each species of larval fish were also accumulated across depths and across log transformed total plankton volume and copepod numbers (+1).

If the physical oceanographic factor (or plankton) and the abundance of the fish larvae are independent, then cases and abundances should accumulate across the environmental gradient at the same rate. If the fish larvae show a pattern of distribution which is associated with the environmental factor, then the larvae should occur more frequently than expected (given the occurrence of samples) along some parts of the factor, and less frequently along other parts of the factor. The total number of larvae or zooplankton observed was very large for some species, so the critical distance for a two-sample Kolmogorov-Smirnov test (the appropriate test statistic for such comparisons - Conover 1980) would be extremely small (sometimes <0.001) and highly variable among taxa. Instead we fit the 99% confidence interval to the cfd of samples, and evaluated the regions where the cfd of fish larvae or plankton lay outside that confidence interval. This allowed a consistent standard of inference to be applied to the possible associations of all taxa. We used the 99% confidence interval because we were seeking associations between the occurrence of each taxon and several environmental variables. By making each individual comparison at the 0.01 level, we kept the probability of falsely inferring the presence of an association with some hydrographic feature at approximately 0.05 for each species of fish or zooplankton.

Zooplankton and fish larvae abundance indices for the study area (Figure 5-3) by year were calculated as overall mean per m^2. To examine the interannual variations and relationships of oceanographic characteristics, abundance indices of zooplankton, and fish larvae in the study area we contoured areas of hydrographical conditions with occurrence of fish larvae and zooplankton taxa more frequent than expected. Because of variability in coverage and density of sampling sites in different parts of the full time series, only a coarse resolution of hydrographic condition ranges was possible, and analytical contouring approaches were inappropriate. Areas of specific hydrographic conditions were contoured and measured in km² using MapInfo - a geographical information system.

Relationships between the annual extent of oceanographic characteristics, abundance indices (log base 10 transformed) of zooplankton, fish larvae, and year-class strengths were investigated using rank correlations and nonparametric kernel estimators (Rice 1993). Among the reasons for using nonparametric density estimation methods, instead of parametric models, are lack of knowledge of the true distributions of errors and lack of smoothness in the unknown true functional relations. Also, because of the skewed distributions of some zooplankton and larval abundances even after transformation, we considered it more appropriate to measure goodness of fit of the deviations of observations from the pdf median rather than deviations from the pdf mean. The sum of squared deviations from the median is then the least-squares criterion to be minimised in setting the kernal span parameter, but it is not a true variance.

RESULTS

Oceanographic structure

In the initial factor analyses, the information in the eight physical oceanographic variables was captured by three orthogonal trends. The remainder of the variability was noise. In the basic analysis, the major overall pattern of association between temperature and salinity (cold water tends to be fresher, warmer waters tend to be more saline) dominated the results (Table 5-1). Subsequent factors were bipolar, with the variance in stratification scattered among all factors. Orthogonal varimax rotation realigned these trends to highlight a gradient in salinity on the first factor, a gradient of temperature on the second factor, and a gradient of increasing temperature range, as an indicator of relative intensity of stratification, as the third factor (Table 5-1). These rotated factors allow direct investigation of the degree to which the distribution and abundance of fish and zooplankton is sensitive to salinity, to temperature, or to stratification.

The scores for samples on all three factors showed significant variation due to year, transect, depth to bottom, and month (Table 5-2). For all three factors the year effect was dominant, and particularly strong for the temperature factor. For that factor there were no consistent differences among transects, although the interaction terms (temperature varied greatly among transects in some years, but not during others) were significant. For both the salinity and stratification factors there was approximately 1/2 to 2/3 as much consistent variation among transects as there was of each sample relative to the "average" condition across 33 years.

Scores of sample sites on all three factors varied with the depth to the bottom (distance from the coast), particularly if the infrequent samples from locations with depths > 1000 m were included. Over those depths salinity and temperature tended to be slightly higher than for samples at locations with bottom depths less than 1000 m. The trend with depth was the opposite over the shelf and banks, in areas less than 800 m. There both temperature and salinity decreased with increasing bottom depth. For the stratification factor, both the scores and their variance increased across the entire depth range. The patterns with depth should be viewed cautiously. Locations in the deep waters of transects 2 and 3 were sampled with variable intensity during 1950-1984, so differences of the deep water samples from the other samples cannot be unconfounded from differences among years.

Table 5-1.   Results of the factor analyses on 8 physical oceanographic variables: (MTEMP=mean temperature, TMIN=minimum temperature, TMAX=maximum temperature, TRANGE=difference between maximum and minimum temperature, MSAL=mean salinity, SMIN=minimum salinity, SMAX=maximum salinity, SRANGE=difference between maximum and minimum salinity.

```
Number of observations = 550
Initial Factor Method: Principal Components
Factor Pattern
                FACTOR1    FACTOR2    FACTOR3    FACTOR4

MTEMP           0.715      0.618     -0.307      0.037
TMIN            0.735      0.226     -0.580      0.262
TMAX            0.550      0.820     -0.048     -0.138
TRANGE         -0.084      0.738      0.520     -0.422
MSAL            0.866     -0.323      0.376      0.025
SMIN            0.864     -0.445      0.162     -0.168
SMAX            0.789     -0.157      0.553      0.216
SRANGE         -0.390      0.539      0.476      0.576


Variance explained by each factor
   FACTOR1     FACTOR2     FACTOR3     FACTOR4
   3.634       2.267       1.403       0.673


-------------------------------------------------
Rotation Method: Varimax
Orthogonal Transformation Matrix

                1         2         3         4

     1       0.754     0.607    -0.026    -0.252
     2      -0.330     0.604     0.601     0.407
     3       0.558    -0.493     0.508     0.433
     4       0.108     0.157    -0.616     0.764

Rotated Factor Pattern

                FACTOR1    FACTOR2    FACTOR3    FACTOR4

MTEMP           0.168      0.964      0.175     -0.033
TMIN            0.184      0.909     -0.339     -0.144
TMAX            0.103      0.831      0.540      0.070
TRANGE         -0.062      0.073      0.970      0.225
MSAL            0.972      0.149     -0.040     -0.167
SMIN            0.870      0.150     -0.104     -0.457
SMAX            0.978      0.146      0.033      0.142
SRANGE         -0.144     -0.056      0.220      0.963


Variance explained by each factor
   FACTOR1     FACTOR2     FACTOR3     FACTOR4
   2.755       2.521       1.439       1.261
```

Table 5-2. Results of a multi-way ANOVA of hydrographic scores. Model: factor 1 salinity (F1SAL), factor 2 temperature (F2TEM), and factor 3 temperature range (F3TR) = Overall mean + Year (1950-1984) + Transect (1-3) + Depth to bottom (0-1000 by 100) + Month (June-July) + interactions + Err.

Dependent Variable: F1SAL

Source	DF	Sum of Squares	Mean Square	F Value	Pr > F
Model	117	38.4	0.3	14.1	0.0001
Error	355	8.3	0.02		
Corrected Total	472	46.6			

R-Square	C.V.	Root MSE	F1SAL Mean
0.82	-305.6	0.15	-0.05

Source	DF	Type I SS	Mean Square	F Value	Pr > F
Year	32	19.8	0.6	26.6	0.0001
Transect	2	11.7	5.9	252.3	0.0001
Depth to bottom	9	1.0	0.1	4.9	0.0001
Month	1	2.0	2.0	85.2	0.0001
Transect*Year	58	2.5	0.0	1.9	0.0003
Transect*Depth to bottom	15	1.3	0.1	3.7	0.0001

Dependent Variable: F2TEM

Source	DF	Sum of Squares	Mean Square	F Value	Pr > F
Model	117	44.4	0.4	8.1	0.0001
Error	355	16.6	0.05		
Corrected Total	472	61.0			

R-Square	C.V.	Root MSE	F2TEM Mean
0.73	9999.9	0.22	0.00

Source	DF	Type I SS	Mean Square	F Value	Pr > F
Year	32	23.5	0.7	15.7	0.0001
Transect	2	0.1	0.1	1.5	0.2331
Depth to bottom	9	8.1	0.9	19.2	0.0001
Month	1	4.8	4.8	102.2	0.0001
Transect*Year	58	4.3	0.1	1.6	0.0063
Transect*Depth to bottom	15	3.6	0.2	5.1	0.0001

Dependent Variable: F3TR

Source	DF	Sum of Squares	Mean Square	F Value	Pr > F
Model	117	48.4	0.4	8.1	0.0001
Error	355	18.1	0.05		
Corrected Total	472	66.5			

R-Square	C.V.	Root MSE	F3TR Mean
0.73	1148.7	0.23	0.02

Source	DF	Type I SS	Mean Square	F Value	Pr > F
Year	32	17.7	0.6	10.8	0.0001
Transect	2	14.3	7.2	140.3	0.0001
Depth to bottom	9	3.2	0.4	7.0	0.0001
Month	1	0.6	0.6	11.7	0.0007
Transect*Year	58	7.2	0.1	2.4	0.0001
Transect*Depth to bottom	15	5.4	0.4	7.0	0.0001

The trends over time were apparent from the sample scores on each factor. For example, the period in the 1950s tended to be warmer than the 1970s (high values on factor 2 - Figure 5-4b); a period of low salinity was present in all transects in the late 1960s and early 1970s (factor 1 - Figure 5-4a), corresponding to the Great Salinity Anomaly (GSA).

The pattern among transects was also apparent in the factor scores. The samples from transect 1 tended to have higher salinities, whereas samples from transect 3 tended to have lower salinities (Figure 5-4a). The temperatures tended to be warmer in transect 1 than in transect 3 during the warm period in the early part of the series, but not during the subsequent colder period (Figure 5-4b). The middle zone (transect 2) does not have intermediate scores, however. In some years samples from the middle transect have scores which resemble those from the northern transect, whereas samples from other years have scores which resemble those from the southern transect. Although the variance in salinity tended to be high in the middle transect the variance in temperature was often less than on the other transects. Transect 2 was marked by extremely strong stratification in most years, however, whereas the other two transects tended to have markedly weaker stratification (Figure 5-4c).

Patterns of occurrence of fish larvae, zooplankton and oceanographic factors across samples

For all six species of larval fish (Atlantic cod, redfish, Greenland halibut, long rough dab, wolffish, sandeels) examined, the occurrences were distributed non-randomly with regard to at least one of the environmental gradients. Shrimp larvae and zooplankton displacement volume were distributed randomly with regard to all three environmental gradients. Copepod occurrences were distributed non-randomly with regard to mean salinity and mean stratification, but randomly relative to mean temperature. The different fish species differed markedly in the degree of sensitivity to the environmental factors, and in which factor(s) had the strongest association(s) (Figure 5-5a-c; only non-random distributions are presented). Moreover, associations were usually much stronger with biotic variables (occurrences of other plankton) than with the abiotic environmental factors (Figure 5-6a-b; only non-random distributions are presented). All fish larvae and Northern shrimp larvae were distributed non-randomly relative to depth to bottom (Figure 5-7).

Cod larvae were significantly less frequent than expected in samples taken from areas with salinities lower than about 33.4 ‰, temperatures less than 2.0° C., and temperature range less than about 2.0 °C. (Figure 5-5a-c). They occurred more frequently than expected in samples with mean salinity, temperature or temperature range higher than those levels. The difference in distribution relative to expected occurrences is slightly stronger for mean salinity than for mean temperature; the response to temperature range

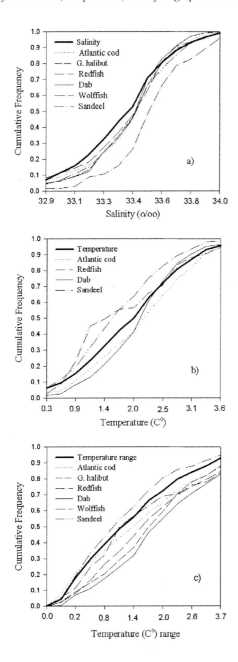

Figure 5-5. Cumulative frequency distributions (cfds) of hydrographic characteristics [salinity (a), temperature (b), and temperature range (c)] in relation to cfds of larvae abundance [$\log_{10}$ (catch in numbers + 1)], 1950-1984. Larval species containing abundance cfd levels inside the 99% confidence interval of the cfd of cases not included, mainly northern shrimp.

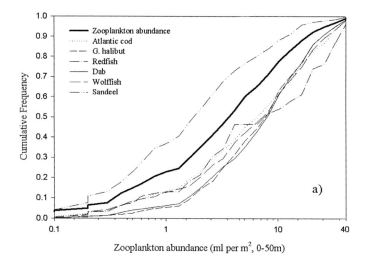

Zooplankton abundance (ml per m², 0-50m)

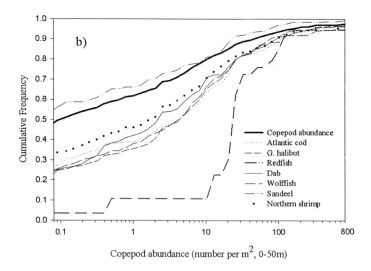

Copepod abundance (number per m², 0-50m)

Figure 5-6. Cumulative frequency distributions (cfds) of zooplankton abundance (displacement volume)(a) and copepod abundance (b) in relation to cfds of larvae abundance [$\log_{10}$ (catch in numbers + 1)], 1950-1984. Total number of cases: (a)=676, and (b)=315. In (a) the cfd of northern shrimp larvae abundance was not significantly (p>0.01) different from the cfd of cases and is therefore not included in the plot.

was substantially weaker. The association of occurrences of cod larvae with total plankton volume was stronger than with any of the physical oceanographic factors, and the association specifically with copepods was stronger yet (Figure 5-6a,b). Cod larvae were more frequent than expected when plankton volumes were greater than about 8 ml per m², when copepod abundance exceeded 5 per m², and at depth to bottom between 50-400 m (Figure 5-7).

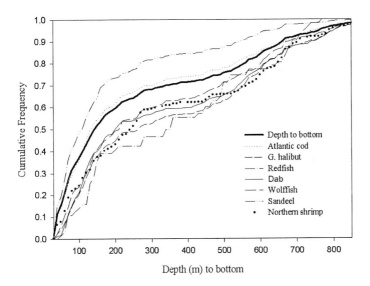

Figure 5-7. Cumulative frequency distribution (cfd) of bottom depths in relation to cfds of larvae abundance ($\log_{10}$ (catch in numbers +1)), 1950-1984. Total number of cases is 618.

Greenland halibut larvae were significantly less frequent than expected in samples taken from areas with salinities lower than about 33.2 ‰, and temperature range less than about 1.4 °C. (Figure 5-5a,c). They occurred significantly more frequently than expected in samples with temperature range higher than 1.4 °C, and marginally more frequently than expected in areas with high salinity (above 33.4 ‰). The occurrence of Greenland halibut larvae along the temperature axis showed a pattern not different from chance. The association of occurrences of Greenland halibut larvae with high total plankton volumes and with high abundance of copepods was much stronger than with salinity or the stratification index (Figure 5-6a,b). Greenland halibut larvae were more frequent than expected at depth to bottom above 400 m (Figure 5-7).

Redfish larvae showed markedly non-random patterns of occurrence with regard to all three environmental factors (Figure 5-5a-c). The association was strongest with the salinity where they were significantly more common than expected in samples with salinity greater than 33.4 ‰ (Figure 5-5a). They were also more frequent than expect in samples with mean temperatures between 0.5 and 2.0 °C, but less frequent than expected on both sides of that range (Figure 5-5b). The association with indices of stratification was weaker, with a tendency to be more common than expected in areas with temperature range above 2.6 °C. Redfish larvae were significantly rarer than expected in areas with total plankton volumes less than 2 ml per m^2, and more common than expected in areas with values greater than 10 ml per m^2 (Figure 5-6a). The relationship with copepods was even more marked, with almost all the redfish larvae found in samples containing more than 10 copepods per m^2 (Figure 5-6b).

Long rough dab larvae showed weak associations with both salinity and temperature, being less common than expected below 33.2 ‰ and 2.0 °C, respectively (Figure 5-5a,b). The association was much stronger with the indices of stratification factor and both biotic variables. They were more common than expected in samples with temperature range greater than 1.4 °C, and in samples with total plankton volumes or copepods more than 5 ml and 0.5 per m^2, respectively (Figure 5-6a,b).

Wolffish larvae were uncommon in most samples, and only weakly associated with salinity (Figure 5-5a). They were significantly less common than expected in samples with temperature range less than 1.4 °C (Figure 5-5c). The associations with total plankton volume were somewhat stronger, occurring more commonly than expected in areas with total plankton volume greater than 8 ml per m^2. The association with copepods was stronger yet, again occurring much more frequently than expected in areas with copepod abundances greater than 5 per m^2. Wolffish larvae are substantially less common than expected in samples from sites shallower than 200 m, and more common than expected in sites deeper than 500 m.

Sandeels larvae showed patterns of occurrence different from all the other species of fish. There were weak but significant associations with the physical oceanographic factors, but the associations with temperature and stratification indices were negative (Figure 5-5b,c). Sandeels were less frequent than expected in areas with temperature warmer than about 1.8 °C, and in samples with temperature range greater than about 2.2 °C. They also show a significant association with total plankton volume (more sandeels than expected when plankton volume is between 0.2 and 4 ml per m^2) and a much weaker but negative association with copepod numbers (more sandeels than expected when copepod numbers are below 10 per m^2) (Figure 5-6a,b). In the case of sandeel larvae the pattern of occurrence with depth was also distinct (Figure 5-7).

Because of the uneven sampling of depths among transects and year, apparent relationships between the occurrence of other fish larvae and depth may contain little information. (It is true that the salinity and temperature ranges were not sampled equally in all years either. However, the interannual differences in occurrences of various physical conditions are a feature of interest, and are part of the analysis and interpretation. Differences across years in occurrences of samples from different depths are a sampling design feature, of little biological interest.) For all the investigated species, occurrence was significantly nonrandom relative to depth, with infrequent occurrences in shallow areas (Figure 5-7). Sandeel larvae were significantly more common than expected in samples from depths less than 150 m, and less common than expected in samples from sites with depths greater than 300 m (Figure 5-7).

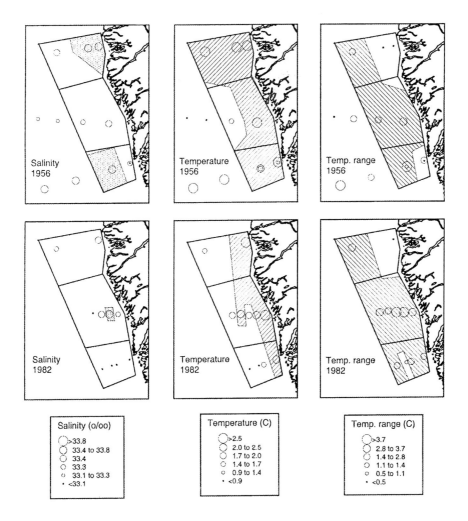

Figure 5-8. Two examples (1956 and 1982) of measuring hydrographic areas (salinity, temperature and temperature range) with above average probability of cod larvae occurrence (salinity>33.4 ‰, temperature>2.0 °C, and temperature range>1.4 °C). The areas were measured using a geographical information system.

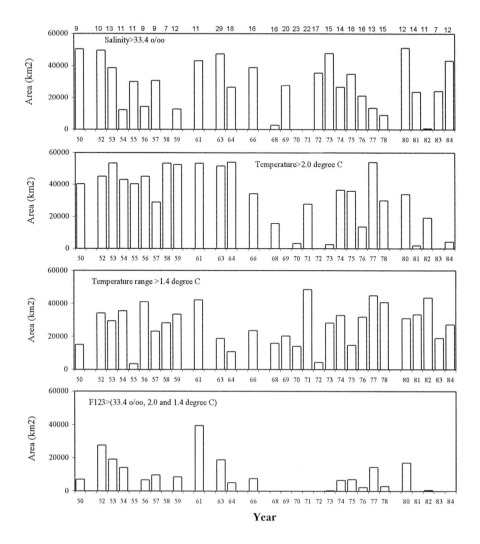

Figure 5-9. Inter-annual variability in hydrographic areas with above average probability of cod larvae occurrence (salinity>33.4 ‰, temperature>2.0 °C, and temperature range >1.4 °C). F123 (lower plot) is the area of overlapping areas with relative higher expectations of cod larvae occurrence. The number of samples is given on the top frame.

Patterns of oceanography, zooplankton, fish larvae, and year-class strength over time

The previous analyses characterize oceanographic conditions associated with high and low abundance (or probability of occurrence) of each species of ichthyoplankton and zooplankton. Over the period 1950 - 1984 the portion of the study area estimated to have conditions characteristic of a higher probability of occurrence or higher density of each species varied greatly (Figures 5-8 and 5-9). For example, the extent of waters with hydrographic conditions characteristic of a higher probability of Atlantic cod larvae occurrence (F1,2,3>33.4‰, 2.0 and 1.4 °C, respectively) was greater in 1950s and early 1960s than in the 1970s and early 1980s (Figure 5-9).

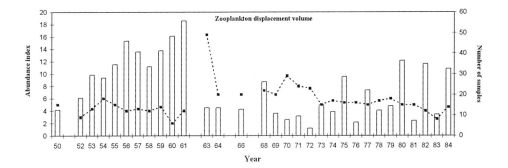

Figure 5-10. Plankton abundance indices (columns) and number of samples (squares), 1950-1984. Breaks in the series these are indicated with spaces on the x-axis.

Zooplankton abundance was generally greater in the period 1953 to 1961 than after 1968 (Figure 5-10), with the intervening years transitional. Zooplankton abundance also fluctuated more between years after 1968. In parts of the zooplankton time series copepods and shrimp larvae have been systematically sorted out and counted from samples (Figure 5-11). Total copepod abundance showed strong fluctuations between years but no clear trend over time. Only in 1976 were no copepods caught. Also *Calanus finmarchicus* and northern shrimp larvae abundance showed strong fluctuations with no clear trend over time (Figure 5-11). However, from 1973 to 1977 no copepods were identified to *C. finmarchicus*. The abundance of copepods, *C. finmarchicus*, and northern shrimp larvae were all exceptionally high in 1982 (Figure 5-11).

There was a trend, albeit weak, of generally higher abundance of Atlantic cod larvae in the period 1950 to 1961 and again in the years 1982, 1983, and 1984 compared to the period 1963 to 1981 (Figure 5-12). The former period also shows a generally lower percentage of stations with nil catch compared to the latter period. Greenland halibut larvae abundance showed strong fluctuations over the years with no clear trend over time (Figure 5-12). Only a low and infrequent number of redfish larvae were caught in the study area over the years and only in the southern part (Figure 5-12). Abundance of dab larvae showed strong fluctuations over the years with a weak trend of less frequent occurrence in the samples after 1969 (Figure 5-12). Also wolffish larvae abundance showed fluctuations over the years with no clear trend over time. However, catches in

1980 and 1982 were exceptionally high (Figure 5-12). Sandeel larvae abundance showed clear trends of low larval abundance in the period 1950-1968, high larval abundance in the period 1969-1976, low again in 1977-1981, and relatively high again in 1982-1984 (Figure 5-12). With the exception of sandeel and redfish, the abundance of copepods and other ichthyoplankton were exceptionally high in 1982 (Figures 5-11 and 5-12).

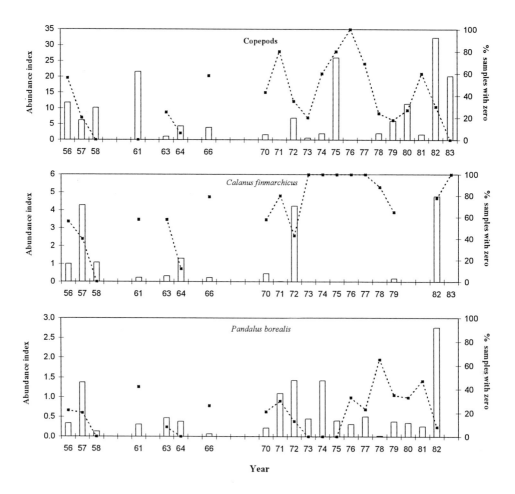

Figure 5-11. Copepod, *Calanus finmarchicus*, and northern shrimp larvae abundance indices (columns) and percentage of samples with nil catch (squares), 1956-1983. Breaks in the series these are indicated with spaces on the x-axis.

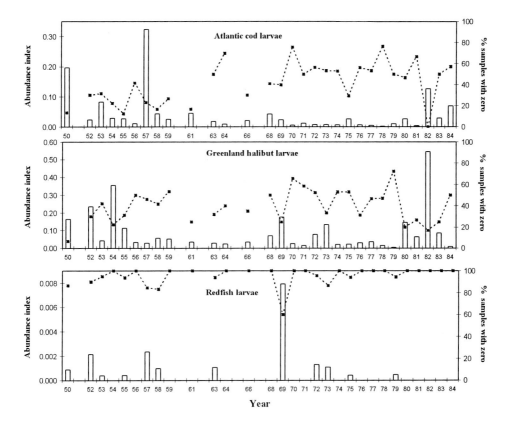

Figure 5-12 (this and next page). Fish larvae abundance indices (columns) and percentage of samples with nil catch (squares), 1950-1984. Breaks in the series these are indicated with spaces on the x-axis.

Of the hydrographic factors, interannual variation in zooplankton abundance was weakly correlated only with mean temperature (n=557, r=0.28, p<0.01). Copepods, *C. finmarchicus*, and northern shrimp larvae abundance showed no relationship over time with any of the hydrographic factors. Of larval fish, only abundance of sandeel and redfish larvae showed significant correlations over time with the extent of specified hydrographic conditions (Table 5-3, Figures 5-8 and 5-9). Sandeel larvae abundance was correlated with the occurrence of low temperature (n=30, r=0.59, p<0.01) and redfish larvae abundance was correlated to the extent of occurrence of high salinity areas (n=30, r=0.37, p=0.05). The kernel estimators did not improve or increase the number of relationships between hydrographic factors and different zooplankton species.

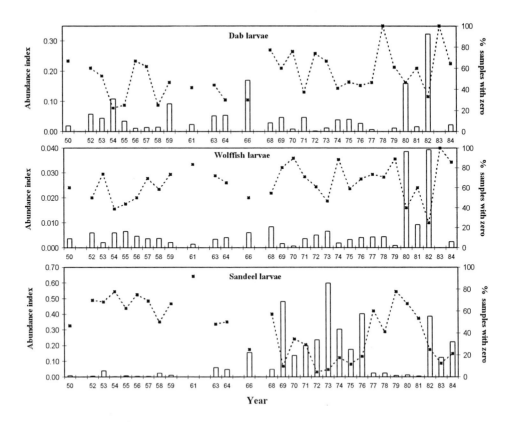

Figure 5-12 (cont'd.). Fish larvae abundance indices (columns) and percentage of samples with nil catch (squares), 1950-1984. Breaks in the series these are indicated with spaces on the x-axis.

Inter annual variation in cod larvae abundance was positively related to biological factors i.e. abundance of zooplankton, copepods and *C. finmarchicus* (Table 5-4). Of other larvae only the abundance of dab larvae and zooplankton showed a positive relationship (n=31, r=0.37, p=0.04). Ordinary linear regression explained 27% of the total variation between abundance indices of cod larvae and zooplankton (or *C. finmarchicus*) while the non-parametric analysis explained much less (15% and 0.02% respectively, Table 5-4). This indicates that the regression model may over-predict the relationships, due to the effect of few outliers or lack of a smooth, continuous relationship. Indices of cod year class strength (numbers at age 3) were correlated to the annual extent of areas with appropriate temperatures, abundance of zooplankton, and abundance of cod larvae (Table 5-4). In all cases the kernel analysis explained less than ordinary linear regression.

Table 5-3. Hydrographic conditions with higher probability of occurrence of fish larvae or copepods.

Species/group	Salinity ‰	Temperature (°C)	Temperature (°C) Range
Atlantic cod	>33.4	>2.0	>1.4
Greenland halibut	>33.2	-	>1.4
Redfish	>33.4	-	>1.4
Long rough dab	>33.2	>2.0	>1.4
Wolffish	>33.4	-	>1.4
Sand eel	>33.4	<2.0	<1.4
Copepods	>33.4	-	>1.4
Calanus finmarchicus	>33.4	-	>1.4

Table 5-4 . Larvae and year class (age 3) abundance indices of Atlantic cod in relationship with areas (km²) of hydrographic conditions with relative higher probability of larval occurrence and zooplankton abundance indices. (-) indicate that the kernel analysis or ogive method did not converge.

		Relationship method			
		Regression		Ogive	
Hydrographic/biological factor	No. obs.	r^2	p	Best window	% reduction
Cod larvae abundance index					
Salinity >33.4 o/oo	30	0.06	0.21	21500	0.03
Temperature >2.0 °C	30	0.00	0.77	-	-
Temperature range>1.4 °C	30	0.01	0.55	-	-
F123	30	0.03	0.41	-	-
Zooplankton abundance index	31	0.27	0.00	0.15	0.15
Copepod abundance index	20	0.23	0.03	0.7	0.23
Calanus finm. abundance index	17	0.27	0.02	2.5	0.02
Cod year class (age 3) abundance index					
Salinity >33.4 o/oo	30	0.12	0.06	1500	0.18
Temperature >2.0 °C	30	0.29	0.00	6750	0.22
Temperature range>1.4 °C	30	0.00	0.99	-	-
F123	30	0.22	0.00	4550	0.09
Zooplankton abundance index	31	0.25	0.00	0.14	0.24
Copepod abundance index	20	0.00	0.99	-	-
Calanus finm. abundance index	17	0.03	0.78	-	-
Cod larvae abundance index	31	0.16	0.03	0.41	0.11
Cod larvae abun. (excl. 1982) ind.	30	0.30	0.00	0.09	0.21

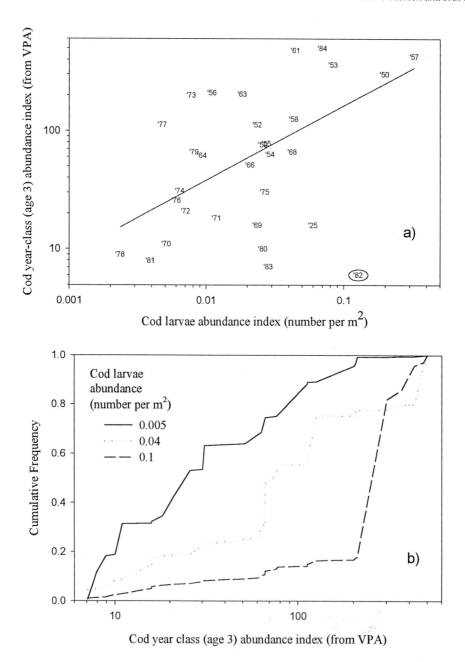

Figure 5-13. Scatterplot of Atlantic cod year-class abundance indices versus larvae abundance indices (a). Solid line shows the fit of a linear regression model (1982 was omitted). Ogives of Atlantic cod year-class indices for three trial larvae abundance indices (1982 was omitted) (b).

Excluding 1982 data in the relationship between the abundance indices of cod year-class and cod larvae increased the variation explained by both linear regression and kernel analyses (Table 5-4, Figure 5-13a,b). At a cod larvae abundance index of about 0.1 the ogives from the kernel analysis showed a probability of less than 0.2 of getting a year-class abundance index below 200 (Figure 5-13b).

DISCUSSION

To eventually address the magnitude and causes of change in the large marine ecosystem of the Greenland Sea, these statistical results need to be synthesized systematically. First we address what interpretations are possible for the individual patterns that were found. Then we can address the changes over time in the patterns of individual species, and how patterns are interrelated among hydrographic and biological time series. The assemblage dynamics built from those patterns provide the framework to allow partial evaluation of the role that various processes have played in the changes in the West Greenland LME.

Interpreting absolute preferences or avoidances

Where taxa are non-randomly distributed, we refer to areas of "preference" or "avoidance". However, these analyses measure only statistical patterns, and not behaviours associated with showing a true preference or avoidance of particular water masses. A species may be more common than expected by chance in a sample due to passive aggregating mechanisms, to actively seeking and maintaining itself in specific conditions, to actively avoiding other conditions, or to higher survivorship under those conditions. It may be rarer than expected because of dispersal away from a sampling locality, active avoidance of specific conditions, or lower survivorship in that area. Moreover, differences in spawning areas of adults of various species will give larvae different probabilities of being found in different parts of the area sampled. Hence it is necessary to consider usual spawning locations when interpreting patterns of occurrence of larvae.

Factors affecting distribution of individual species

Atlantic cod
Spawning sites - Spawning sites of cod have changed substantially over the period of this study (Hansen 1949; Buch *et al.* 1994). For the entire time series there has been spawning in fjords along the West Coast of Greenland, particularly between latitudes 64 and 67°N. These spawning groups introduce eggs and larvae directly into nearshore waters of Davis Strait, where they may move offshore. In addition, between the years 1925 to 1984 large numbers of eggs and larvae from spawning grounds off Iceland

probably were carried by the strong Irminger Current west and south along the southern tip of Greenland, and into Davis Strait from the south (Harden Jones 1968; Dickson and Brander 1993; Buch et al., 1994). Over this period large spawning populations built up to the east and southwest of Greenland along the shelf at depths 50-200 m. Products of these spawners were also carried west and south of Cape Farewell and up into Davis Strait as far north as about 70°N. The latter source of eggs and larvae was drastically reduced by 1968 due to depletion of the offshore spawning stock of cod (Buch et al. 1994).

None of the associations between cod larvae and hydrographic features were strong, and were weakest with the stratification factor. For the middle decades many cod larvae entered Davis Strait via the southern transect, which had relatively low salinity scores. In addition, cod in the fjords annually introduced eggs and larvae into nearshore areas of the middle transect. This is a highly stratified transition area where salinities resemble more southerly waters in some years and more northerly waters in other years. However, the preferences of cod did not correspond to the waters where the spawning products were introduced. Overall, cod larvae were less abundant than expected in waters of lower salinity, and the relationship with stratification suggested a tendency for low abundance of cod larvae in poorly stratified waters, rather than a strong preference for strongly stratified waters. Together these patterns indicate that periods when cod larvae were abundant were likely to be periods when the southern transect was more saline than its long-term average condition, or when the middle transect was neither atypically fresh nor poorly stratified.

The associations with zooplankton volume and copepod abundance were much stronger than associations with any of the hydrographic features. This suggests either abundance of cod was more strongly determined by system productivity than by hydrographic conditions, or that both cod larvae and plankton were aggregated and dispersed by the same mechanisms. In fact, the observed weak preferences for stratified, higher salinity waters might be a secondary consequence of the higher productivity of larval food associated with hydrographic fronts (Cushing 1989; Munk et al. 1995). The extent and strength of these fronts may be the important factor in cod productivity.

Greenland halibut
Spawning Sites - Greenland halibut spawning takes place in January to March, in two major concentrations. One is at 1000 - 3000 m in southern Davis Strait, between approximately 62 and 65°N (Jensen1935; Smidt 1969; Riget and Boje 1989; Jørgensen 1997). The other is at similar depths and latitudes between east Greenland and Iceland (Sigurdsson 1979; Sigurdsson and Magnusson 1980). Eggs and larvae produced by the first spawning group remain in Davis Strait. Depending on details of annual conditions they may be transported either inshore or offshore, and either northward or in a gyre first to the southwest and then back to the northeast. Spawning products from the second

group may be transported by the Irminger Current south around Cape Farewell to enter Davis Strait and the southwest Greenland fjords.

By summer Greenland halibut larvae are widely dispersed throughout the sampling area. The distribution of well-stratified areas with salinity greater than about 33.4 ‰ varied greatly from year to year. In some years the areas characteristic of low Greenland halibut abundance, with salinity below about 33.2 ‰ or temperatures below about 1°C, were widespread, particularly in the southern transect. The relationship with plankton volume was much stronger than the relationship with stratification. This suggests that in addition to whatever transport mechanisms may aggregate both Greenland halibut larvae and plankton in stratified and relatively high salinity waters, the Greenland halibut either survive better in the high productivity areas, or actively maintain themselves in areas of high plankton abundance.

Redfish
Spawning sites - Redfish extrude their larvae over an extended period from April to August, at 300 to 600 m. Spawning occurs in a large area southeast of Greenland and southwest of Iceland in the Irminger Sea along the Reykjanes Ridge (Kotthaus 1965; Anon. 1983, 1984; Pavlov *et al.* 1989; Sibanov *et al.* 1995). The larvae and fry are transported to East and West Greenland waters (Pedersen 1990; Pedersen and Kanneworff 1995; Magnusson and Johannesson 1995). To varying degrees they are carried by currents into the Davis Strait.

There is a strong association of redfish with high salinity waters of intermediate temperature and strong stratification. Almost all larvae were in samples with very high abundance of copepods. Furthermore, all redfish larvae in the sampling area are transported in from the south, and spawning is late enough that they are unlikely to be transported throughout Davis Strait by the time sampling was conducted. In most years the southern transect had low salinity waters and were weakly stratified. Therefore, the patterns of occurrence suggest that conditions favourable to survivorship of redfish larvae occurred infrequently over the sampling period, and coincided with periods of very high productivity. The conditions could be either years of high salinity and stratification in the southern zone, or years of very early redfish spawning and rapid transport, so large numbers of larvae could reach the central and northern sampling areas.

Long rough dab
Spawning sites - Available evidence suggests that long rough dab spawning occurs during March to July, in coastal areas, fjords, and along a wide latitudinal band off the West Greenland coast between 62 and 66°N, with a smaller occurrence off Iceland (Wells 1968; Muus *et al.* 1981; Jørgensen 1992). Whereas there probably is considerable transfer of eggs and larvae by the West Greenland current, it appears that once the fish have settled on the bottom movement is minimal (Pitt 1969).

Dab larvae showed a preference for well-stratified waters, and areas with moderate to high concentrations of zooplankton. The middle transect, adjacent to the suspected spawning areas, is generally most strongly stratified during the sampling period. This pattern would be consistent with some retention of eggs and larvae near the presumed coastal spawning sites, although by the sampling period in June and July dab larvae were much more common in areas with depths greater than 500 m than in shallower, coastal sampling sites, indicating at least movement offshore.

Wolffish
Spawning sites - Almost nothing is known concerning the propagation in Greenland waters of the three *Anarhichas* species (Hansen 1968). From other areas it is known that all three species deposit their eggs in lumps at considerable depths among stones on the bottom. It is believed that adult *Anarhichas minor* live and spawn chiefly in the southern part of West Greenland waters. The larvae which seek the surface are carried by the current to northwest Greenland waters where they go to the bottom and live until they grow to maturity and migrate to the southern spawning grounds (Hansen 1968).

Wolffish larvae occurred generally infrequently, with catches widespread over deeper waters. They do not show close associations with any of the hydrographic factors, beyond some avoidance of extremely weakly stratified sampling areas.

Sandeels
Spawning sites - Two species of sandeel occur at depths between 10 and 80 m off West Greenland, *Ammodytes dubius* and *A. marinus* (Muus *et al.* 1981; Andersen 1985; Horsted 1991). Neither species is common inshore (Horsted 1991) and larvae are scarce in the fjords (Smidt 1979). Little is known of sandeel spawning times, but *Ammodytes dubius* in Newfoundland spawn principally during the period November-January (Winters 1983).

Unlike all other fish larvae, sandeels occurred in water masses that were colder and weakly stratified. The plankton densities characteristic of high abundance of sandeel larvae were lower than the densities characteristic of high abundance of other fish larvae, with sandeels less frequent than expected in samples with high plankton and copepod densities. Sandeel densities were highest in samples from shallower sites than were preferred by other fish larvae as well. These patterns indicate sandeels spawn on the banks in coastal areas, and larvae reach high densities only in relatively shallow areas, which are cold and well mixed.

Northern shrimp
Spawning sites - *P. borealis* in West Greenland spawn in offshore waters starting in July and ending in late August or early September (Horsted 1978). The egg mass is carried by females until spring (March-April) when females move to shallow water to release the first pelagic larval stage (Horsted 1978; Shumway *et al.* 1985). The planktonic larvae are

thought to drift more or less passively during five zoeal stages (approximately 4 months) and settle to the bottom during the sixth (megalopa) stage (Horsted *et al.*, 1978; Shumway *et al.*, 1985).

Shrimp showed no association with specific hydrographic characteristics, being widespread throughout deepwater sites of moderate to high plankton productivity suggesting a wide larval releasing and feeding area. A distribution study of *Pandalus* shrimp larvae in stage 3 and 4 in 1996 support the latter (Pedersen 1998).

Assemblage patterns

Sandeels showed a distinct preference for cold and well mixed waters in relatively shallow areas. Larvae of no other fish species were common under such conditions. Instead, all other species of fish larvae show relatively weak differentiation of water mass preferences. Temperature preferences were weaker than salinity preferences. For wolffish and Greenland halibut abundance was independent of temperature. There was evidence of a gradient of occurrence with regard to salinity, but it was not a partitioning of distributions among species. Rather, the salinity at which all four most common species of fish became less common than expected was very similar; cod and redfish were less common than expected by chance in waters fresher than 33.4 ‰ whereas Greenland halibut and Long rough dab were less common in waters fresher than 33.2 ‰.
  In waters more saline than 33.4 ‰ larvae of all six fish species were more common than expected from the distribution of samples.

Cod, redfish and sandeels larvae showed less preference for well-stratified waters than wolffish, Greenland halibut and dab. However, distributions of redfish and cod larvae overlapped on the stratification gradient, and positions at which tolerance changed were weakly defined. Greenland halibut, long rough dab, and wolffish larvae were more common than expected only when the thermal stratification exceeded 1.4°C. Aside from sandeels, all fish species reached highest abundances in samples also containing relatively high densities of zooplankton. In all cases the association was stronger with copepods than with total plankton volume. Redfish showed the strongest association with abundance with copepods, and also had the highest threshold for occurrence. Redfish larvae were less common than expected until copepod densities reached about 10 per m^2. Cod, Greenland halibut, wolffish and Northern shrimp larvae became consistently more common than expected when copepod abundances exceeded around 3 per m^2, and dab at slightly lower copepod abundances. Long rough dab only became more abundant than expected when total plankton volume (including copepods) exceeded 8 ml per m^2, whereas for Greenland halibut the threshold was 3 ml per m^2, the same as their threshold for copepods alone. Shrimp abundance was only related to copepod density, independent of hydrography conditions and total zooplankton volume (marginally).

Taken together, these patterns do not suggest a suite of fish species whose larvae are adapted to different oceanographic conditions and different ecological niches. Rather, for all species but sandeels, June and July abundances are likely to be higher when water masses are relatively saline and moderately stratified, and when plankton, especially copepods, are abundant. The differentiation is with regard to how saline, how dense in plankton and copepods, and how broad a thermal gradient is required before high abundances become more likely than expected by chance. (Or, comparably, how fresh, how prey poor, and how mixed waters are before species become rarer than expected.) Such an assemblage is consistent with the view of the ecological community of fish larvae as opportunistic feeders on an unpredictable food resource, but a food supply that can be abundant relative to their ration requirements.

The patterns also suggest environmental conditions that are likely to bring about shifts in the living component of the West Greenland large marine ecosystem. Periods of strong inflow of fresh, cold Polar waters are unfavourable for most species of fish larvae. Several mechanisms can be involved causally. For many species, the Irminger Current can transport large numbers of larvae into West Greenland waters from the south around Cape Farewell, if spawning biomasses are adequate in East Greenland or Icelandic waters. This source of larvae may be important for Atlantic cod, redfish, and Greenland halibut. A strong Polar Current would oppose high inflow of the Irminger current, and reduce this source of larvae. The Irminger Current also may act indirectly, bringing the more saline conditions associated with higher abundances of most fish larvae except sandeels. The Polar Current may also transport larvae and their food sources out of Davis Strait, or out in deeper areas, which were poorly sampled. Finally the physical conditions associated with polar waters may be truly unfavourable for eggs or larvae, leading directly to reduced survivorship and abundance. These associations between abundances of the various larvae and the hydrographic gradients indicate that it may be possible to relate interannual variation in larval abundance and subsequent recruitment of year classes to the distribution within the Strait of water masses with varying characteristics. Those analyses are being undertaken currently (e.g. Pedersen, 1998).

Long term variability in zooplankton and fish larvae abundance

In the study area, there was a decreasing trend in the zooplankton abundance indices from the late 1950s and early 1960s to the 1970s (Figure 5-10). Both temperature and indices of zooplankton abundance decreased after 1968 and fluctuate thereafter at lower levels, possibly signalling a general decrease in the overall zooplankton productivity and food for fish predators. This is supported by reduced weight-at-age and recruitment of the West Greenland offshore cod stock after 1968 (Brander 1995; Riget and Engelstoft 1998).

In northern Icelandic waters in spring from the early 1960s to 1970, Astthórsson et al. (1983) also found a decreasing trend in plankton abundance indices and especially of the copepod C.

finmarchicus. According to Astthórsson *et al.* (1983) a reduced influx of Atlantic water to the areas north of Iceland in the late 1960s probably delayed the onset of the spring primary production and thus the zooplankton production. These changes in primary and secondary productivity probably caused a reduction in the productivity of fish stocks in Icelandic waters (Stefansson and Jakobsson 1989).

Similar events may have occurred in Southwest Greenland during the years 1969-1970 due to what has been refereed to as the "Great Salinity Anomaly" (GSA) in the northern North-west Atlantic (Dickson *et al.* 1988; Mertz and Myers 1994). Mertz and Myers (1994) found no significant temperature anomalies south of Greenland corresponding to the GSA and weak evidence for an impact from the GSA on food chain coupling of environment to recruitment. We found evidence of the GSA in salinity, temperature, and stratification measures in the years 1969 and 1970, and zooplankton indices were relatively low during those years (Figures 5-9, 5-10 and 5-11). Zooplankton indices did not recover quickly after the physical signals of the GSA had passed. However indices of many fish larvae were actually average or high from 1969 to 1971, especially for sandeels and redfish. As Mertz and Myers concluded, the ecological consequences of the GSA are not as simple as conjectured in some earlier publications.

The increased sandeel abundances in the study area after 1970 correspond to a general increase in the inflow of the East Greenland Polar Current, bringing cold fresh waters into the survey area. Unexpectedly, redfish larvae also occurred in some of the years of a strong inflow of the East Greenland Polar Current e.g. 1969 (Figure 5-12). Of the fish larvae only the abundance of Atlantic cod larvae showed weak but significant associations with zooplankton, copepod, and *C. finmarchicus* annual abundance indices (Table 5-4). *C. finmarchicus* may be of crucial importance for the cod larvae survival (Bainbridge and McKay, 1968; Cushing, 1995a), although the direct link to interannual variability in fish stock recruitment has not been confirmed (Miller, 1995).

In 1982, the relatively high mean number of Atlantic cod larvae caught in the stramin net hauls on the West Greenland sampling sections 1-3 indicated a good prospect for a large cod recruitment (Figure 5-12). However, the 1982 year-class became poor (Figure 5-13). It has been assumed that the failure of the 1982 year-class was caused mainly by the extremely low winter temperatures in West Greenland during 1982-1984 (Rosenørn *et. al.* 1985). The failure of the 1982 year-class of Atlantic cod and the inability to relate the catches of larvae to subsequent recruitment were the main reasons that put the end to the time-series of zooplankton collections off West Greenland in 1985.

We found weak but significant associations between cod year-class strength (given number at age 3) and areas of both high temperature and cod larvae abundance. Hansen and Buch (1986) also found the year-class strength (given at number at age 3) of Atlantic cod off West Greenland to be positively correlated with the mean water temperature (surface to 45 m over Fyllas Bank in June) and larval abundance.

The field sampling for these analyses concluded in 1984, making it impossible to apply these fully analytical approaches to more recent trends in fish and plankton abundances. However, many aspects of the explanations appear to hold up more recently. For example, the abundance of shrimp has been stable throughout the 1990s (Carlsson et al. 1998), despite substantial changes in hydrographic conditions (Stein 1998a, Anderson 1999). This is consistent with the analytical results indicating larval shrimp abundance was not strongly associated with particular hydrographic conditions, at least within the range recorded historically. This implies (but does not prove) the current continued high abundance of shrimp may be associated with low abundance of predator stocks.

A multispecies trawl survey for groundfish also began in 1982. Unfortunately there are insufficient years of overlap to calibrate the two series, the spatial coverage of the groundfish survey includes less of West Greenland waters but extends into East Greenland, and the hydrographic measure currently available is bottom temperature, which was not one of the attributes measured as part of the historic series. However, the emerging patterns of that series (Rätz 1998) are consistent with the analyses of the larval series. The only pulse of Atlantic cod appeared mostly in West Greenland immediately following a marked influx of Irminger Sea water there in 1984 (Hovgård and Buch 1990), which appeared as a major warming event between 1984 and 1985 in Rätz (1998). Both of the pulses of older redfish, *S. marinus* through the 1980s to 1990, and *mentella* since 1995, show up only in East Greenland, consistent with the very rare recruitment events for redfish in West Greenland. A recruitment pulse of redfish juveniles in East Greenland waters is strong in 1993, and modest from 1995 onward in Cape Farewell and southeast Greenland waters in 1992 and 1993, following a marked temperature minimum in that area in 1992. Long rough dab abundance declines consistently through the 1980s, and recovers weakly, and only in the East, after 1995. The declining temperatures through the 1980s, particularly in the northern part of the area following the single warming event in 1984, would go along with the preference found for warmer waters in the analyses of long rough dab larvae in the earlier years. We stress, though, that these inferred similarities require further investigation, particularly because the hydrographic attributes we analyzed were for the upper 50 m of the water column, and Rätz (1998) reports only bottom temperatures.

CONCLUSIONS

Our analyses found only weak coupling of biological and physical variables with larval abundance, and little differentiation among the fish species in the factors associated with high abundance. Larvae of many fish species were most abundant when zooplankton and *Calanus* were also at high density. We note there were several design factors which could weaken our ability to detect relationships in the time series of this study. These include the low number of sampling stations, the mixture of oblique and depth integrated hauls in the upper 50 m, and coarse spatial extrapolations from a low number of oceanography sampling stations to large areas. However, these concessions were necessary in order to use one of the longest time series available for studying the effects of physical and biological factors on variation in fish larval abundance and year classes in a boreal system. The general patterns found in these analyses are likely to be sound, and although more intensive studies certainly can contribute greatly to understanding the mechanisms behind the processes, they are unlikely to alter our estimations of the strengths of the relationships. Moreover, they would almost certainly have to be on more local spatial and shorter time scales, and sacrifice some of the scope captured in the data sets analysed here.

With regard to major changes in the West Greenland ecosystem, a noticeable minimum in all three oceanographic factors - salinity, temperature, and stratification - occurs in the late 1960s and early 1970s, corresponding to the well-known GSA. The decline in scores of all three oceanographic factors began well before this minimum, as did declines in zooplankton abundance indices and catches of Atlantic cod and redfish. However, fish larval abundance was, in fact, higher during the minimum than during adjacent years for most species. Only for sandeel and *Pandalus*, though, did larval abundance show a major shift, and remain high after 1970. For other fish species and for zooplankton, abundances were average to below average in the 1970s, while the annual scores on oceanographic factors showed average to above average conditions. Only in the early 1980s did fish larval and zooplankton abundance increase markedly again, following a change to relatively cold but highly variable saline conditions, with moderate to strong stratification.

The difficulty in identifying clear regime shifts in this system may not be surprising, given the overall similarities in preferences shown for the various fish species. Cod, Greenland halibut, redfish, dab, and wolffish appear to have similar responses to gradients in oceanographic conditions, and strong affinities for high plankton densities. Only sandeels and *Pandalus* appear to be out of phase. Hence, as the influences of the two major oceanographic features, the East Greenland Polar Current and Irminger Current from the south, increase or decrease, groundfish are either favoured (strong saline Irminger Current) or not (strong cold fresh Polar Current). Although the time series of samples ends in 1984, the continued low abundance of most fish stocks and the strong fishery for *Pandalus* both suggest generally unfavourable conditions have persisted through the past decade. This corresponds to the continued cold sea climate in West Greenland through the 1990s (Stein 1998b,c).

According to Sinclair and Page (1995) there is a need for increased consensus by the scientific community on which of the competing hypotheses on population regulation best capture the critical mechanisms. These results suggest that there may be multiple regulatory factors, even for simply oceanographic processes, individually of only modest strength, and the search for the "best" critical mechanisms may be futile. Moreover, data necessary to evaluate the competing - or augmental - hypotheses of predation and bycatch are only available for years long after the plankton and fish larval sampling was terminated. We hope the present paper will stimulate and add direction to future studies on pelagic ecology, allowing more complete investigations of these factors and recruitment mechanisms for fish and shellfish stocks in West Greenland marine waters.

ACKNOWLEDGEMENTS

The majority of the analyses in this paper were conducted while the second author was a guest professor of the Danish Academy of Sciences, in residence at the Danish Institute for Fisheries Research/University of Copenhagen. We acknowledge the many people involved over the years at the *Danish* and *Greenland Institutes for Fisheries Research*, especially E.L.B. Smidt for making the zooplankton data available for our analyses. Also thanks to Director Klaus Nygaard (*Greenland Institute of Natural Resources*) for support. Finally, we thank Rasmus Nielsen, the ICES Secretariat and Karen L. Nielsen for help with the hydrographical data. During part of the work with this paper the Commission of Scientific Research in Greenland and Danish National Research Council project no. 9803018 funded the first author. The authors appreciate the financial support of IUCN, to allow attendance at the LME Symposium.

REFERENCES

Aebischer, N.J., J.C. Coulson, and J.M. Colebrook. 1990. Parallel long-term trends across four marine trophic levels and weather. Nature 347, no. 6295:753-755.

Andersen, O.G.N. 1985. Forsøgsfiskeri efter tobis i Vestgrønland 1978, del. 1 og 2. (A trial fishery for sandeel at West Greenland). Report Greenland Fisheries and Environmental Research Institute, Series III. 54 p.

Anderson, P.J. 1999. Pandalid shrimp as indicator of ocean climate regime shift. NAFO SCR Doc. 99/80. No. 4152. 14 p.

Anon. 1983. Report on the joint NAFO/ICES study group on the biological relationships of the West Greenland and Irminger Sea redfish stocks. ICES CM 1983/G:3.

Anon. 1984. Report on the joint NAFO/ICES study group on the biological relationships of the West Greenland and Irminger Sea redfish stocks. ICES CM 1984/G:3.

Anon. 1985. SAS User's Guide: Statistics Version 5. Edition. SAS Institute Inc., Raleigh, North Carolina.

Anon. 1995. NAFO Statistical Bulletin, Supplementary Issue, Fishery Statistics for 1960-90. Dartmouth, Nova Scotia, Canada. 156 p.

Anon. 1998a. NAFO Scientific Council Reports 1997. Dartmouth, Nova Scotia, Canada. 274 p.

Anon. 1998b. Report of the North-western Working Group. ICES CM 1998/ACFM:19. 350 p.

Astthórsson, O.S., I. Hallgrimsson, and G.S. Jónsson. 1983. Variation in zooplankton densities in Icelandic waters in spring during the years 1961-1982. Rit Fiskideildar 7(2):73-113.

Bainbridge, V. and J. Corlett. 1968. The zooplankton of the NORWESTLANT surveys. Int. Comm. Northwest Atl. Fish. Spec. Publ. 7 (I):101-122.

Bainbridge, V. and B.J. McKay. 1968. The feeding of cod and redfish larvae. Int. Comm. Northwest Atl. Fish. Spec. Publ. 7 (I):187-217.

Belsley, D.A., E. Kuh, and R.E. Welsch. 1980. Regression Diagnostics: Identifying Influential Data and Sources of Colinearity. John Wiley: New York.

Brander, K.M. 1995. The effect of temperature on growth of Atlantic cod (*Gadus morhua* L.). ICES J. Mar. Sci. 52:1-10.

Buch, E. 1990. A monograph on the physical environment of Greenland waters. Greenland Fisheries Research Institute Report. 405 p.

Buch, E. 2000. Oceanographic Investigations off West Greenland, 1999. NAFO SCR Doc. 00/1., Serial No. N4213. 15 p.

Buch, E., S.Aa. Horsted, and H. Hovgård. 1994. Fluctuations in the occurrence of cod in Greenland waters and their possible causes. ICES Mar. Sci. Symp. 198:158-174.

Carlsson, D., O. Folmer, C. Hvingel, P. Kanneworff, M. Pennington, and H. Siegstad. 1998. A review of the trawl survey of the shrimp stock off West Greenland. NAFO SCR Doc. 98/114. 20 p.

Colebrook, J.M. 1982. Continuous plankton records: seasonal variations in the distribution and abundance of plankton in the North Atlantic Ocean and the North Sea. J. Plankt. Res. 4: 435-462.

Colebrook, J.M. 1985. Sea surface temperature and zooplankton, North Sea, 1948 to 1983. J. Cons. Int. Explor. Mer., 42:179-185.

Conover, W.J. 1980. Practical nonparametric statistics 2ed. John Wiley and Sons, New York. 493 p.

Cushing, D.H. 1989. A difference in structure between ecosystems in strongly stratified waters and in those that are only weakly stratified. J. Plankt. Res. 11:1-13.

Cushing, D.H. 1995a. Population production and regulation in the sea: A Fisheries perspective. Cambridge University Press. 354 p.

Cushing, D.H. 1995b. The long-term relationship between zooplankton and fish. ICES J. Mar. Sci. 52:611-626.

Dickson, R.R., J. Meincke, S.-A. Malmberg, and A.J. Lee. 1988. The "Great Salinity Anomaly" in the northern North Atlantic 1968-1982. Progr. Oceanogr. 20:103-151.

Dickson, R.R. and K.M. Brander. 1993. Effects of a changing windfield on cod stocks of the North Atlantic. Fish. Oceanogr. 2:124-153.

Francis, R.C. and S.R. Hare. 1994. Decadal-scale regime shifts in the large marine ecosystems of the North-east Pacific: a case for historical science. Fish. Oceanogr. 3:279-291.

Grünwald, E. and F. Köster. 1994. Feeding habits of Atlantic cod in West Greenland waters. ICES CM 1994/P:5. 19 p.

Hachey, H.B., F. Hermann, and W.B. Bailey. 1954. The waters of the ICNAF convention area. Int. Comm. Northwest Atl. Fish., Ann. Proc. 4:67-102.

Hansen, H. and E. Buch. 1986. Prediction of year-class strength of Atlantic cod (*Gadus morhua*) off West Greenland. NAFO Sci. Coun. Stud. 10:7-11.

Hansen, P.M. 1949. Studies on the biology of the cod in Greenland waters. Conseil Permanent International pour l'Exploration de la Mer: Rapports et Procés-verbaux des Réunions volume 123:1-77.

Hansen, P.M. 1968. Report on wolffish larvae in West Greenland waters. Int. Comm. Northwest Atl. Fish., Spec. Publ. 7 (I):183-185.

Harden Jones, F.R. 1968. Fish migration. Edward Arnold. Ltd., London. 325 p.

Hermann, F., P.M. Hansen, and S.Aa. Horsted. 1965. The effect of temperature and currents on the distribution and survival of cod larvae at West Greenland. Int. Comm. Northwest Atl. Fish., Spec. Publ. 6:389-395.

Horsted, S.Aa. 1978. Life cycle of the shrimp, *Pandalus borealis* Krøyer, in Greenland waters in relation to the potential yield. Int. Comm. Northwest Atl. Fish., Sel. Papers No. 4:51-60.

Horsted, S.Aa. 1989. Some features of oceanographic and biological conditions in Greenland waters. In: L. Rey and V. Alexander (eds.). Proceedings of the sixth conference of the comité arctique international 13-15 May 1985. E.J. Brill. Leiden. 456-476.

Horsted, S.Aa. 1991. Biological advice for and management of some of the major fisheries resources in Greenland waters. NAFO Sci. Coun. Stud. 16:79-94.

Horsted, S.Aa., P. Johansen, and E. Smidt. 1978. On the possible drift of shrimp larvae in the Davis Strait. ICNAF (NAFO) Res. Doc. 78/XI/93, Serial No. 5309. 13 p.

Horsted, S.Aa. and E. Smidt. 1965. Remarks on effects of food animals on cod behaviour. Int. Comm. Northwest Atl. Fish. Spec. Publ. No. 6:435-437.

Hovgård, H. and E. Buch. 1990. Fluctuation in the cod biomass of the West Greenland Ecosystem in relation to climate. In: K. Sherman, L.M. Alexander, and B.D. Gold (eds.). Large Marine Ecosystems: Patterns, Processes, and Yields. AAAS Press. Washington. 36-43.

ICNAF. 1968. NORWESTLANT Surveys - Text (Part I), Atlas (Part II), Physical and Chemical Oceanographic Data (Part III), Biological Data (Part IV). Int. Comm. Northwest Atl. Fish Spec. Publ. 7; four volumes.

Jensen, A.S. 1925. On the fishery of the greenlanders. Meddelelser fra Kommissionen for Havundersøgelser. Serie: Fiskeri, Bind VII, Nr. 7. København. C.A. Reitzel, Boghandel. 39 p.

Jensen, A.S. 1935. The Greenland halibut (*Reinhardtius hippoglossoides* (Walb.)) its development and migration. Kongelige Danske Videnskabelige Selskabs Skrifter.9rk. 6 (4):1-32.

Jensen, A.S. 1939. Concerning a change of climate during recent decades in the Arctic and subarctic regions, from Greenland in west to Eurasia in the east, and contemporary biological and geophysical changes. Det Kgl. Danske Videnskabernes Selskab. Biologiske Meddelelser XIV,8. København. (Munksgaard. 75 p. with 2 charts.)(in English).

Jørgensen, O.J. 1992. Forsøgsfiskeri efter håising ved Paamiut. (A trial fishery for long rough dab at Paamiut, West Greenland). Report Greenland Fisheries Research Institute (in Danish). 16 p.

Jørgensen, O.J. 1997. Movement patterns of Greenland halibut *Reinhardtius hippoglossoides* (Walbaum), at West Greenland, as inferred from trawl survey distribution and size data. J. Northw. Atl. Fish. Sci., Vol. 21:23-37.

Kotthaus, A. 1965. The breeding and larval distribution of redfish in relation to water temperature. Int. Comm. Northwest Atl. Fish. Spec. Publ. 6:417-423.

Lee, A.J. 1968. NORWESTLANT Surveys: Physical Oceanography. Int. Comm. Northwest Atl. Fish. Spec. Publ. 7 (I):31-54.

Ludwig, D. and C. Walters. 1985. Are age-structured models appropriate for catch-effort data? Can. J. Fish. Aquat. Sci. 42:1066-1072.

Mann, K.H. 1993. Physical oceanography, food chains, and fish stocks: a review. ICES J. Mar. Sci. 50:105-119.

Magnusson, J.V. and G. Johannesson. 1995. Distribution and abundance of 0-group redfish in the Irminger Sea and at East Greenland in 1970-94 and its relation to *Sebastes marinus* abundance index from the Icelandic groundfish survey. ICES CM 1995/G:39.

Mertz, G. and R.A. Myers. 1994. The ecological impact of the Great Salinity Anomaly in the northern North-west Atlantic. Fish. Oceanogr. 3:1-14.

Meyer, A. 1968. 1967 ein erfolgreiches Jahr für die deutsche Grönlandfischerei. Hansa 16:2-4. (in German).

Miller, C. 1995. TransAtlantic Studies of Calanus (TASC) Working Group. U.S. GLOBEC NEWS No. 8 – March 1995.

Munk, P., P.O. Larsson, D. Danielsen, and E. Moksness. 1995. Larval and small juvenile cod (*Gadus morhua*) concentrated in the highly productive areas of a shelf break front. Mar. Ecol. Prog. Ser. 125:21-30.

Muus, B., F. Salomonsen, and C. Vibe. 1981. Grønlands Fauna., Fisk-Fugle-Pattedyr. (The fauna of Greenland, Fish-Birds-Mammals). Gyldendalske Boghandel, Nordisk Forlag A/S Copenhagen. 464 p. (in Danish).

Pavlov, A.J., A.S. Gorelov, and I.A. Oganin. 1989. Larval distribution, elimination indices and spawning stock assessment for beaked redfish (*Sebastes mentella* Travin) in the Irminger Sea in 1984-1988. ICES CM 1989/G:16.

Pavshtiks, E.A. 1968. The influence of currents upon seasonal fluctations in the plankton of the Davis Strait. Sarsia 34:383-392.

Pavshtiks, E.A. 1972. Biological seasons in the zooplankton of Davis Strait. Akad. Nauk. SSSR, Zool. Inst., Explor. Mar. Fauna. 12(20). Israel Prog. Sci. Transl., Jerusalem 1975:200-247.

Pedersen, S.A., 1990. Rødfisk (*Sebastes spp.*) ved Grønland. (Redfish (*Sebastes spp.*) in Greenland waters). Rapport til Grønlands Hjemmestyre fra Grønlands Fiskeri-undersøgelser. January 1990. 103 p. (in Danish).

Pedersen, S.A. 1994a. Multispecies interactions on the offshore West Greenland shrimp grounds. ICES C.M. 1994/P:2.

Pedersen, S.A. 1994b. Shrimp trawl catches and stomach contents of redfish, Greenland halibut and starry ray from West Greenland during a 24-hour cycle. Polar Res. 13:183-196.

Pedersen, S.A. 1994c. Mortality on Northern shrimp (*Pandalus borealis*) and species interactions on the offshore West Greenland shrimp grounds. Ph. D. thesis, University of Copenhagen - Five papers in English and a summary in Danish of 19 p.

Pedersen, S.A. 1995. Feeding habits of starry ray (*Raja radiata*) in West Greenland waters. ICES J. Mar. Sci. 52:43-54.

Pedersen, S.A. 1998. Distribution and lipid composition of Pandalus shrimp larvae in relation to hydrography in West Greenland waters. J. Northw. Atl. Fish. Sci. 24: 39-60.

Pedersen, S.A. and P. Kanneworff. 1995. Fish on the West Greenland shrimp grounds, 1988-1992. ICES J. Mar. Sci. 52:165-182.

Pedersen, S.A. and F. Riget. 1993. Feeding habits of redfish (*Sebastes spp.*) and Greenland halibut (*Reinhardtius hippoglossoides*) in West Greenland waters. ICES J. Mar. Sci. 50:445-459.

Perry, R.I. and S.J. Smith. 1994. Identifying habitat associations of marine fishes using survey data: an application to the Northwest Atlantic. Can. J. Fish. Aquat. Sci., 51:589-602.

Pitt, T.K. 1969. Migration of American plaice on the Grand Bank and in St. Mary's Bay, 1954, 1959, and 1961. J. Fish. Res. Bd. Can. 26:1301-1319.

Rätz. H.-J. 1998. Abundance, biomass, and size composition of dominant demersal fish stocks and trends in near-bottom temperature off West and East Greenland 1982-97. NAFO SCR Doc, 98/21. 22 p.

Rice, J.C. 1993. Forecasting abundance from habitat measures using nonparametric density estimation methods. Can. J. Fish. Aquat. Sci. 50:1690-1698.

Riget, F. and J. Boje. 1989. Fishery and some biological aspects of Greenland halibut (*Reinhardtius hippoglossoides*) in West Greenland waters. NAFO Sci. Coun. Stud.13:41-52.

Riget, F. and J. Engelstoft. 1998. Size-at-age of Cod (Gadus morhua) off West Greenland, 1952-92. NAFO Sci. Coun. Stud. 31:1-12.

Rosenørn, S., J. Fabricius, E. Buch, and S.Aa. Horsted. 1985. Record-hard winters at West Greenland. NAFO SCR Doc. 85/61, Serial No. N1011. 17 p.

Shumway, S.E., H.C. Perkins, D.F. Shick, and A.P. Stickney. 1985. Synopsis of biological data on the pink shrimp, *Pandalus borealis* Krøyer, 1838. NOAA Technical Report NMFS 30 (FAO Fisheries Synopsis No. 144). 57 p.

Sibanov, V.N., A.P. Pedchenko, and S.P. Melnikov. 1995. Peculiarities of formation of oceanic *Sebastes mentella* spawning aggregations in the Irminger Sea. ICES CM 1995/G:23.

Sigurdsson, A. 1979. The Greenland halibut *Reinhardtius hippoglossoides* (Walb.) at Iceland. Hafrannsoknit, 16, 31 p.

Sigurdsson, A. and J.V. Magnusson. 1980. On the nursery grounds of the Greenland halibut spawning in Icelandic waters. ICES CM 1980/G:45.

Sinclair, M. and F. Page. 1995. Cod fishery collapses and North Atlantic GLOBEC. U.S. GLOBEC NEWS No. 8 -- March 1995.

Smidt, E.L.B. 1969. The Greenland halibut, *Reinhardtius hippoglossoides*, biology and exploitation in Greenland waters. Meddelelser fra Danmarks Fiskeri- og Havundersøgesler. N.S. 6 (4):1-79.

Smidt, E.L.B. 1971. Summary report of the ICNAF NORWESTLANT surveys, 1963. Int. Comm. Northwest Atl. Fish. Redbook 1971, Part III: 275-295.

Smidt, E.L.B. 1979. Annual cycles of primary production and of zooplankton at Southwest Greenland. Meddelelser om Grønland, Bioscience No.1. 53 p.

Smith, S.J., R.I. Perry, and L.P. Fanning. 1991. Relationships between water mass characteristics and estimates of fish population abundance from trawl surveys. Envt. Monit.Assmt. 17:227-245.

Stefánsson, U. and J. Jakobsson. 1989. Oceanographical variations in the Iceland Sea and their impact on biological conditions - a brief review. In: L. Rey and V. Alexander (eds.). Proceedings of the sixth conference of the comité arctique international 13-15 May 1985. E.J. Brill. Leiden. 427-455.

Steele, J.H. 1995. Regime shifts in fisheries management. Fish. Res. 25:19-23.

Stein, M. 1998a. Do hydrographic conditions affect the distribution of shrimp (*Pandalus borealis*) off East Greenland. NAFO SCR. Doc. 98/124. 13 p.

Stein, M. 1998b. Climatic Condition Around Greenland – 1996. NAFO Sci. Coun. Studies, 31:147-154.

Stein, M. 1998c. Integrating fisheries observations with environmental data – towards a better understanding of the conditions for fish in the sea. J. Northw. Atl. Fish. Sci. 23:143-156.

Stein, M., and Lloret, J. 1995. Stability of water masses - impact on cod recruitment off West Greenland. Fish. Oceanogr. 4:230-237.

Sverdrup, H.U., M.W. Johnson, and R.H. Fleming. 1946. The Oceans - Their Physics, Chemistry, and General Biology. New York. Prentice-Hall, Inc. 1060 p.

Tiedtke, J.E. 1988. Qualitative und quantitative untersuchungen des mageninhaltes vom kabeljau (*Gadus morhua* L.) aus westgrönlandischen gewässern. Mitteilungen aus dem Institut für Seefischerei der Bundesforschungsanstalt für Fischerei, Hamburg. Nr. 43. 106 p. (in German).

Wells, R. 1968. Report on American plaice eggs and larvae. Int. Comm. Northwest Atl. Fish Spec. Publ. 7 (I):181-182.

Winters, G.H. 1983. Analysis of the biological and demographic parameters of northern sand lance, Ammodytes dubius, from the Newfoundland Grand Bank. Can. J.Fish. Aquat. Sci. 40:409-419.

Large Marine Ecosystems of the North Atlantic
K. Sherman and H.R. Skjoldal (Editors)
© 2002 Published by Elsevier Science B.V.

6

The U.S. Northeast Shelf Large Marine Ecosystem: Zooplankton Trends in Fish Biomass Recovery

Kenneth Sherman, Joseph Kane, Steven Murawski,
William Overholtz, and Andrew Solow

ABSTRACT

As the application of ecosystem-based management of marine fish and fisheries evolves, it will be necessary to extend the scope of stock assessment methodologies to include indicators of trends in lower level food chain components of ecosystems. Zooplankton as a ubiquitous food chain component can be used as an indicator of changing trends in ecosystem states pertinent to ecosystem effects on fish management actions. Spatial and temporal biomass and species diversity of zooplankton have been examined in a continental shelf ecosystem for indications of variability in abundance levels associated with effort reduction management actions for the recovery and sustainable maintenance of depleted bottom fish stocks. Analyses were conducted on 19,000 zooplankton samples collected seasonally over two decades from the U.S. Northeast Shelf ecosystem (1977 to 1999). Fish stock results were obtained from the Northeast Fisheries Science Center's bottom trawl survey indices of spawning biomass of selected demersal (cod, haddock, yellowtail flounder) and pelagic (herring and mackerel) fish species. The emerging pattern of the zooplankton component of the ecosystem showed biomass values close to or above the 22-year median from the late 1980s through the 1990s, during the recovery of herring and mackerel in the mid-1980s, and the initiation of recovery of demersal stocks as indicated by strong year-classes of haddock and yellowtail flounder in the late 1990s following reductions in fishing effort. The zooplankton component of the ecosystem has not undergone any significant decline during the rebuilding of the fish stocks and is being used as an indicator of stability in the lower trophic levels of the NE Shelf ecosystem food web during the fish-stock recovery period.

BACKGROUND

The U.S. Northeast Shelf ecosystem is one of the 50 large marine ecosystems (LMEs) located around the coastal margins of the world's ocean basins (Figure 6-1). It extends from the Gulf of Maine south to Cape Hatteras, and from the coast seaward to the edge of the continental shelf. The ecosystem is one of the more highly productive LMEs

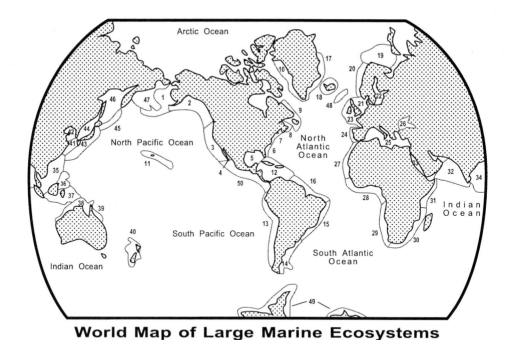

World Map of Large Marine Ecosystems

1. Eastern Bering Sea	14. Patagonian Shelf	27. Canary Current	40. New Zealand Shelf
2. Gulf of Alaska	15. Brazil Current	28. Gulf of Guinea	41. East China Sea
3. California Current	16. Northeast Brazil Shelf	29. Benguela Current	42. Yellow Sea
4. Gulf of California	17. East Greenland Shelf	30. Agulhas Current	43. Kuroshio Current
5. Gulf of Mexico	18. Iceland Shelf	31. Somali Coastal Current	44. Sea of Japan
6. S.E. U.S. Continental Shelf	19. Barents Sea	32. Arabian Sea	45. Oyashio Current
7. N.E. U.S. Continental Shelf	20. Norwegian Shelf	33. Red Sea	46. Sea of Okhotsk
8. Scotian Shelf	21. North Sea	34. Bay of Bengal	47. West Bering Sea
9. Newfoundland Shelf	22. Baltic Sea	35. South China Sea	48. Faroe Plateau
10. West Greenland Shelf	23. Celtic-Biscay Shelf	36. Sulu-Celebes Seas	49. Antarctic
11. Insular Pacific-Hawaiian	24. Iberian Coastal	37. Indonesian Seas	50. Pacific Central
12. Caribbean Sea	25. Mediterranean Sea	38. Northern Australian	American Coastal
13. Humboldt Current	26. Black Sea	Shelf	

Figure 6-1. World map of large marine ecosystems (LMEs).

of the world. The ecosystem has been for the past several decades under considerable stress from overfishing (Murawski *et al.* 1999) and growing problems of degraded water quality, and emerging disease events related to eutrophication of near coastal areas (Epstein 1996).

Effort has been underway since the late 1990s by the New England and Mid-Atlantic Fishery Management Councils to control overfishing and implement a series of management actions leading to the recovery of depleted demersal fish stocks, including cod, haddock and flounder (Sherman *et al.* 1996b). This paper presents the results of a time-series study of the zooplankton component of the ecosystem in relation to fish stock recovery measures enacted by the Fishery Management Councils.

BIOMASS LOSS

During recent years, the public and the scientific communities have signaled concern over the growing number of depleted fisheries that are below their long-term potential yield levels (LTPY). In the United States 30 percent of 275 fish stocks have been classified as below their LTPY (NMFS 1996). Public concern has been registered in newspapers, electronic media, and congressional actions. Scientific concern has moved from the pages of journals to the actions of professional societies, as for example the *Sustainable Biosphere Initiative* of the Ecological Society of America (Lubchenco *et al.* 1991). Responsive actions at the national and international levels have resulted in conventions, protocols and internationally recognized declarations for implementing recovery actions for depleted fish stocks and promoting sustainable marine fisheries globally.

Published commentary on how best to improve the degraded state of fish and fishery resources are not without controversy. While some scientists are concerned with the lack of consistent success in the management of fishery resources (Ludwig *et al.* 1993), others stress the utility of science-based assessments as a key component of marine fishery resource management practices (Rosenberg *et al.* 1993). Given the growing stress from the expanding human population on fishery resources, stewardship institutions cannot wait for science to achieve a full understanding of fish and fisheries within the context of ecosystem structure and function. The best presently available science is needed to monitor and assess changing ecosystem conditions and implement corrective actions to recover lost fisheries biomass and sustain populations at their LTPY levels. In the northeast and northwest Atlantic, scientists have been collecting systematic time-series information for the past 40 years describing the declines in marine fisheries, habitats, and water quality. But it wasn't until the later half of the 1990s that government policies were coupled with actions to accelerate a reversal of overexploitation and environmental degradation. In the United States, among the more forward-looking legislative acts mandating improvements in coastal environments and promoting fisheries sustainability can be found in recent amendments to the Magnuson-Stevens Fishery Conservation and Management Act, and the National Environmental Policy Act (Heinz 2000).

ECOSYSTEM-BASED MONITORING AND ASSESSMENT

In the mid-1960s, in response to public concerns and congressional mandates for improving coastal water quality and fisheries sustainability, NOAA's National Marine Fisheries Service initiated systematic bottom-trawl surveys of the fish inhabiting the

Northeast continental shelf. Oceanographic, plankton, and water quality surveys followed, and the transition from a sector-by-sector approach to resource and environmental monitoring, assessment, and management actions advanced toward an ecosystem-based strategy for recovering lost biomass and improving the health of coastal ecosystems.

The ecosystem-based approach has relevance to the management of large marine ecosystems (Figure 6-1). On a global scale, 50 Large Marine Ecosystems (LMEs) produce 95 percent of the world's annual marine fishery yields and within their waters most of the global ocean pollution, overexploitation, and coastal habitat alteration occur (AAAS 1986, 1989, 1990, 1991, 1993; Sherman *et al.* 1996a; Kumpf *et al.* 1999; Sherman and Tang 1999). The LMEs are regions of ocean space encompassing coastal areas from river basins and estuaries out to the seaward boundary of continental shelves and the outer margins of coastal current systems. They are relatively large, on the order of 200,000 km^2 or greater, characterized by distinct bathymetry, hydrography, productivity, and trophically dependent populations. The theory, measurement, and modeling relevant to monitoring the changing states of LMEs are imbedded in reports on ecosystems with changing ecological states, and on the pattern formation and spatial diffusion within ecosystems (Holling 1973; Pimm 1984; AAAS 1990; Mangel 1991; Levin 1993; Sherman 1994). The concept of monitoring and managing renewable resources from an LME perspective has been the topic of several symposia and published volumes during the past 15 years (Table 6-1).

Temporal and spatial scales influencing biological production and changing ecological states in marine ecosystems have been topics of numerous theoretical and empirical studies. The selection of scale in any study is related to the processes under investigation. Steele (1988) produced a heuristic projection to illustrate scales of importance in monitoring key components of the ecosystem, including phytoplankton, zooplankton, fish, frontal processes, and short-term but large-area episodic effects (Figure 6-2). The LME approach defines a spatial domain based on distinct bathymetry, hydrography, productivity, and trophic linkages and, thereby, provides a basis for focused temporal and spatial ecosystem-based assessment and monitoring efforts in support of management aimed at the long-term productivity and sustainability of marine habitats and resources.

ASSESSMENT METHODOLOGIES

The Northeast Fisheries Science Center (NEFSC) of NOAA's National Marine Fisheries Service has for the past several decades been developing modules of ecosystem-based methodologies for assessing the root causes of changes in the fish and fisheries inhabiting the Northeast Shelf large marine ecosystem and recommending long-term sustainability levels for species biomass yields. The ecosystem extends over 260,000km^2 of the shelf from the Gulf of Maine in the north to Cape Hatteras in the south. Four subareas of the ecosystem have been described (Sherman et al. 1996b): the

Table 6-1. List of 33 LMEs and subsystems for which syntheses relating to primary, secondary, or tertiary driving forces controlling variability in biomass yields have been completed for inclusion in LME volumes.

Large Marine Ecosystem	Volume No.	Authors
U.S. Northeast Continental Shelf	1	M. Sissenwine
	4	P. Falkowski
	6	S. Murawski
U.S. Southeast Continental Shelf	4	J. Yoder
Gulf of Mexico	2	W. Richards and M. McGowan
	4	B. Brown *et al.*
	9	R. Shipp
California Current	1	A. MacCall
	4	M. Mullin
	5	D. Bottom
Eastern Bering Sea	1	L. Incze and J. Schumacher
	8	P. Livingston *et al.*
West Greenland Shelf	3	H. Hovgård and E. Buch
Norwegian Sea	3	B. Ellersten *et al.*
Barents Sea	2	H. Skjoldal and F. Rey
	4	V. Borisov
North Sea	1	N. Daan
Baltic Sea	1	G. Kullenberg
Iberian Coastal	2	T. Wyatt and G. Perez-Gandaras
Mediterranean-Adriatic Sea	5	G. Bombace
Canary Current	5	C. Bas
Gulf of Guinea	5	D. Binet and E. Marchal
Benguela Current	2	R. Crawford *et al.*
Patagonian Shelf	5	A. Bakun
Caribbean Sea	3	W. Richards and J. Bohnsack
South China Sea-Gulf of Thailand	2	T. Piyakarnchana
East China Sea	8	Y-Q Chen and X-Q Shen
Sea of Japan	8	M. Terazaki
Yellow Sea	2	Q. Tang
Sea of Okhotsk	5	V. Kusnetsov *et al.*
Humboldt Current	5	J. Alheit and P. Bernal
Pacific Central American	8	A. Bakun *et al.*
Indonesia Seas-Banda Sea	3	J. Zijlstra and M. Baars
Bay of Bengal	5	S. Dwivedi
	7	A. Hazizi *et al.*
Antarctic Marine	1&5	R. Scully *et al.*
Weddell Sea	3	G. Hempel
Kuroshio Current	2	M. Terazaki
Oyashio Current	2	T. Minoda
Great Barrier Reef	2	R. Bradbury and C. Mundy
	5	G. Kelleher
	8	J. Brodie
Somali Current	7	E. Okemwa
South China Sea	5	D. Pauly and V. Christensen

Vol. 1   Variability and Management of Large Marine Ecosystems (AAAS 1986)
Vol. 2   Biomass Yields and Geography of Large Marine Ecosystems (AAAS 1989)
Vol. 3   Large Marine Ecosystems: Patterns, Processes, and Yields (AAAS 1990)
Vol. 4   Food Chains, Yields, Models, and Management of Large Marine Ecosystems (AAAS 1991)
Vol. 5   Large Marine Ecosystems: Stress, Mitigation, and Sustainability (AAAS 1993)
Vol. 6   The Northeast Shelf Ecosystem: Assessment, Sustainability, and Management (Sherman *et al.* 1996a)
Vol. 7   Large Marine Ecosystems of the Indian Ocean: Assessment, Sustainability, and Management (Sherman *et al.* 1998a)
Vol. 8   Large Marine Ecosystems of the Pacific Rim: Assessment, Sustainability, and Management (Sherman and Tang 1999)
Vol. 9   The Gulf of Mexico Large Marine Ecosystem: Assessment, Sustainability, and Management (Kumpf *et al.* 1999)

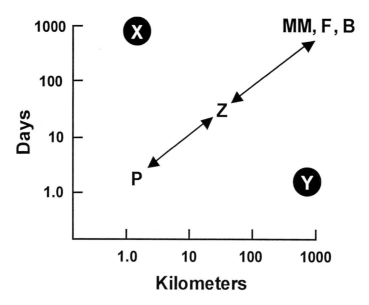

Figure 6-2. A simple set of temporal and spatial scale relations for the pelagic food web. (P) phytoplankton, (Z) zooplankton, (F) fish, (MM) marine mammals, (B) birds. Two physical processes are indicated by (X) scale of variability in predictable fronts and coastal currents, and (Y) weather events occurring over relatively large areas but over short time scales (adapted from Steele 1988).

deep seasonally stratified waters of the Gulf of Maine, the vertically mixed shallow water of Georges Bank, the shoals and seasonally stratified waters of Southern New England, and the estuarine dominated waters of the Mid Atlantic Bight (Figure 6-3). Among the assessment methods used is the systematic monitoring of the zooplankton component of the food web of the ecosystem using continuous plankton recorders and bongo net sampling as an indicator of major changes in stability of the lower levels of the food web and biofeedback responses to oceanographic changes (Sherman *et al.* 1996b). Since the late 1960s the NEFSC has been conducting systematic shelfwide twice a year bottom trawl fish surveys (autumn and spring) and seasonal (4 to 6 times/year) bongo net zooplankton surveys as part of the MARMAP Program (Sherman 1980; Azarovitz 1981; Sherman *et al.* 1998b). During the MARMAP trawling and zooplankton surveys, simultaneous water-column measurements are made of temperature and salinity conditions. In addition, information is available on other components of the ecosystem including pollution and ecosystem health, socioeconomics, and governance (Sherman *et al.* 1996a; Sutinen *et al.* 1998; Juda 1999; and Juda and Hennessey 2001). Fish stock assessments based on survey and catch/effort data are summarized annually (NEFSC 1999).

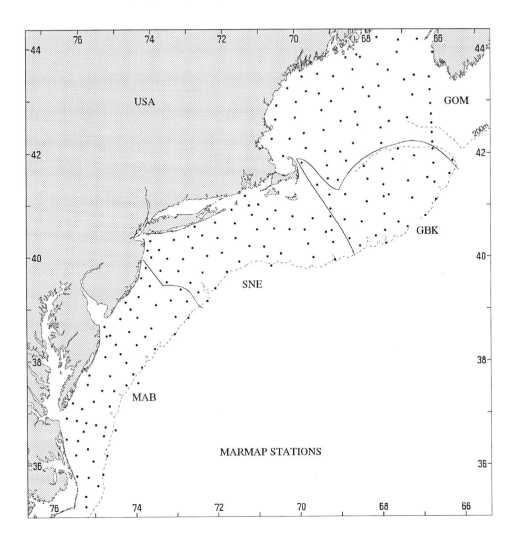

Figure 6-3. The Northeast Shelf ecosystem, extending 260,000km² from the Gulf of Maine in the north to Cape Hatteras in the south. Four subareas are depicted - Gulf of Maine (GOM), Georges Bank (GBK), Southern New England (SNE), and the Mid-Atlantic Bight (MAB). Dots indicate locations of zooplankton towing areas.

BIODIVERSITY AND BIOMASS

The pattern of zooplankton biomass based on mean annual values of volumetric displacement shows the importance of the spring increase in zooplankton on Georges Bank (GB), the principal spawning ground for spring spawning cod and haddock stocks

Table 6-2. Ranked Order of Zooplankton by Taxon (N/100m^3) in Northeast Shelf Ecosystem, 1977-1999.

Rank Order	Taxon	Mean No/100m^3	% Occurrence in Samples
1	*Centropages typicus*	29,534	90.7
2	*Calanus finmarchicus*	14,525	85.3
3	*Pseudocalanus spp.*	13,530	88.5
4	*Temora longicornis*	6,405	52.0
5	*Centropages homatus*	5,914	51.0
6	*Acartia clausi*	4.003	59.5
7	*Paracalanus parvus*	2,765	58.7
8	*Sagitta spp.*	1,780	62.6
9	*Oithona similes*	1,689	73.9

in the Northeast Shelf ecosystem. The seasonal patterns of zooplankton abundance are similar in the Gulf of Maine (GOM) and Southern New England (SNE) subareas with a plateau of biomass extending from May through August. However, the dominant species are different, with the copepod *Calanus finmarchicus* predominant in the Gulf of Maine, and the copepod *Centropages typicus* the dominant species in SNE, along with a wider array of zooplankton biodiversity (Sherman *et al.* 1996b). In the Mid Atlantic Bight (MAB), the annual peak in biomass is reached in early autumn. Biomass is lowest in all subareas during the winter months (Figure 6-4). The principal taxa in the zooplankton are listed in Table 6-2. Biodiversity was dominated by three copepod species: *Calanus finmarchicus, Pseudocalanus spp., and Centropages typicus*. The abundance level for the three dominant species was relatively stable over the 22-year time-series (Figure 6-5a). A slight upward trend in abundance was observed for *Centropages hamatus* from 1977 to 1999. The log mean trends for *Pseudocalanus spp.* and *Calanus finmarchicus* revealed interannual variability but no consistent long-term deviation from the time-series trend line (Figure 6-5b,c). The three species are key prey of larval cod and haddock (Kane 1984), and important prey species of herring (Sherman and Honey 1971; Sherman and Perkins 1971) and mackerel (Bowman *et al.* 1984).

ECOSYSTEM VARIABILITY

Investigators of the zooplankton component of other marine ecosystems have reported major decadal-scale changes in abundance levels for the California Current (Roemmich and McGowan 1995), the Gulf of Alaska (Brodeur *et al.* 1999) and the eastern North Atlantic (Fromentin and Planque 1996). These changes appear related to mega-scale oceanographic regime shifts (Bakun 1999; Lluch-Belda 1999; Taylor, this volume). The principal driving forces affecting zooplankton abundance in the NE Shelf ecosystem

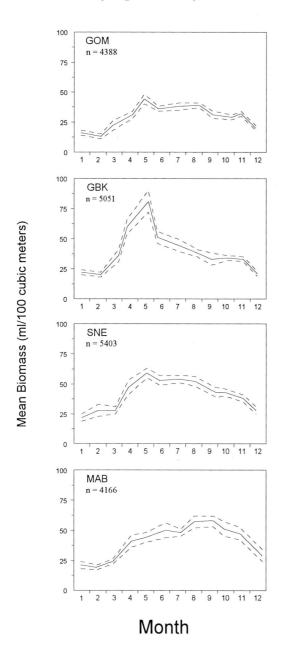

Figure 6-4. Monthly mean values of zooplankton biomass (solid line) and the 95% confidence interval (dashed line) of the mean for 19,008 zooplankton samples from four subareas of the Northeast Shelf ecosystem, 1977-1999.

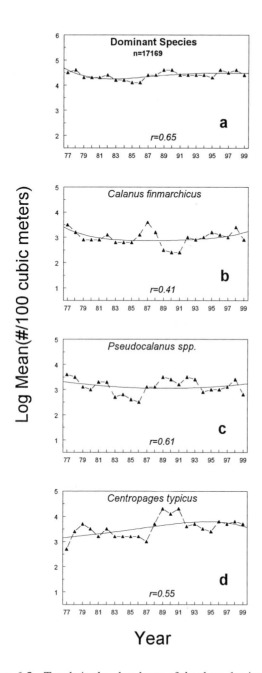

Figure 6-5. Trends in the abundance of the three dominant zooplankton species inhabiting the Northeast Shelf ecosystem (mean annual log number per 100m³ of water), 1977-1999. Total for the three dominant copepod species (**a**), *Calanus finmarchicus* (**b**), *Pseudocalanus spp.* (**c**), and *Centropages typicus* (**d**). The trend line is a fitted polynomial and r value.

are: (1) the within year seasonal pulses in the annual production cycle; (2) annual variability in the changes in the strength of advection of Gulf Stream waters onto the shelf; and (3) the periodic incursion of cold admixture of Labrador and Scotian Shelf water into the Gulf of Maine and Georges Bank subareas of the ecosystem (Mountain *et al*. 2000). The seasonal cold and warm air masses transported westward from central North America and generally warm moist air mass incursions from the Gulf of Mexico produce a wide range of approximately 30° C in average annual water temperatures over the spatial extent of the ecosystem (Hertzman 1996). The long-term bottom temperature signal for the ecosystem shows considerable inter-annual variability (Holzwarth and Mountain 1990; Sherman *et al*. 1998b). Surface temperatures based on measurements made at each of the zooplankton sampling locations show evidence of interannual variation, but no consistent upward or downward trend during the 22-year time-series (Figure 6-6).

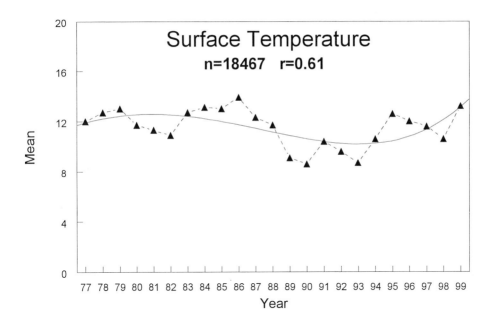

Figure 6-6. Mean surface temperature pattern based on 18,467 measurements taken simultaneously with each of the zooplankton samples, 1977-1999. The trend line is a fitted polynomial and r value.

The mean-annual primary productivity level of the Northeast Shelf ecosystem is 350gCm²/year (O'Reilly and Zetlin 1998). The spatial and temporal pattern of chlorophyll and primary production has not changed significantly since the MARMAP surveys were initiated in the late 1970s (O'Reilly and Yoder 1998), and the seasonal pulse in chlorophyll abundance remains inversely related to the temperature signal in the ecosystem. Water column stratification and the grazing of the zooplankton limits the production of chlorophyll during the summer months (O'Reilly and Zetlin 1998). The dominant zooplankton species appear to have evolved adaptive reproductive strategies that allow for them to sustain their populations at sufficiently high levels to support a highly diverse and abundant fish community (Sherman *et al.* 1998b).

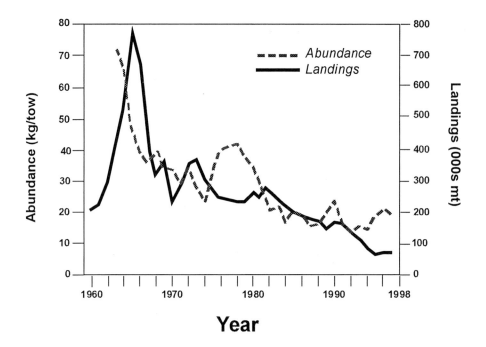

Figure 6-7. Decline of principal groundfish and flounder species of the Northeast Shelf ecosystem, 1960-1998, in relation to biomass (kg/tow) estimated from fishery independent bottom trawl surveys and biomass yield (thousand metric tons) expressed as landings of fish catch (from NEFSC 1999).

FISH BIOMASS DECLINE AND RECOVERY

From the mid-1960s through the early 1990s, the biomass of principal groundfish and flounder species inhabiting the NE Shelf ecosystem declined significantly from overfishing of the spawning stock biomass (NEFSC 1999) (Figure 6-7). In response to the decline, the biomass of skates and spiny dogfish increased from the 1970s through the early 1990s (NEFSC 1999). The impact of the increase in small elasmobranches, particularly spiny dogfish, shifted the principal predator species on the fish component

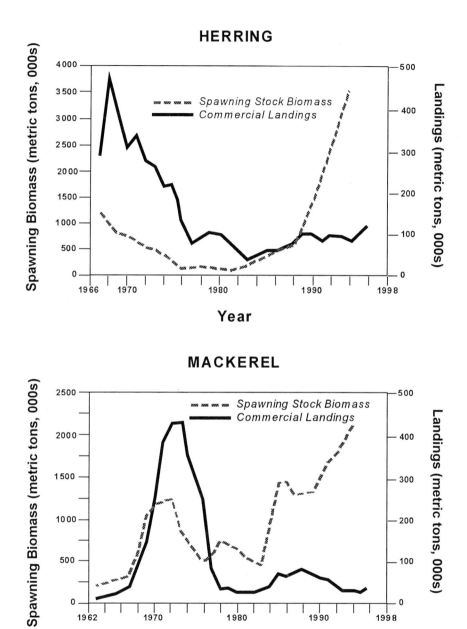

Figure 6-8. Top - Atlantic herring commercial landings and spawning stock biomass, 1967 through 1996 (thousand metric tons). Bottom - Atlantic mackerel landings and spawning stock biomass, 1963 through 1996 (thousand metric tons).

of the ecosystem from silver hake during the mid-1970s to spiny dogfish in the mid-1980s (Sissenwine and Cohen 1991). By the mid-1990s a newly developing fishery for small elasmobranches initiated a declining trend in biomass for skates and spiny dogfish (NEFSC 1999).

Following the secession of foreign fishing on the Georges Bank-Gulf of Maine herring complex and the Atlantic mackerel stock in the late 1970s, and over a decade of very low fishing mortality, both species began to recover to high stock sizes in the 1990s (Figure 6-8). Bottom trawl survey indices for both species increased dramatically, showing over a ten fold increase in abundance (average of 1977-1981 vs. 1995-1999) by the late 1990s (NEFSC 2000a,b; NEFMC 2000). Stock biomass of herring increased to over 2.5 million metric (mm) tons by 1997 and spawning stock biomass (SSB) was projected to increase to well over 3.0 mm tons in 2000 (NEFSC 1999). The offshore component of herring, which represents the largest proportion of the whole complex, appears to have fully recovered from the total collapse it experienced in the early 1970s (NEFSC 2000a). For mackerel, the situation is similar, total stock biomass has continued to increase since the collapse of the fishery in the late 1970s. Although absolute estimates of biomass for the late 1990s are not available, recent analyses concluded that the stock is at or near a historic high in total biomass and SSB (NEFSC 2000b).

Based on historical data, up through the late 1970s, Skud (1982) hypothesized that it would be unlikely for both herring and mackerel to be abundant components of the pelagic fish component of the Northeast Shelf simultaneously. This however, does not appear to be the case, since the recent combined biomass of herring and mackerel appears to be within the "historic range or carrying capacity" of the NE Shelf ecosystem. In addition, recent evidence following substantial reductions in fishing effort indicate that both haddock and yellowtail flounder stocks are responding rather favorably with substantial growth reported in SSB size, since 1994 for haddock and flounder. In addition, in 1997 a very strong year-class of yellowtail flounder was produced, and in 1998, a strong year-class of haddock was produced (Figure 6-9a,b).

ZOOPLANKTON, PRODUCTIVITY, AND RECOVERY TRENDS

At the base of the food web, primary productivity provides the initial level of carbon production to support the important marine commercial fisheries (Nixon *et al.* 1986). Zooplankton production and biomass in turn provide the prey-resource for larval stages of fish, and the principal food source for herring and mackerel in waters of the NE Shelf ecosystem.

Over the past two decades the long-term median value for the zooplankton biomass of the NE Shelf ecosystem has been about 29cc of zooplankton per $100m^3$ of water strained (Figure 6-11) produced from a stable mean-annual primary productivity of 350 gCm^2yr. During the last two decades, the zooplanktivorous herring and mackerel stocks

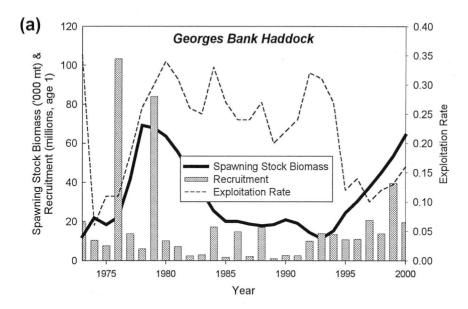

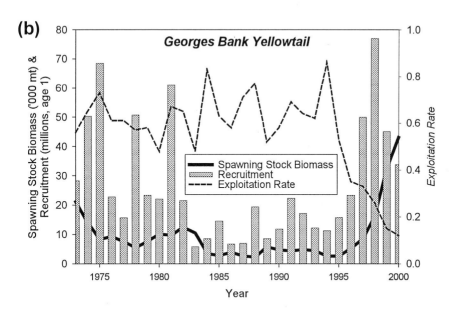

Figure 6-9. Trends in spawning stock biomass (ssb) and recruitment in relation to reductions in exploitation rate (fishing effort) for two commercially important species inhabiting the Georges Bank subarea of the Northeast Shelf ecosystem, haddock **(a)** and yellowtail flounder **(b)**.

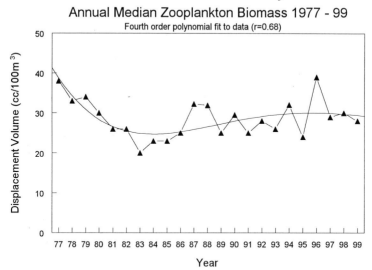

Figure 6-10. Trend in annual median zooplankton biomass of the Northeast Shelf ecosystem (displacement volumes in cc/100m^3 of water), 1977-1999 (n=19,008). The trend line is a fitted polynomial and r value.

underwent unprecedented levels of growth, approaching an historic high combined biomass. This growth is taking place during the same period that the fishery management councils for the New England and Mid-Atlantic areas of the NE Shelf ecosystem have sharply curtailed fishing effort on haddock and yellowtail flounder stocks. Given the observed robust levels of primary productivity and zooplankton biomass, it appears that the "carrying capacity" of zooplankton supporting herring and mackerel stocks and larval zooplanktivorous haddock and yellowtail flounder is sufficient to sustain the strong year-classes reported for 1997 (yellowtail flounder) and 1998 (haddock).

The zooplankton component of the Northeast Shelf ecosystem is in a robust condition at biomass levels at or above the levels of the long-term median values of the past two decades, providing a suitable prey base for supporting a large biomass of pelagic fish (herring and mackerel), while providing sufficient zooplankton prey to support strong year-classes of recovering haddock and yellowtail flounder stocks. No evidence has been found in the fish, zooplankton, temperature, or chlorophyll component that is indicative of any large-scale oceanographic regime shifts of the magnitude reported for the North Pacific or northeast Atlantic Ocean areas.

The robust condition of the plankton components at the base of the food web of the Northeast Shelf ecosystem was important to the relatively rapid rebuilding of zooplanktivorous herring and mackerel biomass from the depleted condition in the early 1980s to a combined biomass in 1999 of an unprecedented level of approximately 5.5 mm tons, following the exclusion of foreign fishing effort and the absence of any significant U.S. fishery on the stocks. The milestone action leading to the rebuilding of lost herring and mackerel biomass was the decision by the United States to extend jurisdiction over marine fish and fisheries within 200 miles of the coastline. Recently the Fishery Management Councils of New England, and the mid-Atlantic coastal states agreed to reduce fishing effort significantly on demersal fish stocks of the NE Shelf ecosystem. With the reduction of exploitation rate, the spawning biomass of haddock and yellowtail flounder increased over a 4-year period and led to the production of large year-classes of haddock in 1998 and yellowtail flounder in 1997.

The Northeast Shelf ecosystem is presently undergoing a significant trend toward biomass recovery of pelagic and demersal fish species important to the economy of the adjacent northeast states from Maine to North Carolina. Although the recovery has not as yet been fully achieved, the corner has been turned from declining overharvested fish stocks toward a condition wherein the stocks can be managed to sustain their long-term potential yield levels. The management decisions taken to reduce fishing effort to recover lost biomass was supported by science-based monitoring and assessment information forthcoming from the productivity, fish and fisheries, pollution and ecosystem health, socioeconomics, and governance modules that have been operational by NOAA's Northeast Fisheries Science Center for several decades in collaboration with state, federal, and private stakeholders from the region. This case study can serve to underscore the utility of the modular approach to ecosystem-based management of marine fish species. In an effort to stem the loss of fisheries biomass in other parts of the world, applications of this modular approach to LME management are presently underway by countries bordering the Yellow Sea, Benguela Current, Baltic Sea, and Guinea Current LMEs (IOC 2000), with financial assistance of the Global Environment Facility, collaborating UN agencies, and the technical and scientific assistance of other governmental and non-governmental agencies and institutions.

ACKNOWLEDGMENTS

We should like to acknowledge with special thanks Dr. Leonard Ejysmont and the staff of the Polish Plankton Sorting and Identification Center for their expert contribution to this study. Scientists and technicians at the Center volumized and taxonomically separated, identified, and enumerated the 19,000 zooplankton samples used in the present study. During the course of a long collaboration between NOAA-NMFS and the Polish Sea Fisheries Institute (SFI) in Gdynia and Szczecin, the Director of SFI, Dr. Tomasz Linkowski, and his staff also provided intellectual support for broadening single fish stock assessment practices toward a more ecologically oriented multispecies ecosystem-based approach to fisheries management practices.

REFERENCES

AAAS. 1986. Variability and Management of Large Marine Ecosystems. AAAS Selected Symposium 99, Westview Press, Inc. Boulder. 319 p.

AAAS. 1989. Biomass Yields and Geography of Large Marine Ecosystems. AAAS Selected Symposium 111, Westview Press, Inc., Boulder. 493 p.

AAAS. 1990. Large Marine Ecosystems: Patterns, Processes and Yields. AAAS Press, Washington, DC. 242 p.

AAAS. 1991. Food Chains, Yields, Models, and Management of Large Marine Ecosystems. Westview Press, Inc., Boulder. 320 p.

AAAS. 1993. Large Marine Ecosystems: Stress, Mitigation, and Sustainability. AAAS Press, Washington, DC. 376 p.

Azarovitz, T.R. 1981. A brief historical review of the Woods Hole laboratory trawl survey time series. In: W.G. Doubleday and D. Rivard (eds.). Bottom Trawl Surveys. Can. Spec. Publ. Fish. Aquat. Sci. Vol. 58. Department of Fisheries and Oceans. Ottawa, Canada. 62-67.

Bakun, A. 1999. A dynamic scenario for simultaneous regime-scale marine population shifts in widely separated large marine ecosystems of the Pacific. In: K. Sherman and Q. Tang (eds.). Large Marine Ecosystems of the Pacific Rim: Assessment, Sustainability, and Management. Blackwell Science, Inc. Malden MA. 2-26.

Bowman, R., J. Warzocha, and T. Morris. 1984. Trophic relationships between Atlantic mackerel and American sand lance. ICES C.M./H:27. 19 p.

Brodeur, R.D. and D.M. Ware. 1995. Interdecadal variability in distribution and catch rates of epipelagic nekton in the Northeast Pacific Ocean. In: R. Beamish (ed.). Climate change and northern fish populations. Can. Spec. Pub. Fish. Aquat. Sci. 121:329-356.

Epstein, P.R. 1996. Emergent stressors and public health implications in large marine ecosystems: An overview. In: K. Sherman, N.A. Jaworski, and T.J. Smayda (eds.). The Northeast Shelf Ecosystem: Assessment, Sustainability, and Management. Blackwell Science, Inc. Cambridge MA. 417-438.

Fromentin, J.M. and B. Planque 1996. *Calanus* and environment in the eastern North Atlantic. II. Influence of the North Atlantic Oscillation on *C. finmarchicus* and *C. helgolandicus*. Mar. Ecol. Prog. Ser. 134:111-118.

Heinz, H. John III Center for Science, Economics and the Environment. 2000. Fishing Grounds: Defining a New Era for American Fisheries Management. Island Press. Washington, DC. 241 p.

Hertzman, O. 1996. Meteorology of the Northeast Shelf. In: K. Sherman, N.A. Jaworski, and T.J. Smayda (eds.). The Northeast Shelf Ecosystem: Assessment, Sustainability, and Management. Blackwell Science, Inc. Cambridge MA. 75-93.

Holling, C.S. 1973. Resilience and Stability of Ecological Systems. Institute of Resource Ecology, University of British Columbia, Vancouver, Canada.

Holzwarth, T. and D. Mountain. 1990. Surface and bottom temperature distributions from the Northeast Fisheries Center spring and fall bottom trawl survey program, 1963-1987; with addendum for 1988-1990. Northeast Fish. Sci. Center Ref. Doc. 90-03. 77 p.

IOC (Intergovernmental Oceanographic Commission). 2000. IOC-IUCN-NOAA Consultative Meeting on Large Marine Ecosystems (LMEs). Third Session, Paris, France, 13-14 June 2000. IOC-IUCN-NOAA/LME-III/3. IOC Reports of Meetings of Experts and Equivalent Bodies, Series 162. 20 p.

Juda, L. and T. Hennessey. 2001. Governance profiles and the management of the uses of large marine ecosystems. Ocean Development & International Law 32:43-69.

Kane, J. 1984. The feeding habits of co-occurring cod and haddock larvae from Georges Bank. Mar. Ecol. Prog. Ser. 16:9-20.

Kumpf, H., K. Steidinger, and K. Sherman (eds.). 1999. The Gulf of Mexico Large Marine Ecosystem: Assessment, Sustainability, and Management. Blackwell Science, Inc. Malden MA. 736 p.

Levin, S.A. 1993. Approaches to forecasting biomass yields in large marine ecosystems. In: K. Sherman, L.M Alexander, and B.D. Gold (eds.). Large Marine Ecosystems: Stress, Mitigation, and Sustainability. AAAS Press, Washington DC. 36-39.

Lluch-Belda, D. 1999. The interdecadal climatic change signal in the temperate large marine ecosystems of the Pacific. In: K. Sherman and Q. Tang (eds.). The Large Marine Ecosystems of the Pacific Rim: Assessment, Sustainability, and Management. Blackwell Science, Inc. Malden MA. 42-47.

Lubchenco, J., A.M. Olson, L.B. Brubaker, S.R. Carpenter, M.M. Holland, S.P. Hubbell, S.A. Levin, J.A. MacMahon, P.A. Matson, J.M. Melillo, H.A. Mooney, C.H. Peterson, H.R. Pulliam, L.A. Real, P.J. Regal, and P.G. Risser. 1991. The sustainable biosphere initiative: an ecological research agenda. Ecology 72: 317-412.

Ludwig, D., R. Hilborn, and C. Walters. 1993. Uncertainty, resource exploitation, and conservation: Lessons from history. Science 260:17-36.

Mangel, M. 1991. Empirical and theoretical aspects of fisheries yield models for large marine ecosystems. In: K. Sherman, L.M. Alexander, and B.D. Gold (eds.). Food Chains, Yields, Models, and Management of Large Marine Ecosystems. Westview Press, Inc. Boulder. 243-261.

Mountain, D., J. Kane, and J. Green. 2000. Environmental forcing of variability in zooplankton abundance and cod recruitment on Georges Bank. ICES C.M./M:100.

Murawski, S.A. *et al.* 1999. New England Groundfish. Our Living Oceans: Report on the Status of U.S. Living Marine Resources, 1999. U.S. Dep. Commer., NOAA Tech. Memo. NMFS-F/SPO-41. Washington DC.

NEFMC. 2000. Atlantic Herring Stock Assessment and Fishery Evaluation Report for the 1999 Fishing Year. New England Fishery Management Council. 42 p.

NEFSC. 1999. G. Atlantic Herring. Report of the 27th Northeast Regional Stock Assessment Workshop (27th SAW). Stock Assessment Review Committee (SARC) Consensus Summary of Assessments. Woods Hole Laboratory Reference Document No. 98-15.

NEFSC. 2000a. Status of Fishery Resources off the Northeastern United States for 1999, Steve Clark, Editor. NOAA Tech. Memo. NMFS-NE-115.

NEFSC. 2000b. D. Atlantic mackerel. Report of the 30th Northeast Regional Stock Assessment Workshop (30th SAW). Stock Assessment Review Committee (SARC) Consensus Summary of Assessments. Woods Hole Laboratory Reference Document No. 00-03.

NMFS. 1996. Our living oceans. Report on the status of U.S. living marine resources, 1995. U.S. Dep. Commer., NOAA Tech. Memo. NMFS-F/SPO-19, 160 p.

Nixon, S.W., C.A. Oviatt, J. Frithsen, and B. Sullivan. 1986. Nutrients and the productivity of estuarine and coastal marine ecosystems. J. Limnol. Soc. Sth. Afr. 12:43-71.

O'Reilly, J.E. and C. Zetlin. 1998. Seasonal, horizontal, and vertical distribution of phytoplankton chlorophyll *a* in the northeast U.S. continental shelf ecosystem. U.S. Dept. Commerce, NOAA Tech. Rep. NMFS 139. 119 p.

O'Reilly, J.E. and J.A. Yoder. 1998. Comparisons between SeaWiFS and CZCS chlorophyll: Northeast US Continental Shelf. American Geophysical Union Fall 1998 meeting, San Francisco, CA. Abstract, page F410.

Pimm, S.L. 1984. The complexity and stability of ecosystems. Nature 307:321-326.

Roemmich, D. and J.A. McGowan. 1995. Climatic warming and the decline of zooplankton in the California Current. Science 267:1324-1326.

Rosenberg, A.A., M.J. Fogarty, M.P. Sissenwine, J.R. Beddington, and J.G. Shepherd. 1993. Achieving sustainable use of renewable resources. Science 262:828-829.

Sherman, K. 1980. MARMAP, A Fisheries Ecosystem Study in the Northwest Atlantic: Fluctuations in Ichthyoplankton-Zooplankton Components and Their Potential for Impact on the System. In: F.P. Diemer, F.J. Vernberg, and D.Z. Mirkes (eds.). Advanced Concepts on Ocean Measurements for Marine Biology. Belle W. Baruch Institute for Marine Biology and Coastal Research, Univ. of South Carolina Press. Columbia SC. 3-37.

Sherman, K. 1994. Sustainability, Biomass yields, and health of coastal ecosystems: An ecological perspective. Mar. Ecol. Prog. Ser. 112:277-301.

Sherman, K. and K.A. Honey. 1971. Seasonal variations in the food of larval herring in coastal waters of central Maine. Rapports et Proces-verbaus des Reunions, Conseil International pour L'Exploration de la Mer 160:121-124.

Sherman, K., N.A. Jaworski, and T.J. Smayda (eds.). 1996a. The Northeast Shelf Ecosystem: Assessment, Sustainability, and Management. Blackwell Science, Inc. Cambridge MA. 564 p.

Sherman, K., M. Grosslein, D. Mountain, D. Busch, J. O'Reilly, and R. Theroux. 1996b. The Northeast Shelf ecosystem: an initial perspective. In: K. Sherman, N.A. Jaworski, and T.J. Smayda (eds.). 1999. The Northeast Shelf Ecosystem: Assessment, Sustainability, and Management. Blackwell Science, Inc. Cambridge MA. 103-126.

Sherman, K., E.N. Okemwa, and M.J. Ntiba (eds.). 1998a. Large Marine Ecosystems of the Indian Ocean: Assessment, Sustainability, and Management. Blackwell Science, Inc., Malden, MA. 394 p.

Sherman, K. and H.C. Perkins. 1971. Seasonal variations in the food of juvenile herring in coastal waters of Maine. Trans. Amer. Fish. Soc. 100:121-124.

Sherman, K., A. Solow, J. Jossi, and J. Kane. 1998b. Biodiversity and abundance of the zooplankton of the Northeast Shelf ecosystem. ICES J. Mar. Sci. 55:730-738.

Sherman, K. and Q. Tang (eds.). 1999. Large Marine Ecosystems of the Pacific Rim: Assessment, Sustainability, and Management. Blackwell Science, Inc. Malden, MA. 465 p.

Sissenwine, M.P. and E.B. Cohen. 1991. Resource productivity and fisheries management of the northeast shelf ecosystem. In: K. Sherman, L.M. Alexander, and B.D. Gold (eds.). Food Chains, Yields, Models, and Management of Large Marine Ecosystems. Westview Press, Inc. Boulder. 107-123.

Skud, B.E. 1982. Dominance in fisheries: A relation between environment and abundance. Science 216:144-149.

III
Insular North Atlantic

Large Marine Ecosystems of the North Atlantic
K. Sherman and H.R. Skjoldal (Editors)
© 2002 Elsevier Science B.V. All rights reserved.

7

Iceland Shelf LME: Decadal Assessment and Resource Sustainability

Olafur S. Astthorsson and Hjálmar Vilhjálmsson

INTRODUCTION

A sentence in an old Icelandic seaman's hymn states "half of our fatherland is the ocean." This is very true, as through the ages the ocean and its resources have been one of the foundations of residence in the country. The settlers of Iceland (c. 870-930) were mainly livestock farmers, but gradually fishing became the main industry (Thor 1997). Since the end of the nineteenth century fishing has been the impetus for economic growth, and seafood products have usually constituted 70-80% of Icelandic merchandise exports. In fact, Iceland is one of few nations in the world today that has been able to build a modern society upon the exploitation of wild animal populations.

During their struggle for existence through the centuries, Icelanders have had their ups and downs. Until the beginning of the nineteenth century, fishing was carried out from rowboats, and weather was therefore the most important factor in determining yearly landings (Jónsson 1994). At the turn of the twentieth century, the fishing industry gradually became more mechanized and concurrently the catches increased. Thus, in addition to the environment affecting the fish stocks and their yield, overfishing or shortsightedness with respect to exploitation became a problem. In recent years the people of Iceland have, however, become more aware that both the changing environment and level of exploitation must be kept in mind and, as far as possible, be taken into account if future exploitation is to be sustainable.

In this paper we present a short description of the main environmental features of the waters around Iceland and how they have varied during recent decades. The exploited fish stocks in the Icelandic ecosystem and their yield during recent decades is also examined, followed by a more detailed discussion of cod (*Gadus morhua*) and capelin (*Mallotus villosus*). These are two of the key exploited species in the Icelandic LME and are also linked through a tight predator-prey relationship (Magnússon and Pálsson 1989, 1991; Vilhjálmsson 1994, 1997). Finally, some examples of how environmental factors have influenced the exploitation of these important stocks are presented, as well as a conceptual model of how changes in hydrography north of Iceland may be transferred up the food chain to influence primary production, zooplankton, capelin, and, eventually, the cod stock.

BOTTOM TOPOGRAPHY

Iceland rises from the crest of the mid-Atlantic ridge and is also part of the system of transverse ridges that extend from Scotland through the Faroes to Greenland (Figure 7-1). These transverse ridges are of importance oceanographically since they separate the relatively warm waters of the northeastern North Atlantic Ocean from the cold Arctic deepwater of the Iceland and Norwegian Seas (Stefánsson 1962). To the south of Iceland lies the deep Iceland Basin, which in the west is separated from the Irminger Sea by the Reykjanes Ridge and in the east from the Norwegian Sea by the Iceland-Faroe Ridge. The Reykjanes ridge is a part of the Mid-Atlantic Ridge.

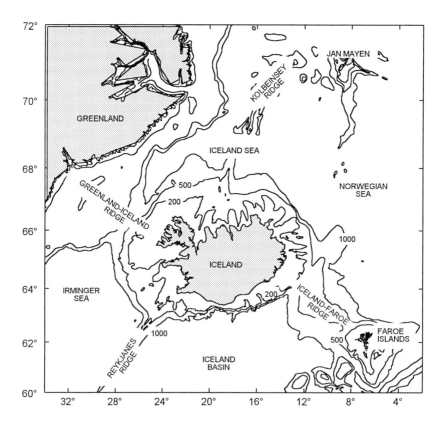

Figure 7-1. Bottom topography around Iceland. The 400 m depth contour is considered to mark the Icelandic shelf area.

The outer boundary of the continental shelf surrounding Iceland roughly follows the 400-500 m depth contour, the total area within the 500 m depth contour being 212 thousand km². The shelf is narrowest off the south coast where in places it extends out only a few nautical miles. From there, the continental slope descends steeply to depths exceeding 1500 m. Off the west, north, and east coasts, however, the shelf is relatively broad and extends 60-90 nautical miles out from the coast (Figure 7-1).

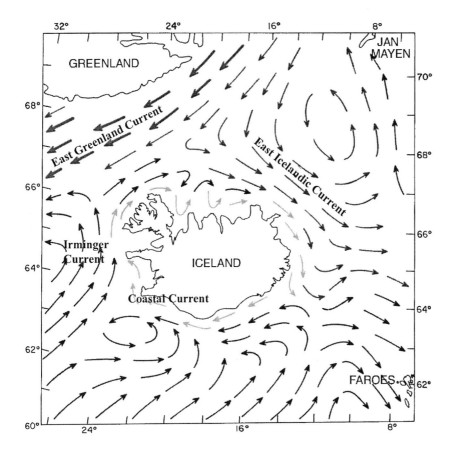

Figure 7-2. Ocean currents around Iceland. Modified from Stefánsson and Ólafsson (1991).

HYDROGRAPHIC FEATURES

The general features of the current system and watermass distribution around Iceland (Figure 7-2) have been described by Stefánsson (1962, 1999) and Valdimarsson and Malmberg (1999). The warm and saline Atlantic water (Irminger Current) flows along the south and west coasts of Iceland and splits into two branches off the Vestfirdir peninsula. The larger branch turns west and southwards, meets the cold East Greenland Current, and forms a cyclonic eddy in the Irminger Sea. The other branch flows eastwards on to the shelf north of Iceland as the North Icelandic Irminger Current. On the north Icelandic shelf, the Atlantic influx is evident as a tongue of relatively warm and saline water, but the proportion of Atlantic water in this east-flowing tongue decreases in the direction of the flow. The cold and low salinity East Icelandic Current, which is a mixture of water derived from both the North Icelandic Irminger Current and the East Greenland Current, flows southeast along the northeast Icelandic continental slope. Close to shore, a freshwater-induced coastal current flows clockwise around Iceland. The deep and bottom currents in the sea around Iceland originate in the overflow of deep cold water from the Nordic Seas and the Arctic Ocean south over the submarine ridges described earlier and into the North Atlantic.

TEMPERATURE AND SALINITY VARIATIONS

Due to the location of Iceland near the boundary between warm and cold currents, i.e. at the oceanic Polar Front in the northern North Atlantic, hydrographic conditions in the shelf area north of the country are highly variable. Consequently, changes in the intensity of the influx of warm Atlantic water and/or variable admixture of polar water may lead to marked fluctuations in temperatures and salinities.

A selected hydrographic section north of Iceland (Siglunes section, 66°16′N, 18°50′W - 67°00′N, 18°50′W) has generally been used to characterize the volume of Atlantic water flowing into the region north of Iceland (Figure 7-3). The period 1952-1964 was characterized by a strong inflow of Atlantic water, while during 1965-1971 the inflow was negligible. Between these periods the difference in the temperature north of Iceland was as much as 4°C. Since 1971 the inflow has been rather variable, often with 3-4 years of marked inflow (*e.g.* 1972-1974, 1984-1987, 1991-1994) and then a few years of limited inflow (*e.g.* 1975-1979, 1982-1983, 1995-1998).

To some extent, these fluctuations can be related to large-scale changes in the atmospheric circulation over the North Atlantic Ocean. Thus, the warm periods coincide to some extent with positive North Atlantic Oscillation (NAO) indices while the cold periods, similarly to some extent, coincide with negative NAO indices (Malmberg *et al.* 1999). However, since the NAO is mainly a measure of the westerly winds blowing from America across the Atlantic at mid latitudes, mainly to the south of Iceland, the relation between climate conditions north of Iceland and the NAO are only indirect. The more detailed hydrographic conditions north of Iceland appear to depend to a greater extent on northerly and southerly winds than on large-scale atmospheric features such as the NAO (Ólafsson 1999).

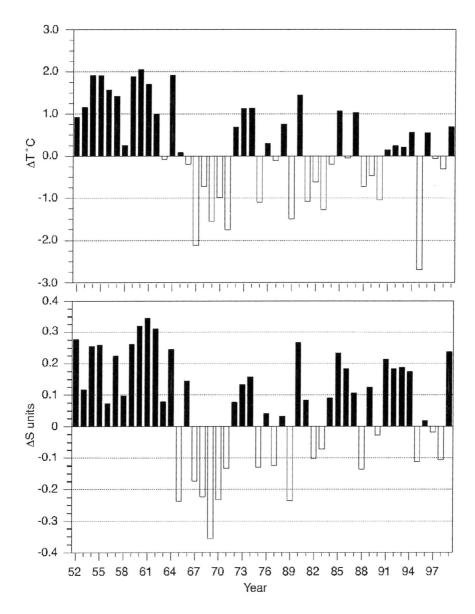

Figure 7-3. Temperature and salinity deviations in late spring from the 1961-1980 average on the Siglunes section to the north of Iceland (the deviation is based on 5 stations between 2 and 46 naut. miles offshore and data from 0, 20, 50, 75, 100, 150, and 200 m. depth). Based on data from Anon (1999).

Atlantic water predominates to the south and west of Iceland. Contrary to the conditions north of Iceland, interannual variations of the main hydrographic features off the south coast are limited and only slight changes in temperature and salinity have been observed (Malmberg 1984; Dickson *et al.* 1988). However, close to shore, hydrographic conditions may vary considerably from one year to another, mainly due to variable timing and magnitude of freshwater run off and variable wind force and direction (Thordardottir 1986; Gislason *et al.* 1994).

PRIMARY PRODUCTION

Extensive primary productivity measurements have been carried out annually in the waters around Iceland for more than four decades (*e.g.* Thordardottir 1976, 1977, 1984, 1986, 1994). The major emphasis has been on spring bloom development in relation to hydrographic features but information has also been collected throughout the growth season.

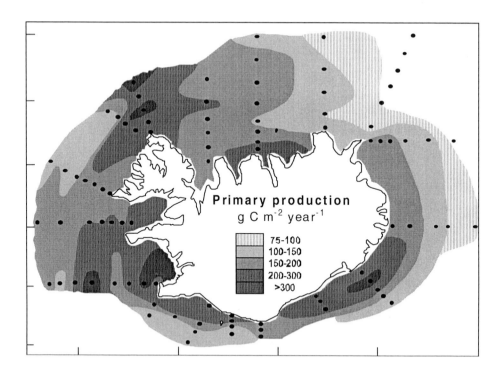

Figure 7-4. Average annual primary production in Icelandic water based on data from the period 1958-1982 (g C m^{-2} year^{-1}). Adapted from Thordardottir (1994).

On the basis of daily production measurements made between 1958-1982, Thordardottir (1994) estimated the average annual primary production for the sea area around Iceland (Figure 7-4). Primary production is generally higher in the warm water off the south and west coasts than in the cold water off the north and east coasts, and it is also higher closer to the land than farther out. On average, the primary production was estimated to be 175 g C m^{-1} year^{-1} or 55 million tonnes C year^{-1} for the combined 314 thousand km^2. However, for the entire 200-mile fisheries zone the annual production is estimated to be 122 million tonnes C year^{-1} (Thordardottir 1994).

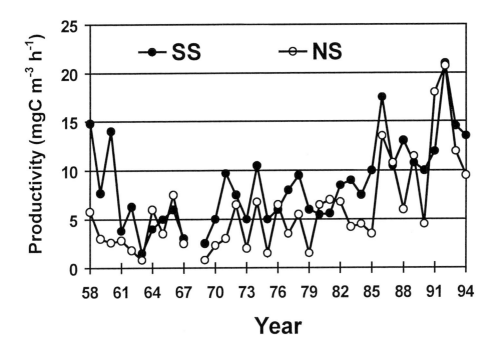

Figure 7-5. Mean productivity (mg C m^{-3} h^{-1}) in the shelf waters (200m depth contour) northeast and southwest of Iceland in spring (16 May to 15 June) for the period 1958-1994. In the northeast region the value for each year is based on 23 measurements per year on average, while for the southwest region on 24 observations per year on average. Redrawn from Gudmundsson (1998).

As pointed out above, the Icelandic LME is located in a frontal area of warm and cold currents and the variable hydrographic conditions result in marked changes in the spring development of phytoplankton from one year to another. Gudmundsson (1998) has recently summarized these changes by using data collected in spring (15 May-14 June) during 1958-1994. Similar to the findings of Thordardottir (1994) with respect to annual production, Gudmundsson (1998) demonstrated that the hourly productivity in spring is generally higher on the southwest shelf region than on the northeast (Figure 7-5). Differences in ambient temperature were believed to partly explain the difference in productivity between the two shelf regions. However, the interannual variations in productivity, observed on the northeast shelf, were ascribed to changes in inflow of Atlantic water and its influence on stratification of the water column and then on spring bloom development. During years of limited Atlantic inflow, the water column stabilizes after the initial spring bloom, nutrients become exhausted and thus the bloom ends abruptly and the biomass remains low throughout the summer (Thordardottir 1977, 1984; Stefánsson and Ólafsson 1991; Gudmundsson 1998). The influence of hydrographic conditions on primary production in spring in the northeast shelf region is further demonstrated in Figure 7-6, where stations with salinity >34.5 are compared to those with salinity <34.5. Clearly, the spring production is generally higher where salinity is higher, *i.e.* where the influence of Atlantic water is more pronounced.

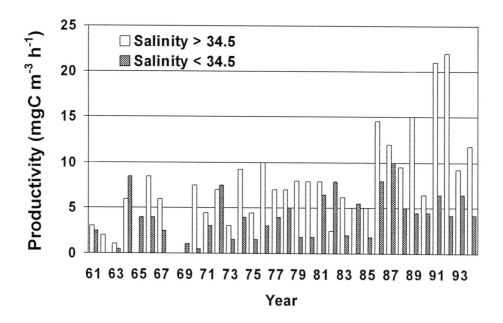

Figure 7-6. Comparison of mean productivity (mg C m^{-3} h^{-1}) in the shelf region northeast of Iceland during spring 1961-1993 at stations with salinity >34.5 and stations with salinty <34.5. Adapted from Gudmundsson (1998).

The average spring productivity in the southwest shelf region also varies considerably between years (Figure 7-5), but there the stratification and onset of growth is believed to depend to a large extent on the combined effect of the amount of runoff from land and the wind regime (*cf.* Thordardottir 1986; Gislason *et al.* 1994).

Figures 7-5 and 7-6 further suggest that from year to year the total annual primary production is likely to vary markedly, depending on climatic and hydrographic conditions, and that those variations will undoubtedly have an influence at higher levels of the food chain.

ZOOPLANKTON BIOMASS

Since the early 1960s, studies on zooplankton biomass and species composition have been carried out on standard transects during late May-June in Icelandic waters (Astthorsson *et al.* 1983; Astthorsson and Gislason 1995). *Calanus finmarchicus* is the dominant member of the plankton community and it commonly constitutes between 60-80% of the zooplankton in spring (Figure 7-7). Therefore interannual variations in biomass mainly reflect variations in this species.

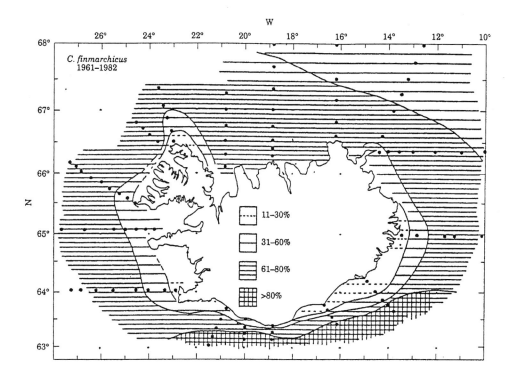

Figure 7-7. Average percentage frequency of *Calanus finmarchicus* at fixed stations around Iceland in May-June 1961-1982. From Astthorsson *et al.* (1983).

The spring zooplankton biomass in the upper 50 m in Icelandic waters generally ranges from ca. 1-10 g dry weight m^{-2}, with an average of 2-4 g dry weight m^{-2}. The lowest biomass is usually observed in the coastal waters off the northeast coast where Atlantic influence is at a minimum, while the highest biomass is found in the frontal area between the coastal and the Atlantic water off the south coast and in the Arctic waters of the East Icelandic Current off the northeast coast (Astthorsson and Gislason 1995). These differences in zooplankton biomass in spring appear to be related both to hydrographic features and composition of the zooplankton, i.e. small-sized species in high numbers in the Atlantic water but large-sized species in low numbers in the Arctic waters.

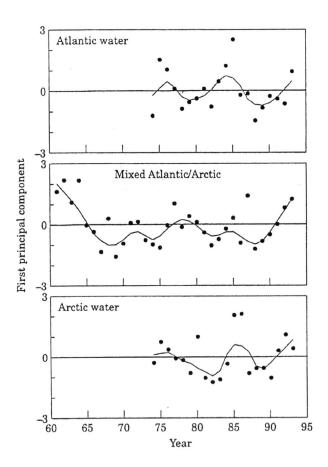

Figure 7-8. First principal component of the fluctuation in zooplankton biomass at three hydrographic regimes (Atlantic, Atlantic/Arctic, Arctic). For each area the principal component has been standardised to show a zero mean and a standard deviation of one. The curved line shows a plain LOWESS smoothing with a span of seven years. From Astthorsson *et al.* (1983).

In order to examine long-term fluctuations in zooplankton biomass in Icelandic waters, Astthorsson and Gislason (1995) grouped data from different hydrographic regimes into three categories, Atlantic water, Mixed Atlantic/Arctic water, and Arctic water, and analyzed it by principal component analysis (Figure 7-8). In the waters of the fluctuating Atlantic/Arctic hydrographic conditions north of Iceland, there was a steady decline in zooplankton biomass from the early 1960s to the late 1960s. The trend then reversed and a positive trend, interrupted by a low during ca. 1973-1975, was observed until a peak was reached in the late 1970s. The zooplankton biomass north of Iceland then declined again. There was a small peak in the mid-1980s, after which a reversal was observed. This positive trend was reversed again during the mid-1990s and since then a downward trend has been observed (Anon. 1999). When these fluctuations were considered in the context of classification into "warm" and "cold" hydrographic conditions north of Iceland, the first principal component in the mixed Atlantic/Arctic water was significantly higher in warm years than in cold years (Astthorsson and Gislason 1994).

The scores of the first principal component for the period during which observations are available in all three hydrographic areas (1974-1993, Figure 7-8) indicate common patterns in the year-to-year changes in biomass. All three areas showed peaks during the mid-1980s, lows during the late 1980s, and a positive trend from then until 1993. Two of the areas (Atlantic and Arctic) showed a peak during the mid-1970s and a decline from then until the early 1980s. The only apparent anomaly is the slight peak observed in the area classified as mixed Atlantic/Arctic during the late 1970s. Furthermore, when data from all areas were combined and compared to fluctuations in the North Atlantic and around the British Isles, as demonstrated by the Continuous Plankton Recorder, very similar fluctuations were apparent (Figure 7-9). This led to the conclusion that variations of zooplankton in Icelandic waters are to a large extent influenced by the same large-scale climatic factors that are believed to affect large-scale plankton dynamics in the northern North Atlantic (Astthorsson and Gislason 1995).

YIELD OF FISH

During the past five decades the catch of demersal fish in the waters around Iceland has been in the range of 450-850 thousand tonnes. The catch was highest in 1954 and 1955, around 850 thousand tonnes, but lowest in 1996 and 1997, 440 thousand tonnes. The average annual catch has been around 650 thousand tonnes (Figure 7-10).

Cod (*Gadus morhua*) is by far the most important of exploited demersal stocks in Iceland. During the past five decades the annual catch of cod has been about 370 thousand tonnes, *i.e.* more than 50-60 % of the total demersal catch. Redfish (mainly *Sebastes marinus*) has yielded the second largest catches and during recent decades the annual catch has been around 90 thousand tonnes on average. The average catch of

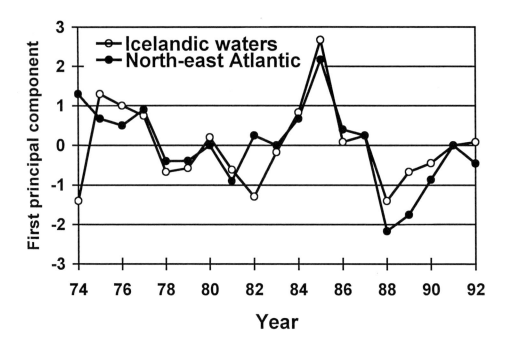

Figure 7-9. First principal component of the fluctuations of zooplankton biomass in Icelandic waters and of the fluctuations of zooplankton abundance in the north-east Atlantic and around the British Isles during 1974-1992 (The values for Atlantic zooplankton were read from Figure 1 in Gamble *et al.* 1993). For both series the principal component has been standardised to show a zero mean and a standard deviation of one. Black dots, Icelandic waters. Open circle, Northeast Atlantic and British Isles. From Astthorsson and Gislason (1995).

saithe (*Pollachius virens*) has been about 70 thousand tones, and of haddock (*Melanogrammus aeglefinus*) about 60 thousand tonnes. The contribution of other demersal species to the total demersal catch, the majority of which are flatfishes [Greenland halibut (*Reinhardtius hippoglossoides*), plaice (*Pleuronectes platessa*), lemon sole (*Microstomus kitt*), witch (*Glyptocephalus cynoglossus*), halibut (*Hippoglossus hippoglossus*), dab (*Limanda limanda*), long rough dab (*Hippoglossoides platessoides*)], has been about 50-60 thousand tonnes during recent decades.

In addition to the demersal fish stocks, a very important fishery for deepwater shrimp (*Pandalus borealis*) has developed during the past four to five decades (Figure 7-11). Initially, this was an inshore fishery yielding a few thousand tonnes per annum, but a significant offshore fishery began during the late 1980s. Since then the total shrimp catch has increased from about 10 thousand to 70 thousand tonnes. However, during the most recent years the offshore shrimp fishery has declined again, mainly due to higher

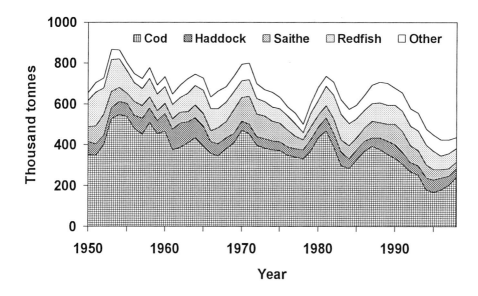

Figure 7-10. The catch of demersal fish in Icelandic waters during 1950-1998.

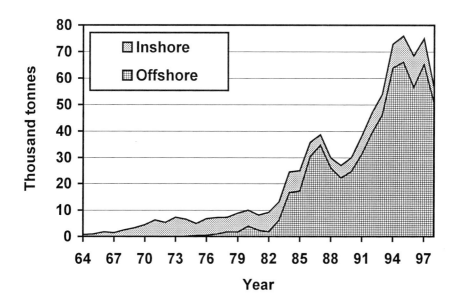

Figure 7-11. The catch of deepwater shrimp (*Pandalus borealis*) in Icelandic waters during 1964-1998.

predatory pressure by the increasing cod stock, but also to some extent because of catches being higher than recommended.

Two pelagic species, herring (*Clupea harengus*) and capelin (*Mallotus villosus*), have been of marked importance to Icelandic fisheries during recent decades (Figure 7-12). During the 1950s and 1960s the herring fishery was mainly based on Atlanto-Scandian herring when feeding north and northeast of Iceland. In addition, fishing for two local Icelandic herring stocks was also of considerable importance during this period. The catch from these three herring stocks in Icelandic waters reached a peak of about 600 thousand tonnes just before their collapse during the late 1960s (Jakobsson 1980, 1985, 1992). During the period 1975-1994 the herring catch in Icelandic waters was entirely based on Icelandic summer spawners. More recently the Icelandic herring fishery has again been based to a considerable extent on migrating Atlanto-Scandian herring. This fishery has, however, been largely conducted in the Norwegian Sea and not on the Icelandic shelf (*e.g.* Misund *et al.* 1998).

Commercial fishing for capelin began in the late 1960s and increased rapidly to a peak of about 1.2 million tonnes in the late 1970s. Stock collapses led to a fishing ban for almost two years in the early 1980s and also to an abrupt reduction in catches about 10 years later. Except for these two occasions, the annual yield of the capelin fishery has remained fairly constant at about 1 million tonnes (Figure 7-12).

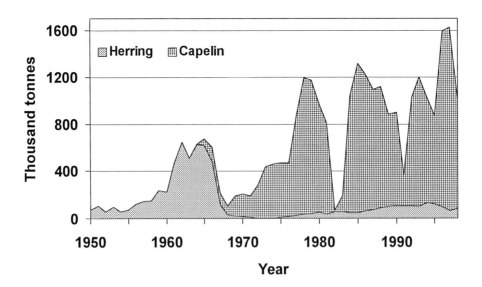

Figure 7-12. The catch of pelagic fish in Icelandic waters during 1950-1998.

THE ICELANDIC COD STOCK

Icelandic cod spawn mainly in late April to early May, relatively close to the coast in the Atlantic waters off the south and southwest coasts of Iceland. The eggs and larvae drift from the spawning grounds with clockwise currents to the nursery and feeding grounds in the waters off the northwest, north, and the northeast coasts. On reaching maturity, the cod migrate back to warmer water for spawning. During some years a considerable fraction of the larvae may also drift to Greenland waters, where they then spend their juvenile years before returning to Icelandic waters as mature individuals to spawn (Astthorsson *et al.* 1994).

Schopka (1994) has recently described in detail the development of the cod fishery during this century. The largest annual catches of about 550 thousand tonnes were taken in the early 1930s and in the mid 1950s. High fishing pressure over several decades, along with changes in recruitment and variations in numbers of immigrants from Greenland, have since reduced the stock. During 1995 the catch of cod was only 170 thousand tonnes, or lower than half a century earlier. Due to strict protection measures in recent years, the stock is now recovering and in 1998 the catch had risen to 242 thousand tonnes (Figure 7-13).

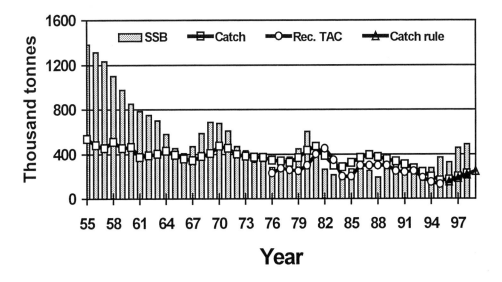

Figure 7-13. Spawning stock biomass, total catch, recommended catch and catch according the catch rule for the Icelandic cod stock during 1955-1998.

The spawning stock declined from about 1400 thousand tonnes in 1955 to less than 300 thousand tonnes in 1982 and remained near that level until 1995; it is about 500 thousand tonnes at present. In 1976, following an extensive debate on the status of the cod stock during the 1970s, the Marine Research Institute began to release annual recommendations on total allowable catch (TAC) (Figure 7-13). Unfortunately this was never fully adhered to. In 1984 a management system consisting of individual transferable quotas (ITQs) was introduced in an attempt to regulate the fishery (Jakobsson and Stefánsson 1998). In spite of this, the yield was consistently above the recommended TAC until the most recent years. In a further attempt to regulate the cod fishery, the Icelandic government introduced a so called "catch rule" in 1995, restricting the cod catch to 25% of the estimated fishable stock biomass as a long-term harvesting strategy. This government decision, which has been strictly adhered to since 1995, is already having an effect toward rebuilding the cod stock (Figures 7-13, 7-14).

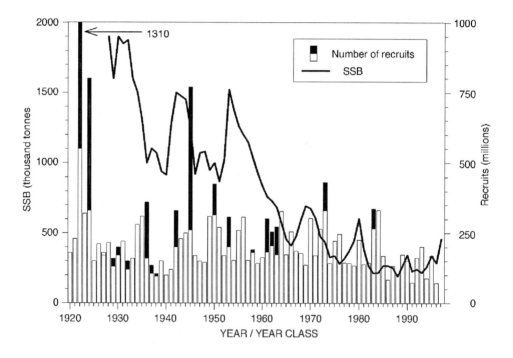

Figure 7-14. Recruitment (at the age of 3 years) and spawning stock biomass of the Icelandic cod stock 1920-1998. The recruits are shown separately for the Iceland (white bar) Greenland (dark bar) components of the stock. Based on Schopka (1994) and additional data.

The year class strength (measured as numbers of 3 year olds) among Icelandic cod has been estimated as far back as 1920, including the component of the year class that consists of immigrants from Greenland to Iceland (Schopka 1994) (Figure 7-14). The 1922 year class is estimated to have been by far the most abundant one during this century (1.3 billion individuals), while the 1924 and 1945 year classes were also very large (almost 800 million). Other strong year classes are from 1936, 1943, 1945, 1950, and 1973. All these year classes include immigrants, and the three largest year classes were to a great extent composed of immigrants from Greenland. Outstanding but pure Icelandic year classes never seem to be stronger than 300-350 million. Very poor year classes, with individuals below 100 million, have only rarely been encountered. It is worth pointing out, however, that during the whole period 1985-1996 recruitment has generally been poor, with the 1991 year class counting only 80 million and being the poorest this century, and the largest year class numbering only some 220 million recruits.

Drift of cod larvae from the spawning grounds at Iceland to Greenland waters extends the nursery area for cod of Icelandic origin. Thus, both climate and fisheries around Greenland directly affect the state of the cod stock at Iceland. The warming of the northern North Atlantic in the early 1920s extended the distribution area of cod along the east, but particularly along the west, coast of Greenland. Combined with the drift of larval cod from Iceland to Greenland and the subsequent return of these fish as mature individuals, this warming is considered to be the main reason for the outburst of the superabundant year classes at Iceland in the early 1920s (Schopka 1994). Schopka pointed out that immigration took place more often prior to 1970 than during the last three decades when it has only been registered twice, i.e. in 1973 and 1984 (Figure 7-14). This is considered to be connected to a much poorer state of the cod stock at Iceland in the last three decades than earlier this century, as well as to there being almost no cod at Greenland during recent years because of their inability to reproduce in the cold marine climate there. Further, there must have been a decline in the frequency of larval drift from the Icelandic ecosystem towards Greenland, possibly caused by changes in the environment of the North Atlantic which took place during the late 1960s (Dickson *et al.* 1988) and/or connected with reduction of the cod stock at Iceland. Dickson and Brander (1993) suggested that increased larval exchange between Iceland and Greenland in warm years was connected to a strengthening of the Irminger/West Greenland Current system.

THE ICELANDIC CAPELIN STOCK

Capelin is a key species in the Icelandic ecosystem. It is by far the largest pelagic species in the system and also the most important food of the largest demersal stock, the cod. Furthermore, capelin is a major food constituent for other predators, such as Greenland halibut, saithe, seabirds, and whales (Vilhjálmsson 1994, 1997).

Capelin spawn off the south and southwest coast of Iceland in February-March. During summer the larvae drift to the nursery and feeding grounds northwest, north, and northeast of the country. Many adult Icelandic capelin migrate north to feed in the deep waters of the Iceland Sea during the months of June-September (Figure 7-15), and then they are out of reach of cod. This situation changes with the return of capelin to the shelf area north of Iceland in late autumn. From then on, and until spawning off the southwest coast, adult capelin are eaten by cod in large quantities (Pálsson 1983; Magnússon and Pálsson 1991).

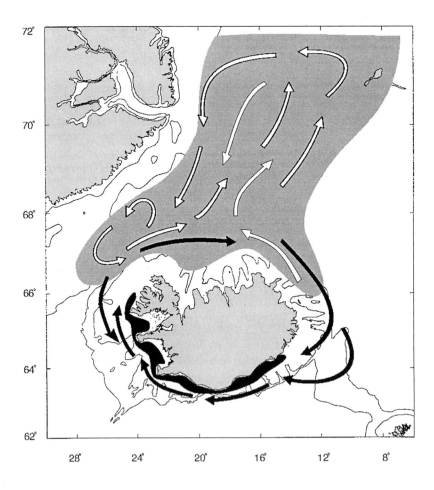

Figure 7-15. Feeding areas and spawning grounds of adult Icelandic capelin. The feeding areas (grey) and spawning grounds (black). White arrows indicate feeding migrations and black arrows spawning migrations. Redrawn from Vilhjálmsson (1997).

No information is available on the stock size of capelin during the period prior to 1970 when Atlanto-Scandian herring migrated across the Norwegian Sea to feed in the waters north and northeast of Iceland. To some extent, the two stocks exploited the same food resource (Astthorsson and Gislason 1998; and unpublished data). It seems possible that if or when the Norwegian spring spawning herring resume their feeding migrations, these species may compete for food or the herring may even prey upon the 0-group capelin. Therefore, we may not see these pelagic stocks very large in Icelandic waters simultaneously.

As stated above, a winter fishery for capelin began on a small scale in the mid-1960s but soon expanded to yield almost 500 thousand tonnes annually. A summer fishery based on feeding concentrations of capelin was begun in 1976, and by 1978 the annual catch had increased from about 400-500 thousand tonnes to more than 1 million tonnes. The adult stock was reduced to a very low level in the early 1980s, and was again reduced to a low level in 1989-1991. Recovery was achieved in 1-2 years on both occasions. Since 1991, the adult stock prior to the fishery has remained between 2-2.5 million tonnes (Figure 7-16).

Since 1980, management of Icelandic capelin has been approached in a multi-species context. Thus, 400 thousand tonnes of capelin have been allowed to spawn each year and the immature part of the stock has been specially protected from fishing. Furthermore, the need of the main predator, the cod, has been taken into account prior to the final decision of the total allowable catch (Vilhjálmsson 1994, 1997).

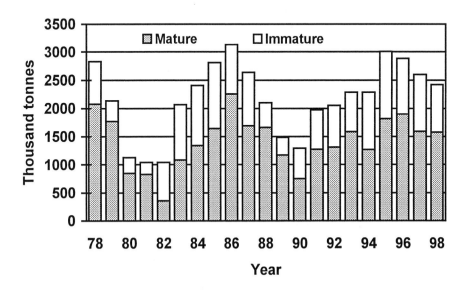

Figure 7-16. Stock biomass of the Icelandic capelin stock on 1 August 1978-1998.

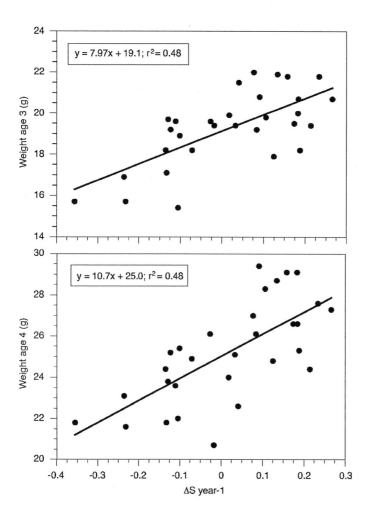

Figure 7-17. Relation between winter weight of 3- and 4-group capelin and salinity deviations in May of preceding year for the period 1970-1998. Redrawn from Vilhjálmsson (1994) and additional data.

RELATION BETWEEN ENVIRONMENT, CAPELIN AND COD

As mentioned earlier, influx of Atlantic water into the North and East Icelandic shelf area leads to increased primary production and zooplankton biomass, while the presence of polar water has the reverse effect. Astthorsson and Gislason (1994) have shown that zooplankton biomass in the waters north of Iceland is on average two times greater during years when Atlantic inflow takes place than during years when it does not occur.

This influence of hydrographic conditions on both primary and secondary production also seems to be clearly reflected in the growth of capelin, which is better during years when the inflow of high salinity Atlantic water is strong (Figure 7-17). Furthermore, Figure 7-18 shows that the total biomass of capelin is generally higher during years when zooplankton biomass is high compared to when it is low. Magnússon and Pálsson (1989, 1991) showed that the Icelandic cod stock had less stomach content, reduced feeding levels, and slower growth rates when the capelin stock was small. The importance of capelin for the condition of the Icelandic cod stock is further demonstrated in Figure 7-19. Thus, it has been observed (Vilhjálmsson 1997) that during the near-collapse of the Icelandic capelin stock in the early 1980s and 1990s, the weight at age in the Icelandic cod stock, in particular among age groups 4-8, decreased by about 25%. The condition of the cod improved again following the quick recovery of capelin in 1983 and 1984, but declined again with reduced capelin abundance in 1989-90, and improved after that in tune with increased capelin abundance. It is quite clear that other available prey species cannot fully substitute for the loss of capelin from the food resources available to cod in the Icelandic area.

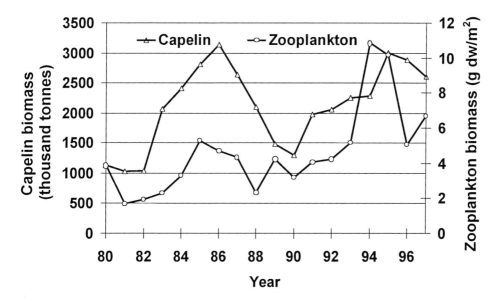

Figure 7-18. Changes in capelin abundance and zooplankton biomass north of Iceland 1980-1997. Redrawn from Astthorsson and Gislason (1994) and additional data.

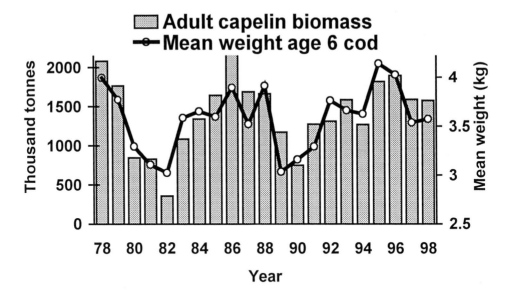

Figure 7-19. Changes in capelin biomass and mean weight of Icelandic cod at age of 6 years. Redrawn from Vilhjálmsson (1997) and additional data.

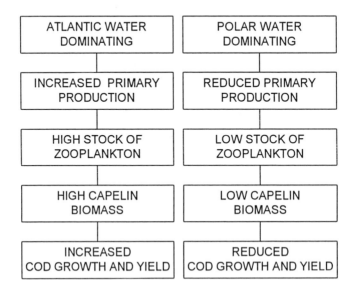

Figure 7-20. A conceptual model of how climatic conditions in Icelandic waters may affect production at lower trophic levels and eventually the yield from the Icelandic cod stock.

CONCLUDING REMARKS

A conceptual model of how climatic factors, during different regimes, may greatly affect the yield of cod in Icelandic waters through the food chain is presented in Figure 7-20. We are well aware of the fact that this is an oversimplification. For example, the model does not consider factors such as recruitment, which is highly variable and to a large extent determined by environmental conditions. However, what it demonstrates is the simplicity of the main trophic links and that oscillations between warm and cold climatic regimes may have sudden and dramatic influences on fish yield in ecosystems bordering on ocean climatic contrasts like the Polar Front.

As pointed out above, the management of Icelandic capelin has been approached in a multi-species context since 1980. Hopefully, Icelandic cod will someday regain its natural abundance; but when it does, the exploitation of Icelandic capelin will probably have to be revised. In the meantime, steps have been taken for obtaining a better understanding of multispecies interactions in the Icelandic ecosystem in general (Anon, 1997).

REFERENCES

Anon. 1997. Fjölstofnarannsóknir 1992-1995 (Multi-species investigations 1992-1995. In Icelandic). Hafrannsóknastofnunin Fjölrit, 57, 411 p.

Anon. 1999. Þættir úr vistfræði sjávar 1997 og 1998 (Environmental conditions in Icelandic waters 1997 and 1998. In Icelandic, English summary). Hafrannsóknastofnun Fjölrit 73, 48 p. (mimeo).

Astthorsson, O.S. and A. Gislason. 1994. Dýrasvif í hafinu við Ísland (Zooplankton in Icelandic waters. In Icelandic). In: U. Stefansson (ed.). Icelanders, the ocean and its resources. Societas Scientarium Islandica, Reykjavik. 89-106.

Astthorsson, O.S. and A. Gislason. 1995. Long term changes in zooplankton biomass in Icelandic waters in spring. ICES Journal of Marine Science, 52: 657-688.

Astthorsson, O.S. and A. Gislason. 1998. On the food of capelin in the subarctic waters north of Iceland. Sarsia, 82: 81-86.

Astthorsson, O.S., A. Gislason, and A. Gudmundsdottir. 1994. Distribution, abundance, and length of pelagic juvenile cod in Icelandic waters in relation to environmental conditions. ICES Marine Science Symposia, 198: 529-541.

Astthorsson, O.S., I. Hallgrímsson, and G.S. Jónsson. 1983. Variations in zooplankton densities in Icelandic waters in spring during the years 1961-1982. Rit Fiskideildar, 7: 73-113.

Dickson, R.R. and K.M. Brander. 1993. Effect of a changing windfield on cod stocks of the North Atlantic. Fisheries Oceanography, 3/4: 124-153.

Dickson, R.R., J. Meinke, H.H. Lamb, S.A. Malmberg, and A.J. Lee. 1988. The "Great Salinity Anomaly" in the Northern North Atlantic 1968-1982. Progress in Oceanography, 20: 103-151.

Gamble, J.C., G.C. Hays, and H.G. Hunt. 1993. The status of the plankton populations in the northwest European shelf seas and northwest Atlantic as determined by the Continuous Plankton Recorder Survey. ICES CM 1993/L:16. 16p.

Gislason, A., O.S. Astthorsson, and H. Gudfinnsson. 1994. Phytoplankton, *Calanus finmarchicus*, and fish eggs southwest of Iceland,1990-1992. ICES Marine Science Symposia, 198: 423-429.

Gudmundsson, K. 1998. Long-term variation in phytoplankton productivity during spring in Icelandic waters. ICES Journal of Marine Science, 55: 635-643.

Jakobsson, J. 1980. The north Icelandic herring fishery and environmental conditions, 1960-1968. Rapport et Proces-verbau Réunion Conseil International Exploration de la Mer, 177: 460-465.

Jakobsson, J. 1985. Monitoring and the management of the Northeast Atlantic herring stocks. Canadian Journal of Fisheries and Aquatic Sciences, 42: 207-221.

Jakobsson, J. 1992. Recent variability in the fisheries of the North Atlantic. ICES Marine Science Symposia,195: 291-315.

Jakobsson, J. and G. Stefánsson. 1998. Rational harvesting of the cod-capelin-shrimp complex in the Icelandic marine ecosystem. Fisheries Research, 37: 7-21.

Jónsson, J. 1994. Fisheries off Iceland 1600-1900. ICES Marine Science Symposia,198: 3-16.

Magnússon, K. and Ó.K. Pálsson. 1989. Trophic ecological relationships of Icelandic cod. Rapport et Proces-verbau Réunion Conseil International Exploration de la Mer, 188: 206-224.

Magnússon, K. and Ó.K. Pálsson. 1991. Predator-prey interactions of cod and capelin in Icelandic waters. ICES Marine Science Symposia, 193: 153-170.

Malmberg, S.A. 1984. A note on the seventies anomaly in Icelandic waters. ICES CM 1984/Gen:14 Mini Symp. 8 p.

Malmberg, S.A., J. Mortensen, and H. Valdimarsson. 1999. Decadal scale Climate and hydrobiological variations in Icelandic waters in relation to large scale atmospheric conditions in the North Atlantic. ICES CM 1999/L:13. 9 p.

Misund, O.A., H. Vilhjálmsson, S.H. Jákupsstovu, I. Röttingen, S. Belikov, O.S. Astthorsson, J. Blindheim, J. Jónsson, A. Krysov, S.A. Malmberg, and S. Sveinbjörnsson. 1998. Distribution, migration and abundance of Norwegian spring spawning herring in relation to temperature and zooplankton biomass in the Norwegian Sea as recorded by coordinated surveys on spring and summer 1996. Sarsia, 83: 117-127.

Ólafsson, J. 1999. Connection between oceanic conditions off N-Iceland, Lake Mývatn temperature, regional and directional wind variability and the North Atlantic Oscillation. Rit Fiskideildar, 16:41-57.

Pálsson, Ó.K. 1983. Feeding habits of demersal fish species in Icelandic waters. Rit Fiskideildar, 7: 1-60.

Schopka, S.A. 1994. Fluctuation in the cod stock off Iceland during the twentieth century in relation to changes in the fisheries and the environment. ICES Marine Science Symposia, 198: 175-193.

Stefánsson, U. 1962. North Icelandic Waters. Rit Fiskideildar, 3:1-269.

Stefánsson, U. 1999. Hafið (The Ocean. In Icelandic). Háskólaútgáfan, Reykjavík. 475 p.

Stefánsson, U. and J. Ólafsson. 1991. Nutrients and fertility of Icelandic waters. Rit Fiskideildar, 12: 1-56.

Thor, J. Th. 1997. Ránargull (Ocean gold, In Icelandic). Skerpla, Reykjavik, 205 pp.

Thordardottir, Th. 1976. The spring primary production in Icelandic waters 1970-1975. ICES CM 1976/L:31. 37 p.

Thordardottir, Th. 1977. Primary production in North Icelandic waters in relation to recent climatic changes. In: M.J. Dunbar (ed.). Polar Oceans, Proceedings of the polar oceans conference held at McGill University, Montreal, May 1974. Arctic Institute of America, Canada. 655-665.

Thordardottir, Th. 1984. Primary production north of Iceland in relation to water masses in May-June 1970-1989. ICES CM 1984/L:20, 17 p.

Thordardottir, Th. 1986. Timing and duration of spring blooming south and southwest of Iceland. In: S. Skreslet (ed.). The role of freshwater outflow in coastal marine ecosystems. Springer-Verlag, Berlin Heidelberg. 345-360.

Thordardottir, Th. 1994. Phytoplankton and primary production in Icelandic waters (In Icelandic). In: U. Stefánsson (ed.). Icelanders, the ocean and its resources. Societas Scientiarum Islandica, Reykjavik. 65-88.

Valdimarsson, H. and S.A. Malmberg. 1999. Near-surface circulation in Icelandic waters derived from satellite tracked drifters. Rit Fiskideildar, 16: 23-39.

Vilhjálmssson, H. 1994. The Icelandic capelin stock. Capelin, *Mallotus villosus* (Müller) in the Iceland - Greenland - Jan Mayen area. Rit Fiskideildar, 13: 1-281.

Vilhjálmssson, H. 1997. Interactions between capelin (*Mallotus villosus*) and other species and the significance of such interactions for the management and harvesting of marine ecosystems in the northern North Atlantic. Rit Fiskideildar, 15: 1-281.

Large Marine Ecosystems of the North Atlantic
K. Sherman and H.R. Skjoldal (Editors)
© 2002 Elsevier Science B.V. All rights reserved.

8

Ecological Features and Recent Trends in the Physical Environment, Plankton, Fish Stocks, and Seabirds in the Faroe Shelf Ecosystem

Eilif Gaard, Bogi Hansen, Bergur Olsen and Jákup Reinert

ABSTRACT

The Faroe shelf water is relatively well separated from the offshore water by a persistent tidal front, which surrounds the islands. The shelf water has neritic phyto- and zooplankton communities, which to a large extent are separated from the surrounding offshore area, although receiving variable influence from the offshore environment. The shelf production of plankton is the basis for production in the higher trophic levels within the ecosystem. The plankton production is interannually variable and in general monitoring data show simultaneous fluctuations at several trophic levels in the ecosystem, including calculated new primary production, fish recruitment, growth and landings, and seabird recruitment and growth. The paper gives an overview of trophic interactions within the Faroe shelf ecosystem and variability in production in the various trophic levels. The production and harvesting potential of the ecosystem is discussed.

INTRODUCTION

Due to an anticyclonic circulation of the water masses on the Faroe shelf and a persistent tidal front that surrounds the shelf at about 100-130 m bottom depth contour (Figure 8-1), the Faroe shelf water is relatively well separated from the surrounding ocean (Hansen 1992; Gaard and Hansen 2000). This current system makes the basis for a small (approximately 8000 km^2) and uniform coastal ecosystem, which is surrounded by an oceanic environment.

The ecosystem of the shelf water is a phyto- and zooplankton habitat that basically, regarding species composition as well as production, is different from the surrounding oceanic environment (Gaard 1996; 1999; 2000). It is also a habitat for benthic fauna and several fish stocks that reproduce and grow within the ecosystem. Furthermore, a large number of seabirds are breeding on the Faroe Islands and feed themselves and their chicks from the shelf water. Hence the Faroe shelf water can be considered an ecosystem with many trophic levels, from phytoplankton through zooplankton and benthic fauna, several trophic levels of nekton, and seabirds. Since the ecosystem is so well defined and geographically uniform, it reacts more or less as one unit towards changes in the environment and is therefore suitable for ecological studies.

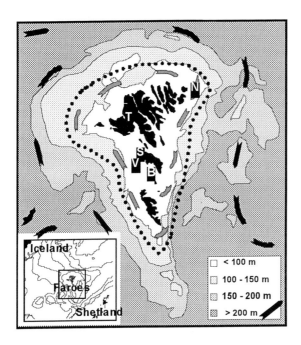

Figure 8-1. The location of the Faroes and bottom topography of the Faroe Plateau and observational sites. The dotted line indicates a typical position of the tidal front that separates the shelf water from the open ocean. The letters refer to stations mentioned in the text.

Landing statistics of the Faroe Plateau cod and haddock throughout the 20th century show that, despite a marked increase in fishing effort during this time, the landings have not increased correspondingly. The long-term average landings of the cod usually have fluctuated between 20,000 and 40,000 tonnes and of haddock between 15,000 and 25,000 tonnes (Figure 8-2). Therefore, catches of these two fish stocks have for a long time reached the limit for long-term production within the ecosystem, and interannual variability in production consequently should be reflected in the cod and haddock catches.

Despite the relative stability in long-term catches, periods occur with high variability. The most pronounced variability was a quite dramatic decrease in catches of these two fish stocks in the beginning of the 1990s. In order to be able to conduct a sustainable and optimal management of the production of the Faroe shelf resources, knowledge about production and regulative key parameters in the various tropic levels is essential.

The aim of this paper is to give an overview of the food web structure and to identify some of the production mechanisms and key organisms in the various trophic levels in the Faroe shelf ecosystem.

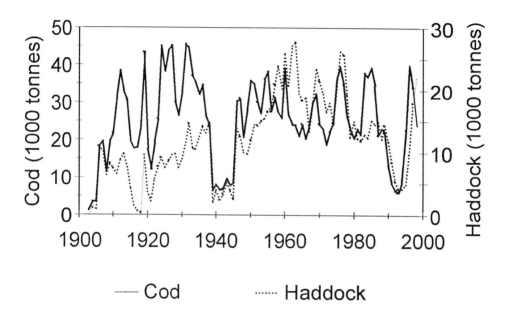

Figure 8-2. Landings of Faroe Plateau cod (left y-axis) and haddock (right y-axis), 1903-1998.

MATERIALS AND METHODS

The materials presented derive from various monitoring activities on the Faroe shelf and ashore. The temperature was measured continuously at station T (Figure 8-1). Samples for nitrate measurements were collected approximately twice a week at a land-based station (marked S on Figure 8-1) and less frequently at two shelf stations (N and V). The nitrate was measured with autoanalyser according to Grasshoff *et al.* (1983). Zooplankton was collected with 200 μm mesh WP2 plankton nets from the upper 50 m throughout the shelf. Cod and haddock stock assessments derive from Virtual Population Analysis (VPA), and the landing statistics derive from ICES (1999) and ICES Bulletin Statistique. Sandeel juveniles were collected as by-catch on annual cod and haddock 0-group surveys during mid summer, using a small pelagic capelin trawl. The trawl opening was about 32 m^2 and the mesh size in the codend was 5 mm. The trawling depth was approximately 25-40 m; the exact depth chosen was based on the recordings from the echo-sounder. Further details about the 0-group surveys are given in Jákupstovu and Reinert (1994). Attending guillemots (*Uria aalge*) have been monitored since 1972 in a study area in the largest guillemot breeding-cliff on the islands (Figure 8-1, site B) and the breeding biology of puffins (*Fratercula arctica*) has been studied on the same island from 1991 to 1995.

THE PHYSICAL ENVIRONMENT

Due to a combined effect of precipitation and reduced water exchange the salinity is slightly lower on the shelf than offshore. The front that separates the shelf water from the offshore water can therefore be identified based on the isohalines (Figure 8-3). It usually follows the 100-130 m bottom contour around the shelf. The salinity usually is lowest in the central and northern part of the shelf while it is somewhat higher in the southern region. This indicates that the separation between these two water masses is highest in the central and northern shelf region - an assumption which is supported by similar isolines of nutrient concentration during summer (Gaard 1996). The average retention time of the shelf water is estimated to about 3 months, however it is variable and 2-monthly oceanic influx averages may in extreme cases fluctuate with a factor of 5 (Gaard and Hansen 2000). The oceanic influx into the shelf water is partly affected by winds (Gaard and Hansen 2000).

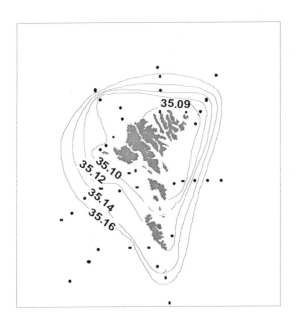

Figure 8-3. Salinity at 50 m depth on the Faroe shelf, 23 June-2 July 1994 (from Gaard 1999).

The separation of the Faroe shelf water mass from the offshore water masses highly affects the species composition and the production potential within the shelf water. It is a prerequisite for keeping the phyto- and zooplankton organisms on the shelf, where they can grow and develop relatively independently from the surrounding oceanic environment (Gaard 1996; 1999; 2000). Another, and maybe even more important, consequence is its ability to keep meroplanktonic eggs and larvae on the shelf. This undoubtedly is a prerequisite for maintaining the benthic fauna as well as the fish populations that inhabit the shelf ecosystem. Hence exchanges of sea water between the shelf area and offshore may affect influx of nutrients and plankton into the shelf and losses of planktonic organisms from the ecosystem. The quantitatively most important planktonic species that is advected into the shelf ecosystem is the copepod *Calanus finmarchicus* (Gaard 1999; Gaard and Hansen 2000) while the most perceptible losses would be of ichthyoplankton (Hansen *et al*. 1994).

The physical environment for plant and animal production on the Faroe shelf therefore is quite different from the surrounding oceans. Due to extreme tidal currents (Hansen 1992) the shallow parts also are well mixed from surface to bottom without any stratification during summer. This affects the primary production, which develops quite independently of the circumstances offshore. Usually it starts earlier in spring on the shelf than offshore (Gaard 1996; 2000).

PHYTOPLANKTON

Due to low irradiation the primary production and phytoplankton biomass is very low during winter and increases during spring and summer (Gaard 1996; 1999). Since the water mass on the shelf and hence the total amounts of nutrients in this water mass are limited, the primary production affects the nutrient concentrations in the shelf water very much, so they in some years may decrease to very low levels (Figure 8-4). A potential new primary production (Dugdale and Goering 1967) is, therefore, limited to the nutrient pool in this water plus the advection of nutrients into the shelf during the summer.

An approximation of relative nitrate assimilation (a potential new primary production) in the shelf water during the high-productive period may be calculated as

Nitrate assimilation = Nitrate decrease + nitrate inflow

The inflow of nitrate can be calculated as

Nitrate inflow = Inflow rate of sea water x ($[NO_3^-]_{offshore} - [NO_3^-]_{shelf}$)

where $[NO_3^-]_{offshore}$ and $[NO_3^-]_{shelf}$ are the nitrate concentrations in the surrounding offshore water and the shelf water respectively during the investigated period.

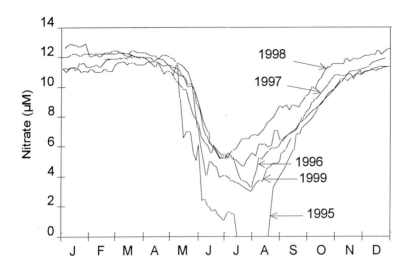

Figure 8-4. Nitrate concentrations on station S, 1995-1999.

Based on an estimated average exchange of 1/90 renewal per day (Gaard and Hansen 2000) and nitrate measurements on the shelf and offshore respectively, an index of a potential new primary production from spring to 26 June, 1990-1999 is calculated (Figure 8-5). The vertical lines show the effect from extremes in renewal rates (Gaard and Hansen 2000).

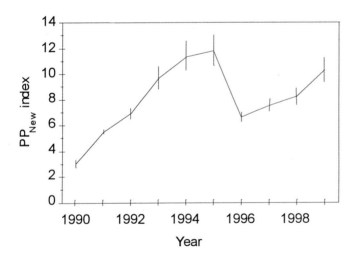

Figure 8-5. Calculated index of a potential new primary production on the Faroe shelf.

The figure shows that the calculated new primary production varied with almost a factor of 4 during the period from which nutrient measurements are available (1990-1999). This interannual variability is so high that it obviously must have affected the higher trophic levels within the ecosystem.

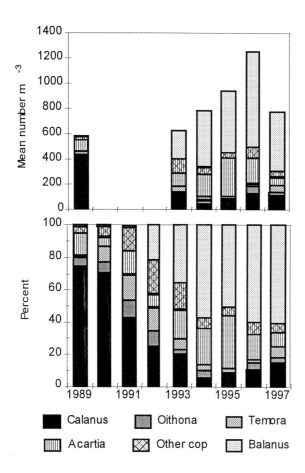

Figure 8-6. Absolute (upper) and relative (lower) abundance of the dominant copepods on stations N and V (see Figure 8-1) in May 1989-1997. For the years 1990-1992 only relative abundance is available.

ZOOPLANKTON

Most years the majority of the zooplankton (by number) on the Faroe shelf consists of neritic species. Most years it is dominated by neritic copepods (mainly *Temora longiremis* and *Acartia spp.*) but during spring it usually also contains high quantities of meroplanktonic larvae (Gaard 1999). Thus, it is basically quite different from the species composition in the surrounding oceanic environment, which mainly consists of the copepod *Calanus finmarchicus*. The offshore zooplankton may also highly affect the shelf communities as *C. finmarchicus* is advected into the shelf and mixed with the neritic species.

Systematic observations of zooplankton on the Faroe shelf have been made since 1989. During this period there has been quite large variability in species composition, abundance and biomass during the summer (Gaard 1999). In 1989 and 1990 the system was dominated by *C. finmarchicus* and there was a low abundance of neritic species. But during the first years of the 1990s the abundance of neritic species increased very much, while *C. finmarchicus* decreased in numbers (Figure 8-6) (Gaard 1999). In 1996 the situation again partly switched towards *C. finmarchicus* (Gaard 1999).

This variability coincides well with the variability in calculated primary production, as shown in Figure 8-5 with neritic dominance in the most productive years and higher *Calanus* advection and dominance in years with low primary production.

FISH

Although the long-term fish landings of Faroe Plateau cod and Faroe haddock have been relatively stable, there indeed have been periods with some quite large fluctuations. With the exception of the World War 2 period, the most pronounced decline in the landings was during the beginning of the 1990s when the landings reached the lowest values ever recorded. However, after that dramatic drop, they increased rapidly to well above the long-term averages of both species (Figures 8-2 and 8-7).

Fishing yield alone may be a somewhat misleading indicator of fish production, since it may also be affected by the effort. However, the biomass fluctuations followed well with the fishing yield for both species (Figure 8-7) and hence the fishing yield can be taken as an approximation for production of these two fish stocks.

With a few exceptions, the cod and haddock recruitment showed roughly the same fluctuations as the biomass and landings (Figure 8-7). Periods with high or low reproduction were reflected in biomass and landings a few years later.

However, the recruitment of the cod and haddock do not always fluctuate simultaneously (Figure 8-7). One main reason for that may be found in the different reproductive strategies of these two species. The main spawning of cod is 2-4 weeks earlier than that of haddock and their spawning areas are also different (Joensen and Tåning 1970; Hansen *et al.* 1990). During the first-feeding period, the larvae of the two species may therefore partly depend on somewhat different food items. However, later

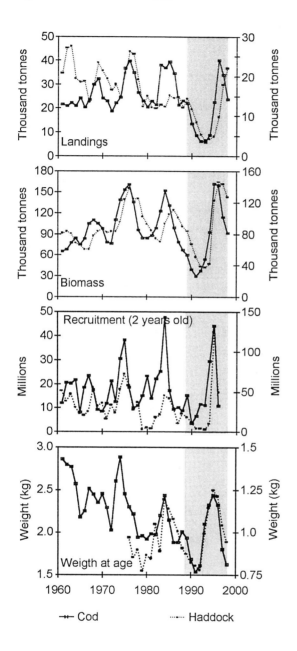

Figure 8-7. Faroe Plateau cod (left y-axis) and haddock (right y-axis) landings, biomass, recruitment as two years old and averaged weights as 2-5 years old during 1960-1998. The grey area shows the period from where comparable environmental data are available.

in the larval and juvenile phases they become mixed on the Faroe Plateau and eat almost the same prey animals (Gaard and Reinert 1996). The nursery areas of the demersal juveniles are also different, as the cod generally migrate into the sublittoral zone while the haddock seek to the bottom on the Plateau and banks (Reinert 1979; 1988). This, in some cases, may also have affected differences in recruitment variability between the cod and haddock.

Although both cod and haddock might be characterised as omnivorous, they seem to prefer some prey items. Normally, adult cod feed mostly on fish, i.e. sandeel, Norway pout, and blue whiting and decapods, supplied by other invertebrates, such as mussels and brittlestars. The main food of haddock is bristle worms, brittlestars, mussels and crustaceans, but at some times of the year and in some areas, sandeel can contribute substantially to their diet (Joensen and Tåning 1970; Rae 1967; Du Buit 1982; unpublished data).

It is well known that temperature highly affects the metabolic rates of poikilotherms in the sea and Brander (1995) has shown a clear relationship between individual growth of several Atlantic cod stocks and their *in situ* temperature. He also found quite good correlation between long-term variability in the Faroe Plateau cod growth rates and the temperature on the Faroe shelf and Weather Station "Mike" in the Norwegian Sea. However, during the 1990s the variability in cod and haddock growth rates (Figure 8-7) did not coincide with the temperature variability in the shelf water (Figure 8-8). On the contrary the annual average temperature tended to decrease during the recovery period in the early 1990s when the average weight at age increased with about 60% for both species during the years 1991-1995. Hence the variability in food availability is a more convincing explanation for the variability in fish growth during that particular period.

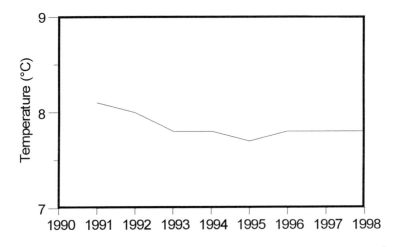

Figure 8-8. Mean annual temperature on the Faroe shelf, 1990-1998.

Another important fish species in the ecosystem is the sandeel. It is not caught commercially and, from an ecological point of view, its production only serves as food supply, mainly for fish, seabirds, and grey seals. Relative abundances of 0-group sandeel, caught as by-catch in the cod 0-group trawling surveys are shown in Figure 8-9. As pointed out earlier, the surveys are not designed for sandeel juveniles, and the abundance indices in Figure 8-9 may therefore not be quantitatively exact. Especially it should be noted that the samples are collected at one depth only (mainly between 25 and 40 m depth), and the index may therefore be sensitive to vertical variations in abundance. However, it may be an indicator of general qualitative trends, which show a low abundance during the late 1980s and the first years of the 1990s and an increase during the early 1990s. The mean length data of the sandeel juveniles are, on the other hand, presumed to be more robust. These data show a general similarity with the weight at age variations of cod and haddock, (Figure 8-7) and also with the plankton variation (Figures 8-5 and 8-6; Gaard 1999).

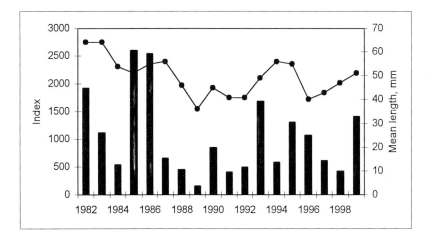

Figure 8-9. Abundance index (bars) and mean length (line) of sandeel juveniles from cod and haddock 0-group trawling surveys on 25-40 m depth, during mid summer 1982-1999.

During the period with abnormally low fish production in the beginning of the 1990s and the recovery during the 1990s (the grey area on Figure 8-7), interannually comparable information of hydrographic and plankton variability is available. This makes it possible to compare the physical and biological environment with the observed trends in landings, growth and recruitment of the Faroe Plateau cod and haddock. The fish recovery coincided very well with the general increase in primary production and the shift towards more neritic zooplankton dominance in the ecosystem.

SEABIRDS

Nearly 2 million pairs of seabirds breed on the Faroe Islands (Bloch and Sørensen 1984; Bloch *et al.* 1996) and take most of their food from the surrounding waters. During the breeding period these birds feed close to the islands and bring food for their chicks. Also a great portion of the immature feed close to the islands during the summer although they are not as restricted as breeding birds. Outside this period the situation is more complicated. Some of the local populations, *e.g.* that of the common guillemot, migrate to other areas while a portion of the Scottish and Icelandic guillemot population spends the winter around the Faroe Islands. Sustained harvest of seabirds and their eggs give an impression of large natural year-to-year variations in the production of seabirds as well as long-term fluctuation in the seabird populations (Nørrevang 1977; Olsen 1992; Reinert unpublished data). Especially during the last two decades, there have been pronounced changes in some of the seabird populations.

The arctic tern *Sterna paradisea* on the Faroes had a period from 1984 to1992 (both years included) where the production of fledged young was very low, and in some of the years no fledged young was seen at all (Olsen 1994). This resulted in a great decline in the breeding population from 9,000-12,000 pairs in 1981 (Bloch and Sørensen 1984) to 2,000 pairs in 1993 (Bloch *et al.* 1996). Since then the population size has been stable or increasing.

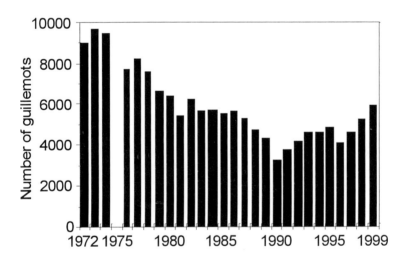

Figure 8-10. Attending guillemots at study site B (Figure 8-1) in early July 1972-1999. No census was made in 1975.

The number of attending guillemots on site B (Figure 8-1) reached its lowest level in 1990 (Figure 8-10) but since then the number has increased to the level of the early

1980s. The most dramatic decline during the 25 years of census occurred from 1987 to 1990, culminating with a 25% decline from 1989 to 1990 that has not been observed before. During these years the production of eggs declined even more and only few chicks survived. The reason for this decline was suggested to be at a level in the food web beneath the predating fishes and auks (Olsen 1991; 1992). The drop and increase in the guillemot population is almost identical to that in the fish production (Figure 8-7) and the recovery during the 1990s coincides with the increase of the calculated primary production index (Figure 8-5).

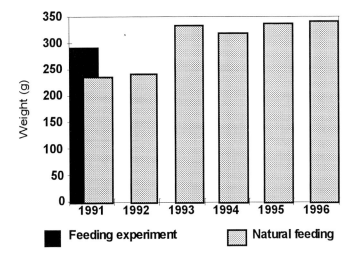

Figure 8-11. Mean weight of puffin chicks with wings longer than 120 mm at study site B (Figure 8-1) during late July-early August 1991-1996.

The harvest of puffins was low in the late 1980s and the early 1990s. Moreover, in 1989, 1990 and 1991 a lot of dead chicks were found in the colonies around the islands and this has never been reported before. In 1991 an experiment with supplementary feeding of the young showed that they were starving. The survival of young fed by their parents alone was less than 50% while all fed experimentally survived. In 1992 and 1993 the survival of young improved to 70% and 98% respectively (Olsen 1994) and remained high at least until 1995. The experimentally fed also reached a higher fledging weight than the controls (Figure 8-11). The fledging weight remained low also in 1992 but after that it was high and stable. The food brought to the young also changed (Figure 8-12). The normal food is 0-group of sandeel (*Ammodytes sp.*), but in 1991 0-group of Norway pout (*Trisopterus esmarki*) and capelin (*Mallotus villosus*) also was a substantial part of the food. The abundance of 0-group capelin, which is a more northern species, was higher in 1991 than usually found on the shelf. In 1992, Norway pout and sandeel dominated. From 1993 to 1996, sandeels were the most common food supplemented by Norway pout. The caloric value of the food was low in 1991 and 1992 as the size of individual fishes was low, but this improved from 1993 to 1995.

Sandeel is an important food during the chick rearing period for most of the Faroese seabirds. Most of the birds can switch to other food items when sandeels are restricted but the three species mentioned above have to carry the food for their young in the beak and they are highly dependent on fish of a certain size at the right time. The terns are especially sensitive, as they are restricted to find their food in the top of the water column. As sandeels are not locally exploited, the seabirds have only to compete with fish and grey seals for the sandeels. Any effect of this competition has, however, not been observed in the seabird populations.

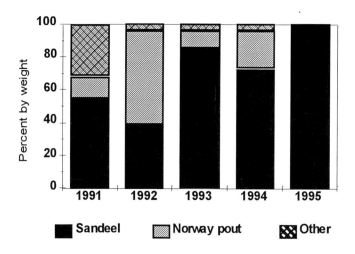

Figure 8-12. Food brought to puffin chicks at study site B (Figure 8-1) during July-early August 1991-1995.

Although the production on the Faroe shelf was low for a period, not all seabird species were negatively affected. This is not conflicting, however, as the seabirds use different feeding strategies. The production of gannet *Sula bassana* young has been increasing since 1989. Breeding gannets can probably forage up to 500 km from the colony for their main food, herring and mackerel (Nelson 1978) and are therefore not dependent only on the production on the Faroe shelf. The production of kittiwakes *Rissa tridactyla* was normal in 1990 and 1991 but failed in 1992, 1997 and 1998. In many areas kittiwakes feed their young on sandeels (Lloyd *et al.* 1991), but at times, when fish is scarce, they will take a wide range of planktonic invertebrates, especially crustaceans (Cramp *et al.* 1974). The fulmar (*Fulmarus glacialis*), the Manx shearwater (*Puffinus puffinus*) and the storm petrel (*Hydrobates pelagicus*), which forage over a wide area and store the food for their young as a fatty oil in their stomach, were not affected

(Olsen 1991). The shag (*Phalacrocorax aristotelis*) and the black guillemot (*Ceppus grylle*), which take their food close to land, also seem to have had normal production.

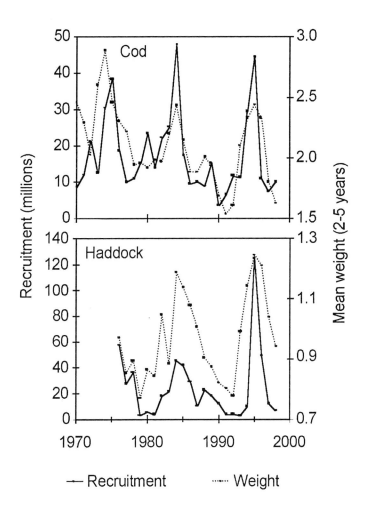

Figure 8-13. Relationship between recruitment of 2 years old cod and haddock and the mean weight of 2-5 years old cod during 1970-1998 and haddock during 1976-1998.

TROPHIC INTERACTIONS ON THE FAROE SHELF

Long-term relations between cod and haddock recruitment and growth (weight at age) show that periods with high weight at age occur simultaneously with periods of good recruitment (Figure 8-13). Long-term data showed that periods with large fish year-classes generally had higher growth rates than periods with small year-classes, indicating that food production for fish may exceed the increased predation pressure during productive periods. The ecosystem therefore seems to have gone through some periods with low recruitment and production levels and through other periods with higher recruitment and production in several trophic levels.

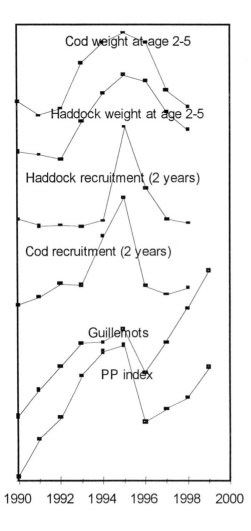

1990   1992   1994   1996   1998   2000

Figure 8-14. Relative variability in calculated new primary production, number of attending guillemots, recruitment of 2 years old cod and haddock, and mean weight of 2-5 years old cod and haddock during 1990-1999.

Systematic monitoring of environmental parameters on the Faroe shelf has only been conducted during the last 10 years. Although this is a short period, it has been very variable regarding productivity and food chain structure, and the variability can be tracked through several trophic levels and parameters in the ecosystem. Some of them are summarized in Figure 8-14. A key question is how the trophic levels may be connected. That question is difficult to answer. However, since the Faroe shelf ecosystem is geographically relatively well defined and homogenous, it is suitable for such ecological studies, and some interpretations can be made.

The years with low primary production coincided with late onset of spring bloom and high abundance of *Calanus finmarchicus* on the shelf and *vice versa*. The calculated new primary production is negatively correlated to the zooplankton biomass and Gaard *et al.* (1998) suggested a top-down effect, where high influx of overwintered *C. finmarchicus* during spring resulted in high grazing pressure on the phytoplankton. It is hypothesised that this may have delayed the spring bloom development. Further studies are needed, however, in order to evaluate this hypothesis.

Egg production of copepods is highly affected by food availability (e.g Kiørboe *et al.* 1990; Kiørboe and Nielsen 1994; Hirche 1996 and references therein; Niehoff *et al.* 1999). This effect is also seen on the Faroe shelf where the seasonal production of phytoplankton reflects development of the copepods (Gaard 1999). A low primary production in spring seems to coincide with low egg production and vice versa. This again may have large consequences for the ichthyoplankton, which during their early feeding stages depend largely on small food items such as copepod eggs and nauplii and small copepodites (Gaard and Steingrund, in press).

A simultaneous low recruitment of cod and haddock on the Faroe shelf during the late 1980s and beginning of the 1990s also leads the attention towards the environment. These species have relatively similar feeding habitats during their pelagic stages and select very much the same prey items (Gaard and Reinert 1996). During these years, the abundance of small-sized prey organisms (recruits of *C. finmarchicus*, small sized neritic zooplankton species and *Balanus* nauplii) seems to have been low in early spring (Gaard 1999). It seems to have increased markedly during the following years, when cod and haddock recruitment also improved markedly. It may therefore be hypothesized that a shortage of small-sized prey organisms during their early larval stages may have contributed to low cod and haddock recruitment during those years. This development could be strengthened by the decline in spawning stock biomass and quality of fish. Another (maybe additional) possibility is that strong winds may have blown the fish eggs out of the current system that maintains the ecosystem, as suggested by Hansen *et al.* (1994).

Detailed knowledge about variability in food consumption of cod and haddock in Faroese waters is poor and not conclusive. Both cod and haddock show diversity in prey items, and predate on benthic fauna as well as fish, with fish being a somewhat more prevalent prey item for cod than for haddock (Rae 1967; Du Buit 1982). Of the fish prey, sandeel seems to be a key species. In deeper areas other species (mainly Norway pout) have been observed to be more important than sandeel as prey item for cod and haddock (Du Buit 1982; Nicolajsen 1993). In shallow areas, *i.e.* inside the tidal front, studies during the years 1949-1962 clearly showed, however, that sandeel formed the principal fish food for the cod (Rae 1967). A comprehensive stomach analysis of cod and haddock was carried out by the Faroese Fisheries Laboratory in 1990 and 1991, and again from 1997 to present. The analysis of these data are not finished yet, but preliminary results show that during 1990 and 1991, sandeel was of little importance, and Norway pout was the principal fish in their diet. In the late 1990s sandeel was their main fish diet in the shallow areas, similar to what was found by Rae (1967), but benthic fauna could also be important. We therefore assume that the interannual variability in production of sandeel has been an important factor affecting the cod and haddock weight at age variability, although the necessary information on stomach content is not available to verify this hypothesis conclusively.

Sandeel also plays a central role in the seabird diet, and for many seabirds, it is the most important food source during the chick-rearing period. The feeding investigations of the puffin chicks, and the coherent recruitment and growth variability partly underline this importance. The fact that the seabirds had problems finding sandeels in sufficient amounts during the end of the 1980s and the beginning of the 1990s strongly indicates that the recruitment of sandeel has failed during that period. During the early 1990 (mainly 1993-1995) this obviously improved. The seabirds took more sandeels, and their recruitment improved markedly. Sandeel hence plays an important role in the ecosystem. It is an important zooplankton predator and an important link between the copepods and higher trophic levels in the shallow regions of the Faroe shelf.

A number of papers have demonstrated an empirical relationship between primary production and fishing yield. (e.g. Iverson 1990; Tatara 1991; Nixon 1992; Nielsen and Richardson 1996). Data from the Faroe shelf shows the relationship in the same order of magnitude as presented by Nielsen and Richardson (1996). Furthermore, the presented data indicate interannual coincidence between primary production, fish growth and reproduction, and seabird abundance and reproduction despite several intervening trophic levels.

The coincidence also demonstrates the limited production and harvesting potential of commercial fish within the ecosystem. Although the variability during the late 1980s and early 1990s obviously is rather extreme, some interannual variability in production is usual (Figure 8-13). The closer the harvesting pressure of the commercial fish is to the upper limit of yield, the clearer will short-term variability in production be tracked in fish landings.

It should be stressed that the available data on new primary production, copepods, fish and seabird variability at present make a basis for general rather than exact relationships. Further detailed studies on (variability in) trophic relationships would be necessary to evaluate more detailed links.

REFERENCES

Bloch, D. and S. Sørensen. 1984. Checklist of Faroese birds. Føroya Skúlabókagrunnur, Tórshavn. 84 p.

Bloch, D., J.-K. Jensen, and B. Olsen. 1996. List of birds seen in the Faroe Islands. Føroya Náttúrugripasavn, Føroya Fuglafrøðifelag and Føroya Skúlabókagrunnur. 14 p.

Brander, K.M. 1995. The effect of temperature on growth of Atlantic cod (*Gadus morhua* L.). ICES J. Mar. Sci., 52: 1-10.

Cramp, S., W.R.P. Bourne, and D. Saunders. 1974. The Seabirds of Britain and Ireland. Collins, London. 287 p.

Du Buit 1982. Essai sur la predation de la morue (*Gadus morhua* L.) L'eglefin (*Melanogrammus aeglefinus* (L.)) et du lieu noir (*Pollacius virens* (L.)) aux Faeroe. Cybium, 6: 3-13.

Dugdale, R.C. and J.J. Goering. 1967. Uptake of new and regenerated forms of nitrogen in primary productivity. Limnology and oceanography, 12: 196-206.

Gaard, E. 1996. Phytoplankton community structure on the Faroe shelf. Fródskaparrit, 44: 59-70.

Gaard, E. 1999. Zooplankton community structure in relation to its biological and physical environment on the Faroe Shelf, 1989-1997. J. Plankton Res., 21(6): 1133-1152.

Gaard, E. 2000. Seasonal abundance and development of *Calanus finmarchicus* in relation to phytoplankton and hydrography on the Faroe shelf. ICES J. Mar. Sci., 57: 1605-1611.

Gaard, E. and J. Reinert. 1996. Pelagic cod and haddock on the Faroe Plateau: Distribution, diets and feeding habitats. ICES CM 1996/L:16, 16 p.

Gaard E., B. Hansen, and S.P. Heinesen. 1998. Phytoplankton variability on the Faroe shelf. ICES J. Mar. Sci, 55: 688-696.

Gaard, E. and B. Hansen. 2000. Variations in the advection of *Calanus finmarchicus* onto the Faroe shelf. ICES J. Mar. Sci., 57: 1612-1618.

Gaard, E. and P. Steingrund. In press. Reproduction of Faroe Plateau cod: Spawning grounds, egg advection and larval feeding. Fróðskaparrit vol. 48.

Grasshoff, K., M. Erhardt, and K. Kremling (eds.). 1983. Methods for Seawater Analysis. Second revised and extended edition. Verlag Chemie, 419 p.

Hansen, B. 1992. Residual and tidal currents on the Faroe Plateau. ICES CM. 1992/C:12, 18 p.

Hansen, B., A. Kristiansen, and J. Reinert. 1990. Cod and haddock in Faroese waters and possible climate influences on them. ICES C.M. 1990/G:33, 23 p.

Hansen, B., E. Gaard, and J. Reinert. 1994. Physical effects on recruitment of Faroe Plateau cod. ICES J. Mar. Sci. Symp., 198: 520-528.

Hirche, H.-J. 1996. The reproductive biology of the marine copepod *Calanus finmarchicus* – A review. Ophelia, 44: 85-109.

ICES 1999. Report of the north-western working group. ICES CM 1999/ACFM:17. 329 p.

Iverson, R.L. 1990. Control of marine fish production. Limnol. Oceanogr. 35(7): 1593-1604.

Jákupstovu, S.H. í and J. Reinert. 1994. Fluctuations in the Faroe Plateau cod stock. ICES Mar. Sci. Symp., 198: 194-211.

Joensen, J.S. and V. Tåning. 1970. Marine and freshwater fishes. Vald. Petersen Bogtrykkeri, Copenhagen, 241 p.

Kiørboe, T., H. Kaas, B. Kruse, F. Møhlenberg, P. Tiselius, and G. Ærtebjerg. 1990. The structure of the pelagic food web in relation to column structure in the Skagerrak. Mar. Ecol. Prog. Ser., 59: 19-32.

Kiørboe, T. and T.G. Nielsen. 1994. Regulation of zooplankton biomass and production in a temperate coastal ecosystem. Copepods. Limnol. Oceanogr., 39: 403-507.

Lloyd, C., M.L. Tasker, and K. Partridge. 1991. The Status of Seabirds in Britain and Ireland. T. & A.D. Poyser, London. 355 p.

Nelson, J.B. 1978. The Gannet. T. & A.D. Poyser, Berkhamsted. 336 p.

Nicolajsen, Á. 1993. A preliminary analysis of stomach data from saithe (*Pollachius virens*), haddock (*Melanogrammus aeglefinus*) and cod *(Gadus morhua)* in the Faroes. Nordic Workshop on predation processes and predation models. Nordiske Seminar- og Arbejdsrapporter 572: 57-63

Niehoff, B., U. Klenke, H.-J. Hirche, X. Irigoien, R. Head, and R. Harris. 1999. A high frequency time series at Wethership M, Norwegian Sea, during the 1997 spring bloom: the reproductive biology of *Calanus finmarchicus*. Mar. Ecol. Prog. Ser., 176: 81-92.

Nielsen, E. and K. Richardson. 1996. Can changes in the fisheries yield in Kattegat (1950-1992) be linked to changes in primary production? ICES J. Mar. Sci., 53: 988-994.

Nixon, S.W. 1992. Quantifying the relationship between nitrogen input and the productivity of marine ecosystems. Proceedings of the Advanced Marine Technology Conference, 57-83.

Nørrevang, A. 1977. Fuglefangsten på Færøerne. Rhodos, 276 p. (In Danish)

Olsen, B. 1991. The Faroese guillemot population in relation to marine biological data. In: F.O. Kapel (ed.). Report of a Nordic seminar on predation and predatory processes in marine mammals and seabirds. Tromsø 25-29 April 1991. Univ. Tromsø: 16-17.

Olsen, B. 1992. Census of guillemots on Høvdin in Skúvoy, 1973 to 1991. Fiskirannsóknir 7: 6-15. (in Faroese with English summary).

Olsen, B. 1994. Faroe Islands. p.18, 100 and 111. In: G.L. Hunt (ed.). Report of the study on seabird/fish interactions. ICES C.M. 1994/L: 3: 1-119.

Rae, B.B. 1967. The food of cod and haddock of Faroese grounds. Mar. Res., 6: 1-23.

Reinert, J. 1979. O-group investigations around the Faroes 1974-1978 and their usefulness in predicting year class strength of cod, haddock, Norway pout and saithe. University of Copenhagen, 1979. (In Danish)

Reinert, J. 1988. Revised indices of the abundance of cod and haddock in the 0-group surveys in Faroese waters 1974-88. ICES Early Life History Symposium, Paper No. 74.

Tatara, K. 1991. Utilization of the biological production in eutrophicated sea areas by commercial fisheries and the environmental quality standard for fishing ground. Marine Pollution Bulletin, 23: 315-320.

IV
Northeast Atlantic

Large Marine Ecosystems of the North Atlantic
K. Sherman and H.R. Skjoldal (Editors)
© 2002 Elsevier Science B.V. All rights reserved.

9

Zooplankton-Fish Interactions in the Barents Sea

Padmini Dalpadado, Bjarte Bogstad, Harald Gjøsæter,
Sigbjørn Mehl, and Hein Rune Skjoldal

ABSTRACT

The Barents Sea is a high latitude ecosystem characterized by high variability. There are large interannual variations in ice cover, spring phytoplankton bloom dynamics, zooplankton stocks, and fish recruitment. The Barents Sea is an important feeding area for commercial fish stocks such as cod (*Gadus morhua*), capelin (*Mallotus villosus*) and herring (*Clupea harengus*). The fish resources in the Barents Sea have shown large changes during the last two decades. The stock size of capelin, which usually is the major planktivorous fish in the Barents Sea, was 4-8 million metric tons during 1973-1983. Due mainly to recruitment failure, a severe decline in capelin biomass (to below 0.5 million tonnes) occurred in the mid-1980s and mid-1990s. Predation by young herring on capelin larvae is regarded as the main cause of the recruitment failure of capelin in the Barents Sea; thus, periods of high abundance of young herring (year classes 1983 and 1991-1992) were followed by periods of low capelin abundance. As capelin are a key species of the Barents Sea ecosystem, the collapse of the stock caused food shortages for higher trophic level populations such as cod, harp seals, and guillemots.

Cod, which preys heavily on capelin, had to switch to macrozooplankton such as krill and amphipods as their main prey when the capelin stock was at extremely low levels. The zooplankton biomass has also shown several-fold variations among years in the central and northern Barents Sea. Predation pressure and climatic fluctuations are possible causes for these variations. Studies carried out in the Barents Sea show clear predator-prey interrelationships among cod, herring, and capelin and between fish and zooplankton. The dramatic changes in fish resources in recent years indicate strong biological interactions in the Barents Sea ecosystem.

INTRODUCTION

The Barents Sea is a large marine ecosystem covering an area of ca. $1.4 \cdot 10^6$ km^2. It is a shallow sea with an average depth of ca. 230 m (Zenkevitch 1963). Three water masses are present in the Barents Sea related to three different currents (Loeng 1989): the

Norwegian Coastal current, the Atlantic current, and the Arctic current systems (Figure 9-1). Climatic variations in the Barents Sea depend mainly on the activity and properties of the inflowing Atlantic water (Midttun and Loeng 1987). Bottom topography strongly influences the circulation and distribution of water masses in the Barents Sea (Skjoldal and Rey 1989).

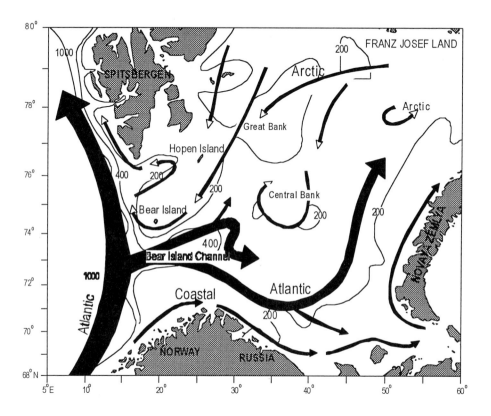

Figure 9-1. Schematic presentation of the major current systems and topographical features of the Barents Sea.

The Barents Sea is an important feeding area for commercial fish stocks. The fish resources there have shown large changes in recent years. The annual fish catch in the late 1970s was around 2.5 million metric tons, while at present the annual catch is around 1.0 million metric tons (ICES 2000a,b). The main reason for the reduction in catch is the complete closure of the capelin (*Mallotus villosus*) fishery during 1994-1998. The capelin stock has started to increase again, with a stock size of ca. 2.8 million metric tons in 1999.

Zooplankton biomass has also shown large variation among years in the Barents Sea (Skjoldal and Rey 1989). Recent studies have shown clear predator-prey interactions among cod (*Gadus morhua*), herring (*Clupea harengus*), and capelin and between fish and zooplankton (Ponomarenko and Yaragina 1979; Dalpadado *et al.* 1994; Dalpadado and Skjoldal 1996; Bogstad and Mehl 1997; Gjøsæter 1998).

Some of the work presented in this manuscript was initiated during the "Pro Mare" program (1984-1989), which dealt with several ecological aspects of the Barents Sea ecosystem. Many institutes and universities from Norway participated in this program. Final results of the proceedings of the Pro Mare Symposium, held in 1990, are published in two volumes in the journal *Polar Research* (Sakshaug *et al.* 1991).

In the present paper we attempt to review studies on zooplankton and fish in the Barents Sea, with emphasis on zooplankton-fish interactions. Some of the results, which have been published in previous papers, are expanded to include recent years. We discuss the large changes in fish stock abundance since the early 1980s in relation to changes in zooplankton biomass and climatic conditions in the Barents Sea.

MAJOR ZOOPLANKTON SPECIES AND THEIR DISTRIBUTION

Crustaceans form the most important group of zooplankton, among which the copepods of the genus *Calanus* play a key role in the sub-Arctic and Arctic ecosystems. *Calanus finmarchicus* (prosome length (PL) up to 3.1 mm) is the most important contributor to the zooplankton biomass of the Barents Sea (Melle and Skjoldal 1998). The mean abundance of this species was about 50, 15 and 3 thousand ind. m^{-2} in the Atlantic, Polar Front, and Arctic waters respectively (Melle and Skjoldal 1998). It has a unique position as the main food for herring, capelin and other plankton-eaters (Lund 1981; Vesin *et al.* 1981; Panasenko 1984; Vilhjálmsson 1994; Huse and Toresen 1996; Astthorson and Gislason 1997). *Calanus glacialis,* which can be up to 4.3 mm (PL), is the dominant contributor to zooplankton biomass of the Arctic region of the Barents Sea. High abundance (11 thousand ind. m^{-2}) of this species was found in the Arctic waters (Melle and Skjoldal 1998). The *Calanus* species are predominantly herbivorous, feeding especially on diatoms (Mauchline 1998).

Krill is another group of crustaceans playing a significant role in the pelagic ecosystem as food for both fish and sea mammals. They appear both in large schools and as continuous layers, often staying deep at daytime and ascending at night. Four krill species are commonly found in the Barents Sea: *Thysanoessa inermis* (total length (TL) up to 35 mm) and *T. longicaudata* (TL up to 25 mm) are the dominant species in the western and central Barents Sea. *T. raschii* (TL up to 35 mm) is more common in the shallow waters of the eastern Barents Sea (Einarsson 1945; Dunbar 1964; Dalpadado and Skjoldal 1996). *Thysanoessa* species penetrate very little into the Arctic water masses in the Barents Sea. The largest of the krill species, *Meganyctiphanes norvegica*

(TL up to 45 mm), is restricted to the warmer Atlantic waters in the west (Dunbar 1964; Dalpadado and Skjoldal 1996).

Three hyperiid amphipod species are found in the Barents Sea: *Themisto abyssorum* (TL up to 17 mm) and *T. libellula* (TL up to 60 mm) are common in the western and central Barents Sea, while *T. compressa* is less common in the central and northern parts of the Barents Sea (Dunbar 1964; Dalpadado *et al.*, in press). This species is restricted to the Atlantic waters of the southwestern Barents Sea (Dunbar 1964). *T. abyssorum* is predominant in the sub-arctic waters. In contrast, the largest of the *Themisto* species, *T. libellula,* is mainly restricted to the mixed Atlantic and Arctic water masses. A very high abundance of this species is recorded close to the Polar Front (Dalpadado *et al.*, in press).

MAJOR FISH SPECIES, THEIR DISTRIBUTION AND DIET

Cod, capelin and herring are commercially and ecologically important fish species in the Barents Sea. Studies carried out in different areas show that zooplankton are a major component of the diet of capelin and herring, whereas to a large extent cod feed on fish (Ponomarenko and Yaragina 1979; Lund 1981; Vesin *et al.* 1981; Panasenko 1984; Vilhjálmsson 1994; Huse and Toresen 1996; Astthorsson and Gislason 1997; Bogstad and Mehl 1997).

The capelin stock spends its whole life in the Barents Sea, spawning along the southern boundary and feeding in the central and northern parts of the sea (Gjøsæter 1998). The krill and amphipod distribution areas overlap with the feeding grounds of capelin, especially in winter to early summer. Distribution areas for cod, capelin and zooplankton species are shown in the following works: Einarsson 1945; Dunbar 1964; Dalpadado *et al.* 1994; Dalpadado and Skjoldal 1996; Bogstad and Mehl 1997. Copepods, krill and amphipods are major prey of capelin, with the relative importance of these preys varying with season, year and capelin size (Lund 1981; Panasenko 1981, 1984; Gjøsæter 1998). The importance of copepods decreases with increasing capelin length, while euphausiids and amphipods are most important for adult capelin (Panasenko 1984).

The Barents Sea is an important nursery area for herring stock. After 2-3 years, the herring move out of the area, to join the adult stock living in the Norwegian Sea (Dragesund *et al.* 1980; Gjøsæter 1995). Capelin mainly utilize the region north of 72° N, while herring is confined to more southern regions. However, there is some overlap between juvenile herring and juvenile capelin. Calanoid copepods were the most important prey of juvenile herring. Appendicularians were the second most abundant prey, and together these two preys made up to 87% of the diet by weight (Huse and Toresen 1996). Krill made up most of the remaining diet.

Atlantic cod is the most important predator in the Barents Sea ecosystem (Bogstad and Mehl 1997). The distribution of cod is confined mainly to the southern Barents Sea (south of 75° N). Cod consume many ecologically and commercially important prey species such as krill, amphipods, shrimps (*Pandalus borealis*), capelin, herring, polar cod (*Boreogadus saida*), redfish (*Sebastes* spp.), cod, and haddock (*Melanogrammus aeglefinus*) (Ponomarenko and Yaragina 1979; Bogstad and Mehl 1997). The distribution area of cod is extended to the north and east in periods of warm climate, while during colder years cod tend to concentrate in the southwestern part of the Barents Sea (Nakken and Raknes 1987; Shevelev *et al.* 1987; Ottersen *et al.* 1998).

TRENDS IN FISH STOCKS

Figure 9-2 shows the stock biomass of capelin (age 1 and older), herring (age 1-3 years) and cod (age 3 and older) in the Barents Sea during 1973-1999. The capelin and herring biomass are acoustic abundance estimates (Gjøsæter and Bogstad 1998; ICES 2000b), while the cod stock biomass is calculated using virtual population analysis (ICES 2000a). The capelin stock biomass varied between 4-8 million metric tons during 1973-1983. In the mid-1980s and 1990s the stock size was at extremely low levels. During periods when the capelin stock was large the herring stock size was very small, and vice versa. The herring estimates before 1990 are believe to be underestimated and therefore cannot be compared to the later years (ICES 1998).

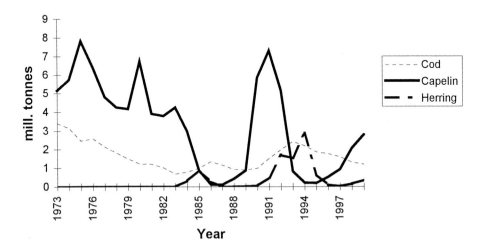

Figure 9-2. Stock biomass of capelin, cod and herring in the Barents Sea during 1973-1999 (Gjøsæter and Bogstad 1998; ICES 2000a, b).

Variation in cod biomass is not as prominent as that for capelin and herring. However, large variations in the individual growth of cod have been observed, and this is related in part to fluctuations in capelin abundance (Mehl and Sunnanå 1991; ICES 2000a). Capelin is a major prey of cod, and in some years cod can consume up to 3 million metric tons of capelin (Bogstad and Mehl 1997).

TRENDS IN ZOOPLANKTON ABUNDANCE COMPARED TO TRENDS IN CAPELIN ABUNDANCE

In Figures 9-3 through 9-8, the variation in various measures of zooplankton abundance during the period 1979-1999 is compared to variation in the capelin abundance during this period.

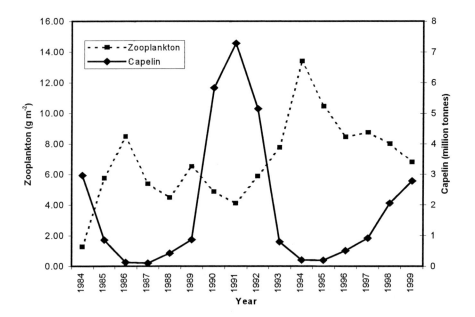

Figure 9-3. Average zooplankton biomass (g m^{-2}) together with total biomass of one year old and older capelin (million tonnes) during 1973-1999, in the Barents Sea (capelin data from Gjøsæter 1998 and Gjøsæter *et al.* 2000).

Figure 9-3 shows zooplankton biomass versus capelin biomass during 1984-1999. Capelin and zooplankton data were restricted to autumn (August to October). The capelin data are extracted from Gjøsæter (1998) and Gjøsæter *et al.* (2000). The sampling gear, coverage and time of year remained similar during the study period for

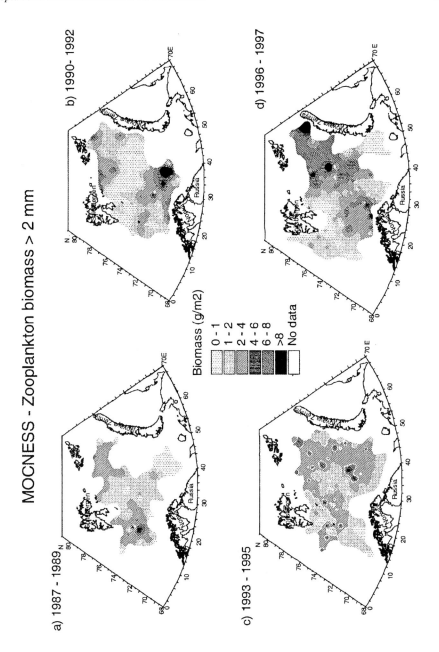

Figure 9-4. The biomass (g m^{-2}) of large zooplankton (>2000 μm) obtained by MOCNESS for four different time periods: a) 1987-89, b) 1990-92, c) 1993-95, and d) 1996-97.

both capelin and zooplankton, except for 1979-1986. During 1979-1986, zooplankton sampling was limited mostly to transects. The zooplankton samples (WP2 and MOCNESS (Anon. 1968; Wiebe *et al.* 1976, 1985)) were used to estimate the average zooplankton biomass for each year. The zooplankton biomass data are size fractionated into 3 categories and the biomass is expressed as g dry weight m^{-2}. The smallest and medium size fractions (below 2000 μm) represent mainly copepods, and the largest size fraction consists mainly of macrozooplankton like krill, amphipods and chaetognaths. The figure shows that during periods with extremely low capelin biomass, zooplankton abundance was high. The highest average zooplankton biomass values of close to 14g/m^2 were observed when the capelin biomass was at extremely low levels. When capelin biomass was large the zooplankton biomass was rather low.

The >2000 μm fraction (only MOCNESS samples) is shown in Figure 9-4 for four different time periods; 1987-89 (capelin stock is increasing), 1990-92 (capelin stock is quite high), and 1993-1995 (capelin stock is decreasing), and 1996-97 (capelin stock is very low). These data are from August and September - end of the capelin-feeding season. When the capelin stock was quite high the zooplankton biomass was low in the central Barents Sea (Figure 9-4a, b). During 1990-1992, a low macrozooplankton biomass (0-1 g/m^2) belt was present between 74-77 °N, which is within the main feeding area of capelin. When the capelin stock was at extremely low levels in 1996 and 1997, the zooplankton biomass was generally quite high (6-8 g/m^2).

The mean biomass of *T. inermis* and *T. longicaudata* observed in each year from 1984 to 1992, together with stock size of capelin, is shown in Figure 9-5 (Dalpadado and Skjoldal 1996). This figure is not updated, as we have not finished the analyses of krill samples. The capelin stock collapsed from 1984 to very low levels in 1986. There was a subsequent strong increase in the biomass of both krill species. We can see an earlier increase for *T. longicaudata*, which has the shortest life span (Figure 9-5b). A marked decrease in abundance and biomass was observed with the recovery of the capelin stock from 1989 to very high levels in 1991. The decrease was most pronounced for *T. inermis*.

Length distributions of *T. inermis* are given in Figure 9-6 for three time periods (Dalpadado and Skjoldal 1996): during 1984-1986 when the capelin stock size was decreasing, during 1987-1989 when the stock was at lowest levels, and during 1990-1992 when the stock was very high. During 1987-89, when capelin stock was very low, larger (> 20 mm) and older age groups (3 and 4 years) of *T. inermis* were present. When predation was low, reduced mortality allowed more individuals to grow into older age groups. A higher proportion of smaller individuals was observed during periods of heavy predation (Figure 9-6c).

Figure 9-7 shows mean abundance of *T. libellula* and *T. abyssorum* observed in each year from 1984 to 1996, together with stock size of capelin (from Dalpadado *et al.*, in press). With the decrease in capelin abundance in the mid-1980s and 1990s an increase in amphipod abundance was observed, and vice versa. The increase in abundance was less pronounced in the mid-1990s for *T. libellula*, probably due to higher grazing pressure from cod (Figure 9-7a). The stock size of cod in the mid-1990s was about twice that of the mid-1980s. The late recovery of the *T. libellula* populations could also be due to higher grazing pressure from other predators such as whales, seals and sea birds that are major predators of zooplankton in the Barents Sea ecosystem.

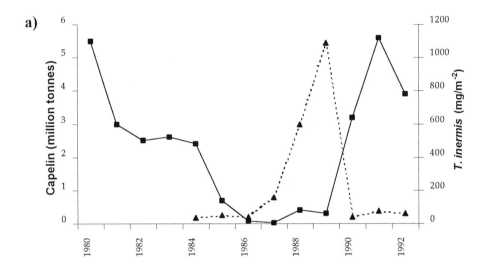

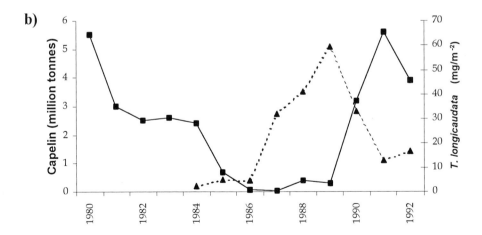

Figure 9-5. Biomass (mg m⁻²) of **a)** *Thysanoessa inermis* and **b)** *T. longicaudata* together with the biomass of one year and older capelin (million tonnes) during 1980-1992 in the Barents Sea. From Dalpadado and Skjoldal (1996).

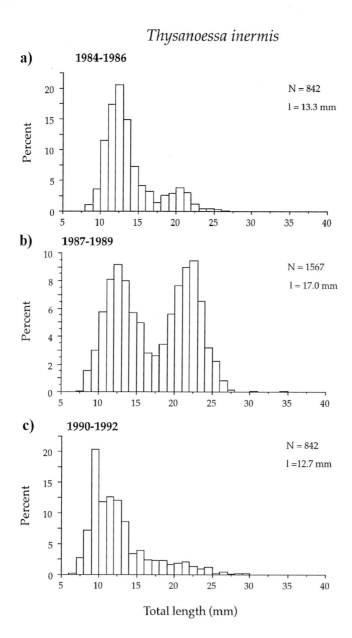

Figure 9-6. Length frequency distributions of *Thysanoessa inermis* for three different time periods, **a)** 1984-1986, **b)** 1987-1989, and **c)** 1990-1992. N = number of krill, 1 = mean length. From Dalpadado and Skjoldal (1996).

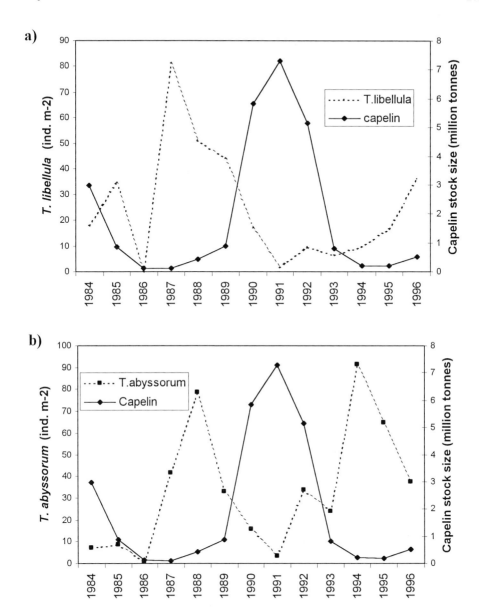

Figure 9-7. Abundance (no. m^{-2}) of **a)** *Themisto libellula* and **b)** *T. abyssorum* together with the total biomass (million tonnes) of one year old and older capelin during 1980-1996 in the Barents Sea. From Dalpadado *et al.* (1994) and Dalpadado *et al.*, in press.

Figure 9-8 shows the size-fractionated plankton biomass and growth of 1-4 year old capelin during the last season (Gjøsæter *et al.* 2000). A statistical analysis showed that the average individual growth of 1 and 2 year old capelin during 1984-1997 was significantly correlated to the abundance of the smallest and medium zooplankton size fractions, mainly copepods (Gjøsæter *et al.* 2000). The growth of age groups 3 and 4 was most correlated with the largest zooplankton size fraction, mainly larger copepods, krill, amphipods, and chaetognaths.

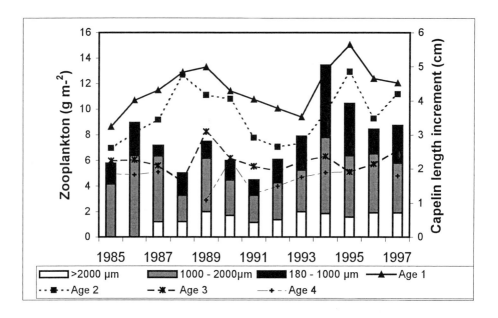

Figure 9-8. Growth of capelin during the last growth season together with the biomass (g m^{-2}) of three size fractions of zooplankton. From Gjøsæter *et al.* (2000).

STATISTICAL ANALYSIS - ZOOPLANKTON BIOMASS VERSUS CAPELIN STOCK SIZE

Regression analysis between mean annual zooplankton biomass (total zooplankton, amphipod, krill biomass) and capelin stock size showed negative slopes, indicating an inverse relationship between zooplankton biomass and capelin stock size (Figure 9-9a-c). Krill and amphipod data from 1986 were excluded from the regression and correlation analyses as the mean biomass/abundance for this year was based on only a few stations, all from a similar locality. The values of the coefficient of determination (r^2) in all three regressions are low, indicating that factors not included in the regression analyses are also of importance for determining zooplankton biomass. The p values for the regression coefficients, though significant at 5% level in Figures 9-9 a,b (p=0.043; p=0.008), may not be precise since autocorrelation has not been taken into consideration.

a)

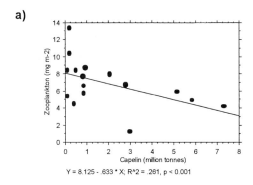

Y = 8.125 - .633 * X; R^2 = .261, p < 0.001

b)

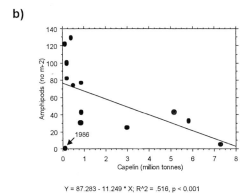

Y = 87.283 - 11.249 * X; R^2 = .516, p < 0.001

c)

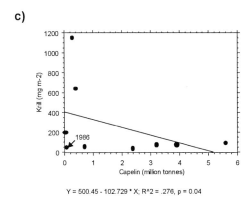

Y = 500.45 - 102.729 * X; R^2 = .276, p = 0.04

Figure 9-9. Regression analysis between mean annual **a)** zooplankton, **b)** amphipod, and **c)** krill biomass versus capelin stock size.

Table 9-1. Correlation analysis between various variables: mean annual - capelin biomass, total zooplankton biomass, amphipod abundance, krill biomass, Kola temperature.

Correlation coefficient
Hypothesized correlation=0

	Correlation	Count	z-value	P-value	95% lower	95% upper
Capelin, zooplankton	-0.511	16	-2.035	**0.0419**	-0.803	-0.021
Capelin, amphipods	-0.719	12	-2.714	**0.0066**	-0.915	-0.246
Capelin, krill	-0.491	8	-1.202	0.2296	-0.888	0.327
Capelin, temperature	0.708	15	3.059	**0.0022**	0.307	0.895
Zooplankton, amphipods	0.375	12	1.183	0.2368	-0.253	0.781
Zooplankton, krill	0.414	8	0.985	0.3245	-0.41	0.866
Zooplankton, temperature	-0.255	15	-0.905	0.3655	-0.679	0.296
Amphipods, krill	0.515	8	1.273	0.2029	-0.298	0.895
Amphipods, temperature	-0.608	12	-2.119	**0.0341**	-0.876	-0.053
Krill, temperature	0.046	8	0.104	0.9172	-0.681	0.727

Correlation analysis was carried out for several variables. This analysis showed that total zooplankton biomass and amphipods were respectively significantly negatively correlated at 5% level (p= 0.041; p=0.006) to capelin stock size (Table 1). The relationship between krill and capelin stock size, though negatively correlated, was not significant (P=0.192). The biomass of capelin was significantly (5% level) positively correlated to the mean annual Kola temperature (p=0.002). Zooplankton and amphipod biomass was negatively correlated to the temperature, though significantly only for amphipods (p= 0.034).

COD AND ZOOPLANKTON

Cod is a major predator in the Barents Sea ecosystem. Cod consume many ecologically and commercially important prey species such as krill, amphipods, shrimps, capelin, polar cod, herring, redfish, cod, and haddock. Figure 9-10 shows the diet of cod during 1984 - 1999. Capelin is one of the most important preys of cod, and in some years cod remove a considerable amount of the total capelin production. Bogstad and Mehl (1997) calculated the annual consumption of capelin by cod. The series has been updated by ICES (2000b) and covers the period 1984-1999. In 1987 cod consumed ca. 200,000 metric tons of capelin. The consumption of capelin then increased to ca. 3 million metric tons of capelin in 1993, before falling to 0.6 million metric tons in 1996. A shift in the diet of cod was observed when capelin stock size was at very low levels. Krill (*Thysanoessa* spp. and *Meganyctiphanes norvegica*) and amphipods (mainly *Themisto libellula*) increased in importance in the diet of cod during 1986-1988 and 1994-1997.

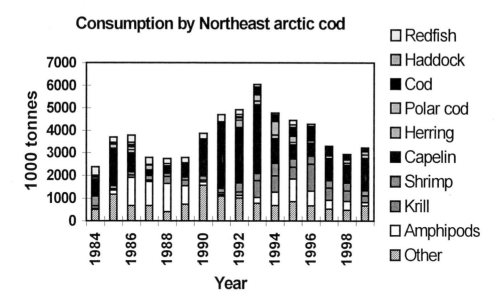

Figure 9-10. Annual food consumption by Northeast arctic cod (in thousand tonnes) in the Barents Sea during 1984-1999. From Bogstad and Mehl (1997); updated by ICES (2000b).

Cod cannibalism also increased to a high level (200,000-500,000 metric tons) in 1993-1997. Herring made up only a small percentage of the cod stock's annual consumption. During 1987-1991 the amount of herring consumed by cod was low (< 50,000 metric tons). In 1992 the consumption of herring by cod increased to about 300,000 metric tons before falling again in subsequent years.

DISCUSSION

Over the last two decades, there have been large changes in the fish stocks of the Barents Sea. For capelin, the decrease in the mid-1980s and 1990s has been explained as due mostly to recruitment failure (Gjøsæter 1998). The Barents Sea capelin stock is potentially the largest capelin stock in the world, with its biomass in some years reaching 4-8 million metric tons (Gjøsæter 1998). During the mid-1980s and 1990s, though the abundance of capelin larvae was high, further recruitment to older age groups was poor. Predation by herring on capelin larvae is regarded as the main cause for the poor capelin recruitment (Hamre 1994; Huse and Toresen 1996; Gjøsæter and Bogstad 1998).

The inflow of 0-group herring to the Barents Sea is dependent on both the spawning stock biomass and the survival of eggs and larvae, and is extremely variable. The variability in year class strength of herring observed in the Barents Sea is further related to variability in cod predation (Hamre 1988; Barros 1995). Increased predation pressure on herring by cod is related to the decline in capelin stock, the major prey of cod (Hamre 1988; Barros 1995). Recruitment of herring is positively correlated to temperature (Toresen and Østvedt 2000).

In some years it appears that the estimated consumption of capelin by cod is higher than the capelin biomass. The capelin biomass estimates are from September/October, which is at the end of the main feeding season of capelin. Therefore, the biomass of capelin shows the standing stock size. It is the capelin production that is available to cod, not the standing stock size. Bogstad *et al.* (2000) compared the estimates of removal of capelin through natural mortality to the estimates of consumption of capelin by cod and other predators. They found that in years of low capelin abundance consumption of capelin by predators was higher than what is available for predation.

No estimates of total zooplankton biomass are currently available, but it would be of interest to compare such estimates (of standing stock in autumn) to estimates of consumption of zooplankton by capelin, herring, cod, marine mammals, and other predators. Zooplankton production is not only governed by the production of phytoplankton and the grazing by predators on the zooplankton, but it is also partly determined by the overwintering stock from the previous year. Slagstad and Støle-Hansen (1991) showed that variability in the overwintering stock of *Calanus* would be an important factor in determining total production the following summer.

A shift in diet was forced upon cod for its survival when its major prey, capelin, decreased to extremely low stock levels. Cod switched to other available prey like krill, amphipods, and herring. The growth rate and average weight at age of cod decreased (Ponomarenko and Yaragina 1979; Bogstad and Mehl 1997) when it was deprived of its main prey, capelin, which is regarded to be rather rich in lipid content (Jangaard 1974; Yaragina and Marshall 2000). These results indicate that the development of herring, cod and capelin stocks are strongly linked biologically.

Results presented in this paper demonstrate that zooplankton biomass in the central Barents Sea in the period 1979-1998 has shown several fold variations among years. Possible reasons for these large variations are the differences in advective transport and predation pressure. Most of the work carried out on advection are modeling studies (Tande and Slagstad 1992; Slagstad and Tande 1996; Giske *et al.* 1998). These studies show that the advection of *Calanus* spp. from the Norwegian Sea to the Barents Sea might govern the amount of *Calanus* in the Barents Sea. Another hypothesis is that major inflow during winter contains no copepods but major inflow in summer will be rich in copepods, especially *C. finmarchicus* (Skjoldal and Rey 1989; Giske *et al.* 1998). This is due to overwintering of *C. finmarchicus* in deeper waters of the Norwegian Sea during winter (Østvedt 1955; Melle and Skjoldal 1998). Unfortunately, not much work has been done on the advection of larger macrozooplankton such as krill and amphipods. The younger developmental stages (drifting stages) of macrozooplankton may be advected by the inflow of Coastal and Atlantic currents from the Norwegian Sea to the Barents Sea.

Characteristics of the interactions between zooplankton and other trophic levels differ between water masses and biological regions. In the northern and central Barents Sea, clear predator-prey interrelationships between planktivorous capelin and their major prey, krill and amphipods, seem to exist. Though statistical analyses validate the interdependencies between capelin and zooplankton biomass, these analyses also show that other factors may be of importance for determining zooplankton biomass. Heavy predation by cod on amphipods and krill in periods of low capelin abundance will most likely modify the interactions between capelin and zooplankton. These variations could also be related to climatic variations in the Barents Sea. Skjoldal and Rey (1989) and Sakshaug *et al.* (1994) relate dramatic changes in plankton and fish resources to the transition from cold to warm periods associated with inflow events.

Several studies have shown that climatic conditions, e.g. temperature, have pronounced effects on recruitment to Barents Sea fish stocks (Sætersdal and Loeng 1987; Ottersen *et al.* 1994; Ottersen and Sundby 1995) as well as individual growth (Loeng 1989; Loeng *et al.* 1995; Ottersen and Loeng 2000). It has been shown that for cod and herring growth through the early pelagic stages is temperature dependent. It has also been shown that the length of cod as 0-group is positively correlated to the year-class strength both as 0-group and as three-year-olds. Climatic fluctuations will also influence both primary and secondary production and thus the food and feeding conditions for planktivorous fish. The feeding distribution of cod and capelin is also known to change with climatic conditions in the Barents Sea, with more easterly and northerly distributions noted in warm years than in cold years (Loeng 1989).

The high latitude Barents Sea ecosystem is characterized by a food web with few dominant species: e.g. diatom→krill→capelin→cod. However, interactions between fish and zooplankton as well as among fish stocks have led to marked changes in trophic pathways and in fish and zooplankton abundance. These changes demonstrate strong biological interactions and some degree of trophic flexibility in the Barents Sea ecosystem, which operates under the strong influence of variations in the physical environment. The case of the Barents Sea ecosystem provides a clear example of a highly variable sysem with a simple structure. The changing nature of marine fisheries requires management approaches that recognize and include ecosystem and environmental effects (Hofmann and Powell 1998).

FOCUS ON FUTURE INVESTIGATIONS

Detailed studies on the diet of capelin and cod, as well other major predators such as mammals and birds, should be investigated in a multispecies context. In the Barents

Sea, the harp seal (*Phoca groenlandica*), minke whale (*Balaenoptera acutorostrata*), and the most common of the sea birds, the Brünnich guillemot (*Uria lomvia*) are major predators of capelin and zooplankton (Mehlum and Gabrielsen 1995; Folkow *et al.* 2000; Nilssen *et al.* 2000). During the period when the capelin stock was small, these predators switched to alternative prey such as krill, amphipods, and herring. This type of information/data should be incorporated in future investigations of predator-prey interactions.

Observed year-to-year variation in both fish and zooplankton has been attributed to variability in climate. Climatic variations in the Barents Sea have been explained by variable inflow of Atlantic water (Ådlandsvik and Loeng 1991). Much work has been done on the advection of copepods to the Barents Sea (Tande and Slagstad 1992; Slagstad and Tande 1996; Giske *et al.* 1998). Future work should also focus on the advection of large macrozooplanktonic organisms. We have evidence that both krill and amphipods reproduce in the central and northern Barents Sea (Einarsson 1945; Zelikman 1958; Dunbar 1964; Dalpadado and Skjoldal 1996), but we do not know the quantity of these organisms that is advected from the Norwegian Sea.

Distribution of species will also vary according to climatic conditions in the Barents Sea, e.g. on the amount of warm Atlantic water inflow. As discussed above, temperature has great influence on the distribution, growth, and recruitment of major fish species in the Barents Sea. The size of fish stocks will determine predation pressure exerted on zooplankton populations. Climatic parameters, such as temperature, will also have an influence on zooplankton abundance and distribution. More research is needed to clarify these relationships.

Modeling studies are of importance to simulate distribution and migration patterns of plankton and fish under different climatic scenarios. These model results are often not verified using field observations. Better temporal and spatial coverage of zooplankton and fish are needed to validate the conditions described in models.

Several attempts have been made to model fish-fish and also marine mammals-fish interactions in the Barents Sea, as described by Bogstad *et al.* (1997), Hamre and Hatlebakk (1998), and Tjelmeland and Bogstad (1998). However, plankton is modeled in a very coarse way in these models and thus a good description of the plankton dynamics cannot be expected. Similarly, in models focusing on zooplankton, phytoplankton and oceanography (see overview in Giske *et al.* 1998), the modeling of fish stocks does not capture the main dynamics of these stocks. Linking models for plankton and oceanography to models of fish and marine mammal stocks is not a simple task, however, as these models are based on quite different modeling concepts and operate on different spatial and temporal scales.

ACKNOWLEDGEMENTS

We thank our colleagues at the Institute of Marine Research, Bergen, Norway, for their contributions to our studies of the Barents Sea ecosystem. We are grateful to Drs. J. Giske, H. Loeng and G. Ottersen for constructive comments on the manuscript. This work was partly financed by the EU project (FAIR: PL 95 817) - Barents Sea Impact Study (BASIS).

REFERENCES

Ådlandsvik, B. and H. Loeng. 1991. A study of the climatic system in the Barents Sea. In: E. Sakshaug, C.C.E. Hopkins, and N.A. Øritsland (eds.). Proceedings of the Pro Mare Symposium on Polar Marine Ecology, Trondheim, 12-16 May 1990. Polar Res. 10(1). 45-49.

Anon. 1968. Smaller mesozooplankton. Report of Working Party No. 2. In: D.J. Tranter (ed.). Zooplankton sampling (Monographs on oceanographic methodology 2), UNESCO, Paris. 153-159.

Astthorson, O.S. and A. Gislason 1997. On the food of capelin in the subarctic waters north of Iceland. Sarsia 82(2):81-86.

Barros, P.C. 1995. Quantitative studies on recruitment variations in Norwegian spring-spawning herring (*Clupea harengus* Linnaeus 1758), with special emphasis on the juvenile stage. Dr. Scient. thesis, Department of Fisheries and Marine Biology, University of Bergen, Norway. 143 p.

Bogstad, B., T. Haug and S. Mehl. 2000. Who eats whom in the Barents Sea? NAMMCO Sci. Publ. 2: 98-119.

Bogstad, B., K. Hiis Hauge and Ø. Ulltang. 1997. MULTSPEC- A Multi-Species Model for Fish and Marine Mammals in the Barents Sea. J. Northw. Atl. Fish. Sci. 22: 317-341.

Bogstad, B. and S. Mehl. 1997. Interactions between Atlantic cod (*Gadus morhua*) and its prey species in the Barents Sea. Forage Fishes in Marine Ecosystems. Alaska Sea Grant College program. AK-SG-97-01: 591-615.

Dalpadado P., N. Borkner, B. Bogstad, and S. Mehl. In press. Distribution of and predation on *Themisto* (Amphipoda) spp. in the Barents Sea . ICES J. Mar. Sci.

Dalpadado, P. and H.R. Skjoldal. 1996. Abundance, maturity and growth of krill species *Thysanoessa inermis* and *T. longicaudata* in the Barents Sea. Mar. Ecol. Prog. Series 144: 175-183.

Dragesund, O., J. Hamre, and Ø. Ulltang 1980. Biology and population dynamics of the Norwegian spring spawning herring. Rapp. P.-v. Réun. Cons. int. Explor. Mer, 177:43-71.

Dunbar, M.J. 1964. Serial Atlas of the Marine Environment. Folio 6 –Euphausiids and Pelagic Amphipods. Distribution in the North Atlantic and Arctic waters. American Geographical Society. 17 p.

Einarsson, H. 1945. Euphasiacea 1. Northern Atlantic species. Dana Rep.27:1-185.

Folkow, L.P., T. Haug, K.T. Nilssen, and E.S. Nordøy. 2000. Estimated food consumption of minke whales *Balaenoptera acutorostrata* in Northeast Atlantic waters in 1992-1995. NAMMCO Sci. Publ.2: 65-80.

Giske, J., H.R. Skjoldal, and D. Slagstad. 1998. Ecological modelling for fisheries. In: T. Rødseth (ed.). Models for multispecies management. Physica-Verlag.11-68.

Gjøsæter, H. 1995. Pelagic fish and the ecological impact of the modern fishing industry in the Barents Sea. Arctic 48(3): 267-278.

Gjøsæter, H. 1998. The population biology and exploitation of capelin (*Mallotus villosus*) in the Barents Sea. Sarsia 83:453-496.

Gjøsæter, H., P. Dalpadado, A. Hassel, and H.R. Skjoldal. 2000. Growth of the Barents Sea capelin (*Mallotus villosus* Müller). In: H. Gjøsæter, Dr. Philos. thesis, Department of Fisheries and Marine Biology, University of Bergen, Norway. Paper IV.

Gjøsæter, H. and B. Bogstad. 1998. Effects of the presence of herring (*Clupea harengus*) on the stock-recruitment relationship of Barents Sea capelin (*Mallotus villosus*). Fisheries Research 38: 57-71.

Hamre, J. 1988. Some aspects of the interrelation between the herring in the Norwegian Sea and the stocks of capelin and cod in the Barents Sea. ICES CM 1988/H:42. 15 p.

Hamre, J. 1994. Biodiversity and exploitation of the main fish stocks in the Norwegian-Barents Sea ecosystem. Biodiversity and conservation 3:473-492.

Hamre, J. and E. Hatlebakk. 1998. System Model (Systmod) for the Norwegian Sea and the Barents Sea. In: T. Rødseth (ed.). Models for multispecies management. Physica-Verlag. 93-115.

Hofmann, E.E. and T.M. Powell. 1998. Environment variability effects on marine fisheries: four case histories. Ecological applications. 8(1) supplement: 23-32.

Huse, G. and R. Toresen. 1996. A comparative study of the feeding habits of herring (*Clupea harengus*, Clupeidae L.) and capelin (*Mallotus villosus*, Osmeridae, Müller) in the Barents Sea. Sarsia 81:143-153.

ICES. 1998. Report of the Northern Pelagic and Blue Whiting Fisheries Working Group. ICES CM 1998/ACFM:18. 276 p.

ICES. 2000a. Report of the Arctic Fisheries Working Group. ICES CM 2000/ACFM:3. 312 p.

ICES. 2000b. Report of the Northern Pelagic and Blue Whiting Fisheries Working Group. ICES CM 2000/ACFM:16. 226 p.

Jangaard, P.M. 1974. The capelin (*Mallotus villosus*). Biology, distribution, exploitation, utilization and composition. Bulletin 186:1-70.

Loeng, H. 1989. The influence of temperature on some fish population parameters in the Barents Sea. Journal of North West. Atlantic Fishery Science 9:103-113.

Loeng, H., H. Bjørke, and G. Ottersen. 1995. Larval fish growth in the Barents Sea. Can. Spec. Publ. Fish. Aquat. Sci. (121): 691-698.

Lund, A. 1981. Ernæring hos lodde *Mallotus villosus villosus* Müller, i Barentshavet. Cand real thesis. Institute of Fisheries Biology, University of Bergen, Norway. 128 p. (In Norwegian).

Mauchline, J. 1998. The biology of calanoid copepods. Advances in Marine Biology 33, Academic Press, London. 710 p.

Mehl, S. and K. Sunnanå. 1991. Changes in growth of Northeast Arctic cod in relation to food consumption in 1984-1988. ICES Mar. Sci. Symp. 193: 109-112.

Mehlum, F. and G.W. Gabrielsen. 1995. Energy expenditure and food consumption by sea bird populations in the Barents Sea. In: H.R. Skjoldal, C. Hopkins, K.E. Erikstad, and H.P. Leinaas (eds.). Ecology of Fjords and coastal waters. Elsevier Science. 457-470.

Melle, W. and H.R. Skjoldal 1998. Reproduction, life cycles and distributions of *C. finmarchicus* and *C. hyperboreus* in relation to environmental conditions in the Barents Sea. In: W. Melle, Dr. Scient. thesis, Department of Fisheries and Marine Biology, University of Bergen Norway. Paper II, 32 p.

Midttun, L. and H. Loeng 1987. Climatic variations in the Barents Sea. In: H. Loeng, (ed.). The effect of oceanographic conditions on distribution and population dynamics of commercial fish stocks in the Barents Sea. Proceedings of the third Soviet-Norwegian Symposium, Murmansk, 26-28 May 1986. Institute of Marine Research, Bergen, Norway. 13-27.

Nakken, O. and A. Raknes. 1987. The distribution and growth of Northeast Arctic cod in relation to bottom temperatures in the Barents Sea, 1978-1984. Fish. Res. 5: 243-252.

Nilssen, K.T., O.P. Pedersen, L.P. Folkow, and T. Haug. 2000. Food consumption estimates of Barents Sea harp seals. NAMMCO Scientific Publications 2: 9-27.

Ostvedt, O.J. 1955. Zooplankton investigations from weathership M in the Norwegian Sea 1948-49. Hvalrådets Skrifter, 40:1-93.

Ottersen, G. and H. Loeng. 2000. Covariability in early growth and year-class strength of Barents Sea cod, haddock and herring: The environmental link. ICES J. Mar. Sci. 57: 339-348.

Ottersen, G., H. Loeng, and A. Raknes. 1994. Influence of temperature variability on recruitment of cod in the Barents Sea. ICES Mar. Sci. Symp. 198: 471-481.

Ottersen, G., K. Michalsen, and O. Nakken. 1998. Ambient temperature and distribution of north-east Arctic cod. ICES J. Mar. Sci. 55: 67-85.

Ottersen, G. and S. Sundby. 1995. Effects of temperature, wind and spawning stock biomass on recruitment of Arcto-Norwegian cod. Fish. Oceanogr. 4(4): 278-292.

Panasenko, L.D. 1981. Diurnal rhythms and rations of capelin feeding in the Barents Sea. ICES C.M. 1981/H:26. 14 p.

Panasenko, L.D. 1984. Feeding of the Barents Sea capelin. ICES CM 1984/H:6. 16 p.

Ponomarenko, I. Ya. and N.A. Yaragina. 1979. Seasonal and year to year variations in the feeding of the Barents Sea cod on Euphausiacea in 1947-1977. ICES CM 1979/ G:17. 20 p.

Sætersdal, G. and H. Loeng. 1987. Ecological adaptation of reproduction in Northeast Arctic Cod. Fish. Res.(5):253-270.

Sakshaug, E., A. Bjørge, B. Gulliksen, H. Loeng, and F. Mehlum 1994. Structure, biomass distribution, and energetics of the pelagic ecosystem in the Barents Sea: A Synopsis. Polar Biology 14:405-411.

Sakshaug, E., C.C.E. Hopkins, and N.A. Øritsland 1991. Proceedings of the Pro Mare Symposium on Polar Marine Ecology, Trondheim, Norway. Polar Research 10:1-662.

Shevelev, M.S., V.V. Tereschchenko, and N.A. Yaragina. 1987. Distribution and behaviour of demersal fishes in the Barents and Norwegian Seas, and the factors influencing them. In: H. Loeng (ed.). The effect of oceanographic conditions on distribution and population dynamics of commercial fish stocks in the Barents Sea. Proceedings of the third Soviet-Norwegian Symposium, Murmansk, 26-28 May 1986. Institute of Marine Research, Bergen, Norway. 181-190.

Skjoldal, H.R. and F. Rey. 1989. Pelagic production and variability of the Barents Sea Ecosystem. In: K. Sherman and L. Alexander (eds.). Biomass Yields and Geography of Large Marine Ecosystems. AAAS Selected Symposium . 241-286.

Slagstad, D. and K. Støle-Hansen. 1991. Dynamics of plankton growth in the Barents Sea. In: E. Sakshaug, C.C.E. Hopkins, and N.A. Øritsland (eds.). Proceedings of the Pro Mare Symposium on Polar Marine Ecology, Trondheim, 12-16 May 1990. Polar Res. 10(1):173-186.

Slagstad D. and K.S. Tande. 1996. The importance of seasonal migration in across shelf transport of *Calanus finmarchicus*. OPHELIA 44:189-205.

Tande, K. and D. Slagstad 1992. Regional and interannual variations in biomass and productivity of the marine copepod *Calanus finmarchicus*, in subarctic environments. Oceanologica Acta 15:309-321.

Toresen, R. and O.J. Østvedt. 2000. Variations in abundance of Norwegian spring spawning herring throughout the 20th century and the influence of climatic fluctuations. Fish and Fisheries 1: 231-256.

Tjelmeland, S. and B. Bogstad. 1998. System Model (Systmod) for the Norwegian Sea and the Barents Sea. In: T. Rødseth (ed.). Models for multispecies management. Physica-Verlag. 69-91.

Vesin, J.P., W.C. Legget, and K.W. Able. 1981. Feeding ecology of capelin (*Mallotus villosus*) in the estuary and Western Gulf of St. Lawrence and its multispecies implications. Can. J. Fish. Aquat. Sci. 38:257-267.

Vilhjálmsson, H. 1994. The Icelandic capelin stock. Journal of the Marine Research Institute, Reykjavík, Iceland. Vol. XIII no. 1. 281 pp.

Wiebe, P.H., K.H. Burt, S.H. Boyd, and A.W. Morton. 1976. A multiple opening/closing net and environmental sensing system for sampling zooplankton. Journal of Marine Research 34: 313-326.

Wiebe, P.H., A.W. Morton, A.M. Bradley, R.H. Backus, J.E. Craddock, V. Barber, T.J. Cowles, and G.R. Flierl. 1985. New developments in the MOCNESS, an apparatus for sampling zooplankton and micronekton. Marine Biology 87:313-323.

Yaragina, N.A. and C.T. Marshall. 2000. Trophic influences on interannual and seasonal variation in the liver condition index of Northeast Arctic cod (*Gadus morhua*). ICES Journal of Marine Science 57: 42-55.

Zelikman, E.A. 1958. On gonad maturation and female productivity in species of euphasians abundant in the Barents Sea. Dokl. Akad. Nauk. S.S.R. 118:201-204.

Zenkevitch, L. 1963. Biology of the Seas of the USSR. George Allen and Unwin, London. 955 pp.

Large Marine Ecosystems of the North Atlantic
K. Sherman and H.R. Skjoldal (Editors)
© 2002 Elsevier Science B.V. All rights reserved.

10

Dynamics and Human Impact in the Bay of Biscay: An Ecological Perspective

Luis Valdés and Alicia Lavín

INTRODUCTION

Geographically, the Bay of Biscay constitutes a geomorphological unit characterised by presenting a break in the north-south direction of the eastern Atlantic's continental margin (Figure 10-1). This break produces an inlet in the coastal topography occupying a surface area of approximately 175,000 km^2. The coastal margin of the Bay of Biscay has been inhabited since prehistoric times and nowadays the region has a population density in the average of the EU (113 inhabitants per km^2) showing a clear increasing trend: e.g. the population density in the French coastal lands (a band of 10 km-wide) has continually increased during the most recent censuses: 111 inhabitants per km^2 in 1962, 126 in 1977, 134 in 1982 and 141 in 1990 (OSPAR Commission 2000).

Living marine resources exploited in the Bay of Biscay include a wide range of organisms, from seaweeds to molluscs and whales. Retrospective analyses on the human uses of the marine resources in the coastal margin of the Bay of Biscay have demonstrated the harvesting of gastropods and other intertidal animals since the Palaeolithic (17,000 BP) [e.g. long records on abundance of intertidal animals were provided by shells deposited in prehistoric middens (Ortea 1986)]. Whale hunting was a common practice all along the north Spanish coast from the Middle Ages (Figure 10-2) until its prohibition in the mid-1980s. Recent uses of marine resources include traditional fisheries of both pelagic and demersal species, and nearly 5000 French and Spanish boats are currently active in the Bay of Biscay. The coastal margin of the Bay of Biscay is also under considerable pressure from industrial activities known to cause pollution (paper milling, petroleum refining, iron and steel working, chemicals, etc.).

Descriptive aspects of the Bay of Biscay, both on the French and Spanish continental margins, have been extensively studied, and it is possible to recognise general patterns in circulation, annual cycles of planktonic communities, and distribution of demersal and pelagic fish populations. In spite of the effort devoted to the description of the assemblages of the marine communities, it is generally recognised that we do not know how species extend and contract their spatial distribution, nor how their abundance increases and decreases. This is due to the complexity of relationships among abiotic

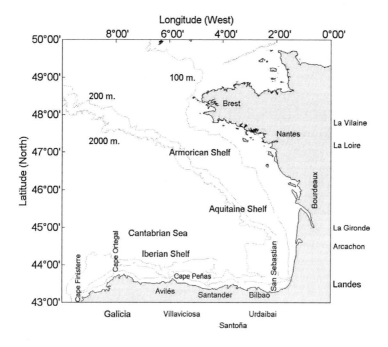

Figure 10-1. Map of the Bay of Biscay showing its main divisions and locations.

Figure 10-2. Middle Age crest of Ondarroa (North Spanish coast).

and biotic properties of the ecosystem, and because marine ecosystems are subject to different sources of variability (both natural and anthropogenic).

Natural variability occurs on a wide range of space and time scales and is inherent to the functioning of marine ecosystems. An ecosystem's variability over time is manifested through its seasonal, inter-annual, decadal, and centennial cycles. Spatial variability is associated with the vertical and horizontal movements of water. Human impacts on ecosystems are many (fishing effects, marine litter, pollutants, etc.) and their effects on the ecosystem usually remain even when their original cause has disappeared. In addition, we should consider the impact of global warming, whose long-term impact on an ecosystem and its species remains unknown.

These cycles and sources of variability interact with the biological cycles of species, producing fluctuations in their abundance that are not always easy to explain and make it very difficult to determine the states of equilibrium of species and communities. The effects of pollution and global change on marine ecosystems are generally perceived by society only when these effects become evident at the upper end of the trophic system. Such effects include alterations in the abundance, distribution and diversity of fish and marine mammals but they also alter the equilibrium of invertebrates like limpets and echinoderms.

All the mentioned factors and our inability to detect emerging environmental problems at an early stage, when remedial measures are still possible, add a high degree of uncertainty to the proper management of marine resources and uses of coastal areas, and in consequence, limit our capacity to plan a sustainable development policy for the coastal areas.

The present paper gives a description of the dynamics of the Bay of Biscay focusing in particular on the factors that are natural sources of variability and on the human activities that cause environmental degradation and alter the marine ecosystem. Some considerations for the scientific actions needed for better future management of marine living resources are discussed in the last section.

THE BAY OF BISCAY AS A LARGE MARINE ECOSYSTEM

Classical approaches to divide the World Ocean into regional ecosystems are based on the association of marine organisms that show a similar spatial distribution within geomorphological or ecological barriers that isolate them from floras and faunas of other areas (Golikov *et al.* 1990; Olson and Hood 1994). Parallel with the search for biogeographic patterns, new proposals have been made for an ecological geography of the oceans based on ecological criteria such as rates of critical physiological processes, physical forcing and plankton properties, timing in seasonal stages, features of the surface circulation, etc. (Nixon 1988; Legendre and Le Fevre 1991; Longhurst 1998). Finally, some recent attempts to classify the pelagic ecosystem into objectively defined geographic compartments include not only ecological criteria but also consider man as part of the ecosystem, interacting in many forms: exploiting resources, impacting the ecosystem, establishing management criteria, etc. Among these, the Large Marine Ecosystem approach (Sherman and Duda 1999) offers a comprehensive division of the coastal regions into 50 LME entities.

The Bay of Biscay lies in two different LMEs, though the scientific literature refers to this area as a single entity. This is supported by biogeographic patterns, comparative ecology and management criteria.

In terms of biogeography, the Bay of Biscay is a region of transition from subtropical to boreal regimes. Its most remarkable characteristic lies in its biological richness in floral and faunal species: at least 800 species of phytoplankton, more than 100 species of copepods, around 400 species of fish, 28 species of cetaceans, etc., were identified in the region. Many species of seaweeds, invertebrates, fish and marine mammals reach into the Bay of Biscay as the southern or northern limits of their European continental margin distribution (Fischer-Piette 1955; Whitehead *et al.* 1984; Quéro *et al.* 1989; Sanchez *et al.* 1995). Good examples of the association of species in the Bay of Biscay are given by seaweeds: e.g. Bretagne and Galicia regions present species adapted to cold waters (*Laminaria saccharina, Himanthalia elongata,* and *Palmaria palmata*); meanwhile, the inner Bay of Biscay presents species adapted to warmer conditions (*Gelidium sesquipedale, Gelidium latifolium* and *Corallina elongata*). Also, the spatial distribution of some commercial pelagic fishes like the mediterranean horse mackerel (*Trachurus mediterraneus*) and anchovy (*Engraulis encrasicholus*) is clearly restricted to the inner Bay of Biscay where they maintain a self-sustaining population; others such as sardine (*Sardina pilchardus*) extend the spatial distribution outside the Bay of Biscay, both to the north (Brest) and to the south-west (Galicia), but maintain main population of adults in both the Spanish and French margins of the bay. Migratory species such as mackerel (*Scomber scombrus*) arrive in the bay each year during the spawning season. Tagging experiments that monitor the migration routes have shown that mackerel stay in the Bay of Biscay until they commence the reverse migration to northern waters where the bulk of tags were recovered (Uriarte *et al.* 1998). Also, large pelagic fish such as bluefin tuna (*Thunnus thynnus*) and albacore (*Thunnus alalunga*), which live in subtropical areas of the western North Atlantic, make annual feeding migrations to the Bay of Biscay, reaching their maximum concentration in the bay during July and August when they are caught by the French and Spanish fleets. Thus, the Bay of Biscay maintains many different living species organised in both self-sustaining populations and seasonal/trophically dependent populations.

The recent review of Longhurst (1998) on the comparative ecology of the sea divides the world oceans into four primary biomes which are subdivided into provinces. According to Longhurst, the North East Shelves Province can be split into primary divisions, one of them being the southern outer shelf from northern Spain to Ushant, including the Aquitaine and Armorican shelves off western France (i.e. the Bay of Biscay). This is supported by the characteristics of regional oceanography and by the

Figure 10-3. Temperature distribution (SST) of the Bay of Biscay (NE Atlantic) on July 7, 2000.

biological and ecological response to the regional environment. One of the most illustrative differences between the Bay of Biscay and the surrounding regions is given by the seasonal heating of surface waters from May to October that produces a shift in the distribution of the isotherms in the Bay of Biscay in such a way that a thermal discontinuity is created with the rest of the Atlantic (Figure 10-3). Because this thermal structure is maintained throughout the growth season, it has major implications for critical biological processes, to the extent that such a discontinuity corresponds with an ecological boundary. Other ecological aspects, such as physical forcing in relation to scales of temporal variability and the environmental risk induced by human activities, will be treated in detail in the following sections.

Management criteria regarding the Bay of Biscay include regulations on several fish species subject to Total Allowable Catches (TACs), such as anchovy, hake, blue whiting, etc. which are shared by France and Spain. Restrictions on catches are discussed in international commissions: ICES (International Council for the Exploration of the Sea), STECF (Scientific, Technical and Economic Committee for Fisheries), etc. France and Spain also have a long history of scientific co-operation and promotion of regional research programmes for the Bay of Biscay (e.g. IEO-IFREMER Conference for the development of an integrated research of the Bay of Biscay ecosystem). Finally, France and Spain are signataries of several international conventions concerning conservation of natural environments comprising the Bay of Biscay.

The critical processes controlling the structure and functioning of biological communities, the stress and degradation on coastal spaces, and the exploitation of fish and other renewable resources, are the same at the regional scale of the whole Bay of Biscay, to the extent that the biological, ecological and management characteristics confirm that the Bay of Biscay is a well defined LME.

DYNAMICS OF THE BAY OF BISCAY MARINE ECOSYSTEM

The main scheme

The Bay of Biscay exhibits the ecological functioning of the temperate seas, whose dynamic is governed by climate and tides. Like the entire north-eastern Atlantic, this region undergoes a seasonal climatic cycle that strongly affects the pelagic ecosystem through three interrelated forcing factors over the year: sunlight exposure, heat input, and mechanical forcing on the surface due to wind. These forces produce a regular pattern in hydrographic conditions characterised by winter mixing of waters, followed by summer stratification. Phytoplankton blooms occur during the transition between both periods. The spring phytoplankton bloom, generally in March-early April, occurs when sunlight exposure is intense and long enough for net photosynthesis and is characterized by a dominance almost exclusively of diatoms (Casas *et al.* 1997). During summer stratification, nutrient concentration drops, phytoplankton biomass decreases to low levels and dinoflagellates are dominant. In winter, water mixing and low irradiance prevent phytoplankton growth in spite of high concentrations of nutrients.

The currents are governed by winds, tides, and density gradients. Wind-driven flow (Sverdrup circulation) induces the subpolar and subtropical gyres in the North Atlantic. The Bay of Biscay is located in the eastern part of these, influenced by the North Atlantic drift to the north and the Azores Current to the south. The currents induced by this general circulation are not very intense in the Bay of Biscay, and its being an oceanic bay further lessens their effects. Its geostrophic circulation is weak, being anticyclonic in the oceanic part (Saunders, 1982; Maillard, 1986) and cyclonic on the continental margin (Pingree, 1993) (Figure 10-4). Tidal currents have an oscillatory component (semi-diurnal) and a long-term component (residual current). The semi-diurnal component is greatest over the north-west Armorican shelf (about 0.5 m s^{-1}, locally reaching 1 m s^{-1}), decreasing towards the south along the Aquitaine shelf. It is not particularly intense over the Iberian shelf areas (10 cm s^{-1}). The residual component tends to be very weak (less than 1 cm s^{-1}) over most of the shelf. Locally, however, it may be one order of magnitude higher; this is the case near the islands of Noirmoutier, Oléron and Ré, on the inner Armorican shelf, where it can reach 10 cm s^{-1} and plays a major role in long-term transport.

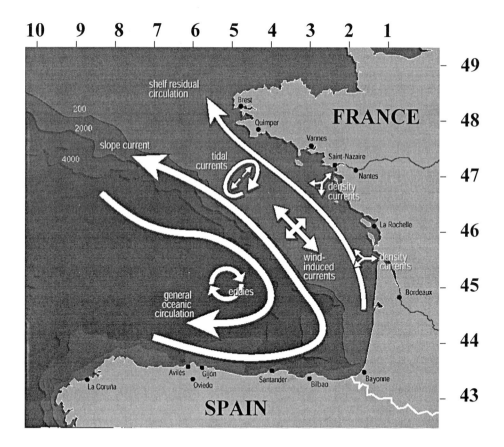

Figure 10-4. Circulation and currents in the Bay of Biscay (from Koutsikopoulos and Le Cann 1996).

Seasonal variability in the vertical structure of coastal and oceanic waters

Both atmospheric events and internal waves modify the steady state of the water column by direct transfer of energy (e.g. wind stress and turbulence), flows of surface water with different properties (coastal run-off and river plumes), and modulation of the depth of the mixed layer.

Coastal run-off and river plumes: The Gironde, the Loire and, to a lesser extent, the Vilaine Rivers provide large volumes of fresh water (which is turbid but rich in nutrients from the mainland) during spring. This results in the formation of dilution plumes at the surface of coastal waters, which drive significant northward currents over the inner Armorican shelf. They are clearly stratified vertically and delimited horizontally by a density front at the mouth of the estuary. Because the tidal currents along the Atlantic coast are much weaker than in the English Channel, these plumes can extend over several hundred kilometres in length, becoming progressively diffuse. With the continual input of nutrients they provide, these river plumes maintain very large 'new' phytoplankton production along the coastal fringe, and sometimes even to the edge of the continental shelf. Due to the haline stratification that maintains the phytoplankton cells in a very thin layer of water, phytoplankton production can start very early in the year. Triggering of this production could be favoured by an anticyclone regime, which is frequent in the region during winter (January to March). Studies have shown that these winter blooms linked to plumes from large rivers are relatively short, because they are quickly limited by phosphorous due to the high N/P ratios of river waters, which are highly unbalanced in favour of nitrogen. A lower volume of river run-off and a much narrower shelf off the Iberian Peninsula act in tandem to make buoyant plumes much less persistent over the Cantabrian coasts than over the Armorican shelf.

Internal waves and tidal fronts: At the Armorican shelf break, when the water column is vertically stratified, tides generate internal waves that propagate both on- and off-shelf from about 5° W to 9° W (*bourrelet froid*). Such internal waves appear to be responsible for significant mixing and upwelling of nutrients. Oscillations involving a sharp interface (e.g. a seasonal pycnocline) may sometimes produce surface signatures, as well as reinforce the barotropic current (due to oscillation of the ocean free surface) or act as forcing mechanisms of long-term phenomena (Sournia *et al.* 1990). This tidal front recurs annually, starting in June or July at the shelf-break where it has its maximum intensity and thus its greatest impact on primary production. Satellite images clearly show this phenomenon. Signs of upwelling on the front include an increase in fluorescence values, zooplanktonic biomass, and eggs and larvae of pelagic fishes, e.g. sardine, mackerel and horse-mackerel (Lago de Lanzós *et al.* 1997). The abundant, sustainable 'new' production makes this 100-km wide fringe, overhanging the continental shelf break, an oasis in the open sea and it is used by pelagic species as spawning grounds to nourish future larvae (Arbault and Boutin 1968; Lago de Lanzós *et al.* 1997).

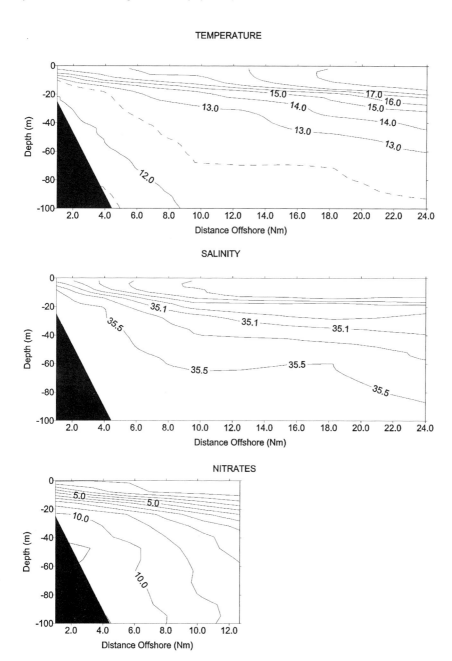

Figure 10-5. Cross-shelf section off Santander showing the signal of temperature, salinity and nitrates during an upwelling event (June 1995).

Interannual variability of mesoscale events

In the ocean, mesoscale motions are typically the most energetic physical processes; consequently, they represent major perturbations. Episodic mesoscale events are common in the Bay of Biscay, showing both high seasonal and inter-annual variability (Koutsikopoulos and Le Cann 1996). They have a dramatic impact on the ecosystem's productivity and in the transport of drifting biological material.

Coastal upwelling events occur on the Spanish and French continental margins. On the Spanish continental margin, upwelling extends from Galicia to the Cantabrian Sea (Molina 1972; Estrada 1982; Botas *et al.* 1990; Ríos *et al.* 1992; Lavín *et al.* 1998). Upwelling is more intense to the west of Cape Peñas and Ortegal, and acts as a mechanism generating spatial variability between the western and eastern zone of the Cantabrian Sea, and between the coastal mixed waters and the neighbouring oceanic stratified areas (Figure 10-5). Upwelling events are highly variable in intensity and frequency and they show a significant variability from year to year. When upwelling is particularly intense, surface signs of upwelling are observed first on the northern coast of Galicia, then off Asturias, and later to the east off Cantabria. On the French continental margin, mainly in summer, weak upwelling events are induced along the Landes coastline by northerly winds.

Navidad. Especially in winter, there is a warm and salty current (called *Navidad*) flowing from the central Atlantic and running from west to east along the slope of Spain's northern coast, then flowing northward along the French continental slope (Frouin *et al.* 1990; Pingree 1994) (Figure 10-6). Mean velocities of 30 cm s^{-1} were recorded at the shelf break in the central Cantabrian Sea during late December 1995 and January 1996 (Díaz del Río *et al.* 1996). Its influence on the area's biology has also been studied by Botas *et al.* (1988), Varela *et al.* (1998), Sánchez and Gil (2000), etc. Although the extent of intrusion is not a well-studied phenomenon, it seems to be more intense in the western zone of the Cantabrian Sea. These saline intrusions present a high space-time variability (Pingree and Le Cann 1990; Varela *et al.* 1995; Moreno-Ventas *et al.* 1997).

Slope Water Oceanic eDDIES (SWODDIES). Quasi-periodically (in some cases associated with the *Navidad* current, whose instability generates meanders that can break off in the form of free eddies) different anticyclonic and cyclonic eddies have been observed in the Bay of Biscay (Howe and Tait 1967; Pingree 1979; Dickson and Hughes 1981; Pingree and Le Cann 1992a; Moreno-Ventas *et al.* 1997). The formation of rings in the Bay of Biscay is frequent due to the currents' interaction with the topography of the continental margins (Pingree and Le Cann 1992a, b; Pingree 1994; Gil 1995). A set of regions with a high incidence of rings has been determined, such as the area between Cape Finisterre and Ortegal (Gil 1995; Porteiro *et al.* 1996; Díaz del Río *et al.* 1996) and the innermost part (*cul-de-sac*) of the bay (Pingree and Le Cann 1992a, b; Pingree 1994; Díaz del Río *et al.* 1996) (Figure 10-7). Although these structures, which have a diameter of approximately 100 km, move slowly (2 km d^{-1}), some persist during more than one year, drifting over oceanic waters without

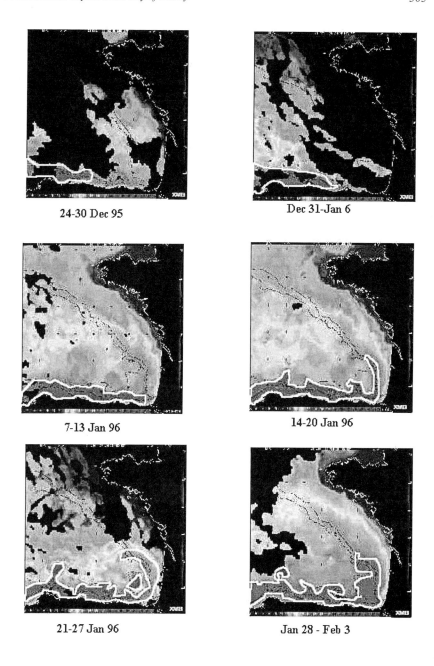

Figure 10-6. Weekly evolution of the poleward current (Navidad) from December 1995 to January 1996 in the Bay of Biscay. For illustration purposes the current is contoured by a white line on SST images (weekly average values) (from Moreno-Ventas *et al.* 1997).

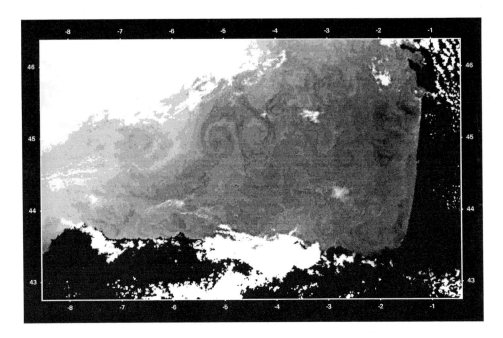

Figure 10-7. AVHRR satellite image showing the formation of eddies in the Bay of Biscay (24 May 1999).

necessarily being associated with the continental margin (Pingree and Le Cann 1992a). In summer, the persistence at the core of these eddies, between 100 and 300 m, brings, by hydrostatic effect, an uprising of isotherms (and therefore, of the thermocline) towards the surface. Due to their formation on the edge of the slope, these rings can trap and transport biological material from the shelf-break area, thus dispersing or accumulating populations (Porteiro *et al.* 1996).

Jet filaments and fronts. It is also necessary to consider fronts as elements that can disrupt the general pattern over the platform or along the shelf break. Fronts have diverse origins (e.g. upwelling filaments, instability of currents, anomalies associated with saline intrusions), but in any case they have great biological interest, as well as playing an essential role as barriers to the transport of eggs and larvae. In many areas of the Atlantic, a close association has been found between spawning areas and shelf-break fronts (Iles and Sinclair 1985; Heath and MacLachlan 1987; Coombs *et al.* 1990). The success of pelagic fish recruitment depends to a great extent on whether spawning occurs in areas where larvae can find the proper prey in size and quality. Therefore, the presence of fronts during the period in which recruitment occurs can be decisive for the success of local fisheries.

Decadal variability at basin-scale

Climate and oceanic circulation interact in basin-scale water transport. Changes in heat fluxes on time-scales of one to several years can have a dramatic impact on standing stocks of plankton and higher trophic levels. The following oscillations have been shown to influence the Bay of Biscay.

NAO (North Atlantic Oscillation). The Bay of Biscay forms part of the eastern sector of the North Atlantic, and therefore it is in the area of influence of the periodic oscillations that act upon it. The most significant oscillation, with a periodicity of approximately 5-7 years, is the North Atlantic Oscillation (NAO) (Hurrell 1995). Related to the NAO, an increase in surface temperatures has been detected near Iceland, off Norway and in the North Sea (Dickson *et al.* 1988; Turrell *et al.* 1997). In the Bay of Biscay, a significant correlation (P< 0.01) has been found between the NAO index and the air temperatures measured in Santander by the INM (Instituto Nacional de Meteorología) for the historic series between 1961 and 1998 (Anon. 1999).

Gulf Stream North Wall. The position of the Gulf Stream's north wall, where it separates from the American continental shelf, is a factor of great importance in climatic studies, since this current is responsible for the greatest heat transport in the Northern Hemisphere. The Gulf Stream latitude index (GULF) has been described by Taylor (1996), and was extrapolated from monthly charts of the latitude of the Gulf Stream's north wall at six longitudes, measured between 65° and 79° W for the period of 1966-1996. Applying a principal component analysis to the resulting correlation matrix, Taylor found a common pattern of variation at the six longitudes. Theatmospheric displacement (changes in NAO index) seems to be able to induce (with an adjustment time) a displacement in the position of the Gulf Stream path. The adjustment time for oceanic circulation has been fixed at two years by Taylor and Stephens (1997) for the three decades studied.

Long-term trends

Accurate records of sea surface temperature in the Bay of Biscay are available for the whole century and data show an oscillatory pattern with first a rising period, from 1920 to 1960, when temperature increased in the order of half a degree; after that there was a phase of cooling which reversed in 1981 (Southward *et al.* 1995) and continues at present. Reasons that explain global warming include solar cycles like the 10-12 years sunspot cycle and the longer harmonics, the 30-40 years cycle in ocean-atmosphere interactions (Stocker 1994), and the greenhouse effect. However, in terms of solar cycle and orbital momentum, the earth is entering into a cooler phase, thus the climatic basis of global warming is not clear and it is suggested that the most recent warming represents the greenhouse effect superimposed on a natural cooling (Southward *et al.* 1995), which remark the importance of the human impact on climate. Modellers (e.g.

Stouffer *et al.* 1994) think that the signal of global warming is stronger than any stochastic variation.

In the Southern Bay of Biscay a warming trend is currently detected down to a depth of 75 m on the shelf and oceanic waters (Lavín *et al.* 1998). The same authors have estimated that in the period 1991-1999 the temperature at 10 m had risen 0.5 °C in the Southern Bay of Biscay (annual rate of 0.06 °C, Figure 10-8). A similar rise was reported off San Sebastián between 1986 and 1990 by Valencia (1993), and throughout the Bay of Biscay by Koutsikopoulos *et al.* (1998). This atmospheric and surface-water warming could have severe consequences for the ecology of the ecosystem. Lavín *et al.* (1998) noted that the water column experienced a higher degree of stratification, which also remained stratified for a longer period of time. Valdés and Moral (1998) have related this higher degree of water-column stratification to a less intense exchange of nutrients with the surface layers, which in the end means a reinforcement of the microbial loop, a decrease in the mesozooplankton biomass, a drop in the number of species per unit of volume, and a shift of the classic trophic chain to a less efficient one.

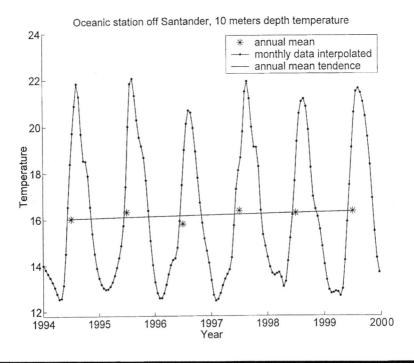

Figure 10-8. Time series of temperature at 10 m depth off Santander in the period 1991-1999. Linear trend shows an increase in annual mean temperature of 0.06 °C.

HUMAN IMPACT IN THE BAY OF BISCAY

The Bay of Biscay has experienced dramatic changes in the coastal use of land and living resources during the last 50 years. Fast growth of population and socioeconomic development have resulted in environmental imbalances; but information and data are scarce, often restricted to inner coastal areas and usually produced by local agencies that strongly limit the access to data and its accuracy. Thus, it is difficult to: i) determine the full range of human impacts in the Bay of Biscay, ii) evaluate the environmental importance and the economic cost of man's uses of the marine ecosystem, and iii) establish social and administrative priorities addressed to rational management and sustainable use of the resources. In the absence of a quantitative scale of the importance of human impacts on the ecosystem in the Bay of Biscay, the effects of human pressure on the marine ecosystem (33 different impacts are described in Anon. 1998) are grouped here by type of human activity (human settlements, extractive activities, industrial activities, and transport, building and maintenance of infrastructures). In general, the disturbances produced by man can be stated as inputs of specific substances, physical disturbances and direct impacts on biological communities and species.

Human settlements on the coast in the last century have resulted in large populations, with all the problems associated with urban agglomerations. Tourism, new urbanisation of coastal areas and recreational uses of beaches and shores have added pressure, including the disposals of by-products of human sewage. Ecological disturbances produced by human settlements include microbiological pollution, eutrophication, land reclamation, loss of habitats and marine litter. Estuaries and coastal lagoons receive the main impact of microbiological contamination from urban origin, which has a strong impact on the quality of bathing waters and shellfish salubrity. The implementation of recent Directives from the EU on water treatment and depuration will result in a diminishing of this kind of risk in the near future. Most of the data on floating debris or litter along the coast (particularly on beaches) refers to glass, plastics, metal, paper, bottles, clothing, foodstuff, wood, rubber, packaging materials, rest of trawl and other fishing gear. In general, plastics constitute about 85% of the accumulations because of their poor degradability. Plastics enter the marine environment as discards by recreational users of beaches, and from ships, sewers and coastal runoff. Although human pressure is strong on the region, the risks derived from human settlements are not severe in the Bay of Biscay, and only in some cases have local imbalances been described.

Extractive activities include fishing, aquaculture and farming. The ecological disturbances include direct impact on target species, overfishing, alterations on the seabed, introduction of non-indigenous species, agriculture sewages, etc. The Bay of Biscay has traditionally been an area of intense fishing activity and nearly 5000 fishing boats operate there. Trawlers and purse seiners are the main fishing vessels used for demersal and pelagic species. Other gears used to a lesser extent are gill nets, lines, dragnets, etc. Fishing activities not only affect the pelagic and demersal fish species.

Many intertidal populations are also subject to exploitation: clams, crabs, octopus and others, including the sea urchin *Paracentrotus lividus,* which is intensively exploited on the north coast of Spain and whose intertidal population has virtually disappeared from wide areas (this species is now restricted to tide pools and made up of small-sized individuals). The aquaculture industry has increased greatly in the last decade and environmental risks associated with this activity are under debate. Society has recently been made aware of the use of GMOs (Genetically Modified Organisms), but the use of such organisms has not yet been reported in the Bay of Biscay.

Since the 1970s the use of DDT in agriculture has been banned in most countries. Nevertheless, residues of DDT and in particular its metabolites, DDE and DDD, are found in the environment in suspended material, sediments and biota; the main places for accumulation are rias, estuaries and coastal lagoons. The inputs usually occur during the run-off periods (Ferreira and Vale 1995). Agriculture is also the primary source of nutrients carried to the coastal water by the rivers.

Industrial activities have traditionally supported the economy of the Bay of Biscay in both France and Spain. Many of these activities are known to be polluting, e.g. paper milling, petroleum refining, iron and steel working, chemicals, etc. Disturbances of the ecosystem include industrial discharges and inputs of specific pollutants (both inorganic and organic compounds). Mercury is associated with papermill industries and is recognised as one of the most important inorganic pollutants. PAHs (Policyclic Aromatic Hydrocarbons) can occur naturally but their concentrations may increase significantly due to human activities: incomplete combustion, marine oil extraction, industrial discharge, oil traffic and handling, etc. Petrochemicals are one of the most significant human-induced problems of contamination.

Transport, building and maintenance of infrastructures include activities such as coastal protection, land reclamation, dredging, and shipping. Sediments are often dredged in harbour areas, estuaries, and navigation channels. The material excavated is usually sand, silt or gravel. The quantities of dredged material vary from year to year according to the patterns of sediment movement and accretion that make recurrent maintenance dredging necessary, as well as to new projects for harbour development requiring capital dredging. On the French coastal and estuarine areas of the Bay of Biscay, the annual quantity of dredged and dumped material represented roughly 10^6 m^3 in 1993, about one percent of this amount being composed of contaminated sediments. The dumping of material produced by maintenance dredging of ports and navigation channels has a potential temporary and long term impact on the bottom and water column of the dump site because of the scale of the dumping and the general contaminated nature of the sediments. The highest levels of contaminants in water, sediments and biota were found in coastal areas, mainly in estuaries, rias and semi-enclosed sites, due to slow water renewal and to higher urban and industrial concentrations. Shipping accidents can produce unintentional pollution; in December 1999 the supertanker *Erika* wrecked on the coast of France and 10,000 tons of oil were spilled in shallow waters. Due to the strong wind in the area, the "black tide" moved to the coast and large expanses of French beaches were contaminated by oil.

EFFECTS OF THE ECOSYSTEM VARIABILITY AND HUMAN IMPACT ON SPECIES AND HABITATS OF THE BAY OF BISCAY

The ecosystem dynamics and human impacts mentioned in the preceding sections interact with the biological cycles of the species, producing fluctuations in their abundance which are not always easy to explain. At high trophic levels our perception of the states of equilibrium of species and communities is even worse because, depending on the biology of the species (e.g. short/long life), natural and anthropogenic variabilities have an impact on future generations. This means that although fluctuations are more common and sharper in short-lived species, they are more persistent over time in those species for which various cohorts co-exist simultaneously in the ecosystem.

Interactions between environmental variability and population dynamics

Small pelagic fishes of commercial value, such as sardine (*Sardina pilchardus*) and anchovy (*Engraulis encrasicolus*), are highly dependent on environmental conditions. The distribution of sardine extends outside the Bay of Biscay, both to the north (Brest) and to the south-west (Galicia). The sardine, like other clupeids, is a relatively short-lived species at the base of the food chain, and is subject to a very high rate of natural mortality. The variability in annual recruitment leads quickly to changes in population abundance, and consequently affects the fishery's productivity. Sardine catches in the Bay of Biscay have diminished in recent years and the species is under biological regulations. Sardine recruitment has been related to the frequency and intensity of upwelling events (Robles *et al.* 1992; Roy 1993; Cabanas and Porteiro 1998). According to Roy (1993), sardine adapts its spawning strategy to upwelling, taking advantage of an environmental window of opportunity to optimise larval survival. Below a certain level of upwelling intensity, survival falls; and above a certain intensity threshold, eggs and larvae are advected towards the ocean, where they are lost to the population.

Sardine recruitment has also been correlated with air temperature (Figure 10-9) (Lavín *et al.* 1997; Cabanas and Porteiro 1998). Sea surface temperature in coastal waters of the southern Bay of Biscay has increased in recent years (Lavín *et al.* 1998). This suggests that the observed rise in temperature reported in these waters has produced an increase in the thermal stratification of the water column. A linear decreasing relationship between water-column stratification and the number of copepod species has been observed in the shelf waters of the southern Bay of Biscay during the period 1991-1996 (Valdés and Moral 1998). Roemmich and McGowan (1995), Southward and Boalch (1994) and Southward *et al.* (1995) also suggest that a warming trend is related to zooplankton decay in the Pacific and in the English Channel, due to a reduction in nutrient supply to the photic layers and to the predominance of microflagellates in the plankton. Thus, the relationship between air temperature and sardine recruitment may be explained by a cascade relationship between water-column stratification and low food availability during warm years. Consequently, when searching for factors forcing the sardine population, we have to take into account not only local conditions, but also those climate changes that are occurring at a basin or global scale.

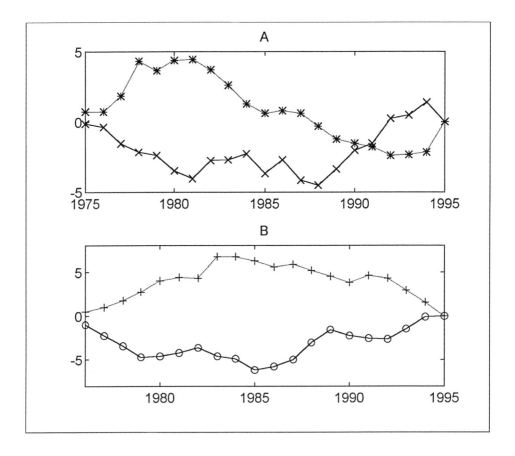

Figure 10-9. **(A)** Cumulative sum of anomalies of NAO (X) and albacore recruitment (*) between 1976 and 1995. **(B)** Cumulative sum of anomalies of air temperature (o) and sardine recruitment (+) between 1976 and 1995 (Lavín *et al.* 1997).

The anchovy is the only pelagic species whose principal population is confined to the Bay of Biscay. The Bay of Biscay anchovy shows dramatic inter-annual recruitment fluctuations that make it difficult to offer scientific advice for the management of the fishery. The species shows a strong connection between its spawning habits and recruitment success and oceanographic features such as coastal run-off, shelf-break fronts and eddies (Motos *et al.* 1996). Recruitment is influenced by the wind regime and Borja *et al.* (1996) found a clear relationship between anchovy recruitment intensity and upwelling intensity on the French and Spanish coast. Junquera (1988) has also suggested a relationship between anchovy recruitment and global warming.

Oceanic pelagic species of tuna found in the Bay of Biscay are the albacore, *Thunnus alalunga,* and the bluefin, *Thunnus thynnus.* Both show oscillations in their abundance and distribution in response to basin-scale shifts in the North Atlantic oceanic current system. The NAO and GULF indices provide a connection between northward displacement of the Gulf Stream current and tuna distribution. Both albacore and bluefin tuna are present in this area from the beginning of spring to mid-autumn, depending on oceanographic conditions, especially temperature. These two species live in subtropical areas of the western Atlantic and make annual feeding migrations to the Bay of Biscay. Tuna schools begin to move in an easterly direction at the beginning of spring, and reach their maximum concentration in the Bay of Biscay during July and August, when they are caught by the French and Spanish fleets. The albacore tuna population present in this area is comprised of juveniles (1-4 years) from the northern Atlantic stock. The annual catches are on the order of 16,000 t in the Bay of Biscay and north-east Atlantic (average data from 1993-1997), with a maximum of 40,000 t in 1960 when baitboat and troll fleets from France and Spain were targeting this species. The bluefin population is also made up, in part, of juveniles from the eastern Atlantic stock.

Cyclical oscillations, such as the NAO, have been related to fluctuations in abundance of albacore and bluefin tuna (Ortiz de Zárate *et al.* 1997; Santiago 1997). Results suggest that during periods of high NAO, when the westerly winds intensify, the recruitment of albacore is low (Figure 10-9). Increased turbulence due to these strong winds could have a negative impact on recruitment. During periods of low NAO, convective activity in the Sargasso Sea is high (Dickson *et al.* 1996), which may provide good recruitment conditions. Bluefin shows opposite patterns to those shown by albacore. With regard to aggregated immature catches of albacore (ages 2 and 3) of the northern stock, correlation is significant and negative with the index of Gulf Stream latitude and mean sea level. These two factors may provide an idea of the influence of the North Atlantic Current on young albacore. The northward displacement of the current could bring these tuna out of the fisheries area.

Environmental risks due to human activities

Fisheries have severe impacts on the ecosystem. Direct impacts include effects on the target species: increase in the mortality rate, reduction of average age and size, and decrease of the fecundity rate, which can result in a depletion of the spawning stock to a level where the sustainability of the resource is threatened. This is particularly the case for species with late maturity and low fecundity. Examples of such impacts on marine fish populations have been reported for *Raja clavata* and *Scyliorhinus canicula* (Quero and Cendrero 1996; Sánchez *et al.* 1997), which have been proposed as key species to monitor changes in marine exploited systems. Post escapement mortality, discards and lost gear also directly affect target and non-target species. Indirect effects on the ecosystem include disturbances of the seabed with impact on benthic communities, by-catches of cetaceans and seabirds, changes in species assemblages and their relative abundances, i.e. reduction in diversity through the elimination of the specialist species with low birth rates, alteration of the natural equilibrium between predator and prey species, etc. (OSPAR Commission 2000).

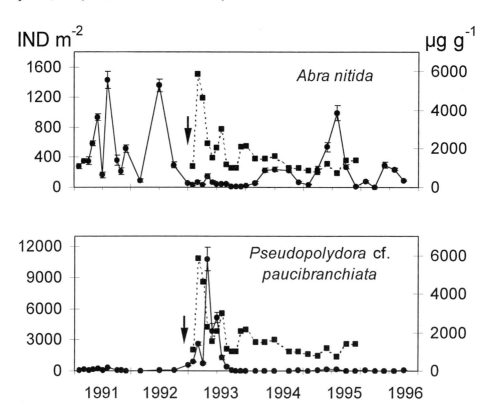

Figure 10-10. Decrease in the abundance of the species *Abra nitida* (upper panel, solid line) and colonisation of substrata by the opportunistic species *Pseudopolydora cf. paucibranchiata* (lower panel, solid line) after the *Aegean Sea* oil spill (dashed lines show oil concentration in sediment, arrows refer to oil spill) (Parra and López-Jamar 1997).

Benthic communities are very sensitive to changes in water properties and human impacts. One example is the toxic effects of the oil from the supertanker "Aegean Sea" (wrecked in La Coruña in 1992), which caused a temporal reduction of the macroinfauna (amphipods, echinoderms, molluscs) with a simultaneous dramatic increase of opportunist polychaetes (mainly *Pseudopolydora paucibranchiata* and *Capitella capitata*) (Parra and López-Jamar 1997) (Figure 10-10). Although the immediate effects of the contamination were dramatic, the abundance and diversity of the meiofauna recovered after three years.

Other impacts on benthic organisms are due to industrial activities and industrial waste, which have led to some alterations in macroalgal assemblages at specific locations; and urban sewage, which increases the level of nutrients and enhances the proliferation of green algae, e.g. *Ulva* and *Monostroma*. Most of these alterations in the Bay of Biscay take the appearance of local imbalances rather than serious alterations in the ecosystem.
Solid waste has become a new form of pollution by the generalised use of plastic. A risk of the impact on cetaceans from marine litter, mainly plastic bags and other debris, was recently reported. In one case, more than 50 kg of plastics were found in the stomach content of a fin whale (*Balaneoptera physalus*) stranded on a beach in Cantabria in November 1997 (García-Castrillo 1998). Also, between 1988 and 1998, plastics were found in the stomach contents of sea turtles (*Dermochelys coriacea)* in 22 of 43 autopsies recorded (Duguy *et al.* 1998). Other causes of mortality presenting a real risk to marine mammal species have arisen, such as ship collisions, noise, anthropogenic disease agents, fishing gear, etc.

Other human activities that cause environmental degradation and local imbalances in the Bay of Biscay include shipping, microbiological pollution, eutrophication, marine litter, and contaminants. TBT has been reported to produce shell calcification anomalies in adult oysters, and declining reproduction in the Bay of Arcachon. TBT also alters the physiology of dogwhelks *Nucella lapillus*, wherein females develop male sexual characteristics. It can also cause sterility and have destructive effects on the population, as reported from industrial bays and some estuaries of NW Spain (OSPAR Commission 2000).

Finally, during the last decades there have been new appearances of introduced species in the coastal areas. The entrance vectors of these alien species are both natural (e.g. transported by currents) and human induced, caused by intensification of communications and shipping, travel in ballast water on commercial vessels or adherence to a ship's keel and hull, or transfer of shellfish. The introduction of non-indigenous species carries the risk of introducing pests and disease, and establishing undesirable ecological effects in relation to existing species. Examples of species introduced in the Bay of Biscay include seaweeds (e.g. *Undaria pinnatifida*, which escaped from farms and disseminated in the natural environment, where it carries out its entire cycle), molluscs (e.g. *Crassostrea gigas* and *Crepidula fornicata*), and crustaceans (*Hemigrapsus penicillatus* and *Elminius modestus*).

Overlapping all these fluctuations and human impacts, there are global climatic processes that affect large areas and influence all populations. These changes tend to operate slowly and can affect: i) the behaviour of species (e.g. changes in migratory routes), ii) their recruitment (due to changes in the environmental conditions in the spawning and/or recruitment areas), and iii) the spatial distribution of species. Generally, these changes mark long-term population trends.

Southward and Boalch (1994) and Southward *et al.* (1995) suggest that if the predicted global warming results in increases in temperature on the order of 2 °C, then there will be considerable changes in marine communities, producing latitudinal shifts in species distribution on the order of 200-400 miles, especially for mobile animals such as fish. In fact, the warming trend seems to be responsible for the appearance of tropical fish species on the south-east shelf of the Bay of Biscay. Several tropical species (*Cyttopsis roseus, Zenopsis conchifer* and *Sphoeroides pachygaster*) have been caught throughout the Portuguese, Spanish and French Atlantic margin (Quéro *et al.* 1997, 1998). The subtropical copepod *Temora stylifera* has also shown a significant increasing trend during the last 15 years in the Bay of Biscay (Villate *et al.* 1997). Sessile intertidal organisms such as limpets and barnacles were also reported to show changes in their spatial distribution towards northern areas (Southward *et al.* 1995).

In addition to the impact of human activities on communities and species, there is also an impact on the different types of coastal habitats. Among the coastal habitats (estuaries, coastal lagoons, rocky cliffs, shingles, rocky shores, sandy and muddy shores), estuaries are among the most productive and dynamic coastal ecosystems because of their role in the transformation and transfer of materials from the land to the sea. The main physical disturbance in the Bay of Biscay is the loss of estuarine habitats - mainly wetlands, saltmarshes and dunes - caused by land reclamation, sand extraction, agriculture, and aquaculture. Moreover, the sediments accumulate many organic and inorganic pollutants from mines, farms and industries. The most heavily polluted of such areas are the relatively small rías of Avilés and Bilbao. Depending on the season, eutrophication is also considerable in Arcachon (OSPAR Commission 2000).

Estuaries are subject to intense human pressure and are also the areas most vulnerable to sea level rise due to global warming. Thus, they are the areas of greatest importance for conservation. The large estuaries of the Gironde and Loire (France) are important areas to protect, as are many marshes of the Spanish northern estuaries, such as those of Villaviciosa, Santoña and Urdaibai, which are important in terms of conservation (these are the most important migratory passes of waders and wintering areas for birds in the north of Spain). Other areas of importance for conservation coincide with the main seabird breeding colonies.

CONCLUDING COMMENTS

The above discussion indicates that we have a certain level of knowledge regarding the time-space variability factors regulating the fluctuations of biological communities. However, we still do not understand the connections between biological and physical processes, the parameterisation of flows, or possible cascade transmissions of variability. In other words, we do not know how the ecosystem is regulated. This lack of knowledge is evidence of the great need for more studies on ecosystem functioning in the Bay of Biscay.

It is also evident that all the pressing environmental problems caused by man's activities and the complexity of ecosystem interactions add a high degree of uncertainty to the proper management of living marine resources and uses of coastal areas; and it makes imperative that the future management of marine living resources be based on a more solid basis of scientific knowledge. Integration of disciplines is necessary to provide comprehensive answers to the central issues of ecosystem management from an holistic approach. Time series programmes on adequate spatial and time scales supported by governmental agencies should be encouraged. There is also a need for improved forecasting of human impacts on the ecosystem. Future integrative projects should give increased attention to the human role in the regulation of the marine ecosystem. Searching for changing trends based on key species, and monitoring the preservation state of selected areas should also be a major priority.

REFERENCES

Anon. 1998. Priorisation of human pressures for the Greater North Sea Quality Status Report 1999. An interactive and iterative Multi Criteria Analyses approach. Results and methodology. Delft, RA/98-331, December 1998.

Anon. 1999. Report of the ICES Working Group on Oceanic Hydrography. ICES CM 1999/C: 8.

Arbault, S. and N. Boutin. 1968. Ichtyoplancton. Oeufs et larves de poissons téléostéens dans le Golfe de Gascogne en 1964. Revue des Travaux Institut Scientifique et Technique des Peches Maritimes, 32(4): 413-476.

Borja, A., A. Uriarte, V. Valencia, L. Motos, and Ad. Uriarte. 1996. Relationship between anchovy (*Engraulis encrasicolus* L.) recruitment and the environment in the Bay of Biscay. Scientia Marina, 60 (2): 179-192.

Botas, J.A., A. Bode, E. Fernández, and R. Anadón. 1988. Descripción de una intrusión de agua de elevada salinidad en el Cantábrico central: distribución de los nutrientes inorgánicos y su relación con el fitoplancton. Investigación Pesquera, 52 (4): 561-574.

Botas, J.A., E. Fernández, A. Bode, and R. Anadón. 1990. A persistent upwelling off the Central Cantabrian Coast (Bay of Biscay). Estuarine, Coastal and Shelf Science, 30: 185-199.

Cabanas, J.M. and C. Porteiro. 1998. Links between the North Atlantic Sardine recruitment and their environment. ICES CM 1998/R: 23.

Casas, B., M. Varela, M. Canle, N. González, and A. Bode. 1997. Seasonal variations of nutrients, seston and phytoplankton, and upwelling intensity off La Coruña (NW Spain). Estuarine, Coastal and Shelf Science, 44: 767-778.

Coombs, S.H., J. Aiken, and T.D. Griffin. 1990. The aetiology of mackerel spawning to the west of the British Isles. Meeresforschung, 33: 52-75.

Díaz del Río, G., A. Lavín, J. Alonso, J.M. Cabanas, and X. Moreno-Ventas. 1996. Hydrographic variability in Bay of Biscay shelf and slope waters in spring 1994, 1995, 1996 and relation to biological drifting material. ICES CM 1996/S: 18.

Dickson, R.R. and D.A. Hughes. 1981. Satellite evidence of mesoscale eddy activity over the Biscay abyssal plain. Oceanologica Acta, 4: 43-46.

Dickson, R.R., P.M. Kelly, J.M. Colebrook, W.S. Wooster, and D.H. Cushing. 1988. North winds and production in the eastern North Atlantic. Journal of Plankton Research, 10(1): 151-169.

Dickson, R.R., J. Lazier, J. Meincke, P. Rhines, and J. Swift. 1996. Long-term coordinated changes in the convective activity of the North Atlantic. Progress in Oceanography, 38: 241-295.

Duguy, R., P. Moriniene, and C. Le Milinaire. 1998. Facteurs de mortalité observés chez les tortues marines dans le Golfe de Gascogne. Oceanologica Acta, 21(2): 383-388.

Estrada, M. 1982. Ciclo anual del fitoplancton en la zona costera frente a Punta Endata (Golfo de Vizcaya). Investigación Pesquera, 46(3): 469- 491.

Ferreira, A.M. and C. Vale. 1995. The importance of runoff to DDT and PCB inputs to the Sado estuary and Ría Formosa. Netherland Journal of Aquatic Ecology, 29 (3-4): 211-216.

Fischer-Piette, E. 1955. Sur les déplacements des frontièrs biogeographiques intercotidales observables en Espagne: situation en 1954-55. Comptes Rendus. Academie des Sciences (Paris), 241: 447-449.

Frouin, R., A.F.G. Fiuza, I. Ambar, and T.J. Boyd. 1990. Observations of a poleward surface current off the coasts of Portugal and Spain during winter. Journal of Geophysical Research, 95 (C1): 679- 691.

García-Castrillo, G. 1998. Informe sobre la ballena (Balaneoptera physalus) de Oriñon. Informe Interno Consejería de Cultura y Deporte, Gobierno de Cantabria.

Gil, J. 1995. Inestabilidades, fenómenos de mesoescala y movimiento vertical a lo largo del borde sur del Golfo de Vizcaya. Boletín Instituto Español de Oceanografía, 11(2): 141-159.

Golikov, A.N., M.A. Dolgolemko, N.V. Maximovich, and O.A. Scarlato. 1990. Theoretical approaches to marine biogeography. Marine Ecology Progress Series, 63: 289-301.

Heath, M.R. and P. MacLachlan. 1987. Dispersion and mortality of yolk-sac herring (Clupea harengus L.) larvae from a spawning ground to the west of the Outer Hebrides. Journal of Plankton Research, 9: 613-630.

Howe, M.R. and R.I. Tait. 1967. A subsurface cold core eddy. Deep-Sea Research, 33: 383- 404.

Hurrell, J.W. 1995. Decadal trends in the North Atlantic Oscillation: Regional temperatures and precipitation. Science, 269: 676-679.

Iles, T.D., and M. Sinclair. 1985. An instance of herring larval retention in the North Sea. ICES CM 1985/H: 43.

Junquera, S. 1988. Changes in the anchovy fishery of the Bay of Biscay in relation to climatic and Oceanographic variations in the North Atlantic. Symposium on Long-term Changes of Marine Fish Populations, Vigo 1986: 543-554.

Koutsikopoulos C. and B. Le Cann. 1996. Physical processes and hydrological structures related to the Bay of Biscay anchovy. Scientia Marina, 60(2): 9-19.

Koutsikopoulos C., P. Beillois, C. Leroy, and F. Taillefer. 1998. Temporal trends and spatial structures of the Sea Surface Temperature in the Bay of Biscay. Oceanologica Acta, 21(2): 335-344.

Lago de Lanzós, A., L. Valdés, C. Franco, A. Solá, and X. Moreno-Ventas. 1997. Bay of Biscay environmental scenary in June and distribution of early fish stages. Plankton Symposium. Galway, Ireland.

Lavín, A., L. Valdés, X. Moreno-Ventas, V. Ortiz de Zárate, and C. Porteiro. 1997. Common signals between physical, atmospheric variables, North Iberian sardine recruitment and North Atlantic albacore recruitment. ICES/GLOBEC Workshop on Prediction and Decadal-scale Fluctuations on the North Atlantic. Copenhagen. Sept 1997.

Lavín, A., L. Valdés, J. Gil, and M. Moral. 1998. Seasonal and interannual variability in properties of surface water off Santander (Bay of Biscay) (1991-1995). Oceanologica Acta, 21 (2): 179-190.

Legendre, L. and J. Le Févre. 1991. From individual plankton cells to pelagic marine ecosystems and to global biogeochemical cycles. In S. Demers, ed. Particle analysis in Oceanography. NATO ASI Series, G 27: 261-300.

Longhurst, A. 1998. Ecological geography of the Sea. Academic Press, San Diego. 398 p.

Maillard, C. 1986. Atlas Hydrologique de l'Atlantique Nord-Est. IFREMER. Brest. 260 p.

Molina, R. 1972. Contribución al estudio del upwelling frente a la costa nordoccidental de la Peninsula Ibérica. Boletín Instituto Español de Oceanografía, 152.

Moreno-Ventas, X., A.Lavín, and L. Valdés. 1997. Hydrodynamic Singularities Observed by Satellite Imagery in the Continental Margin of the Bay of Biscay. 2° Simposium Internacional sobre el Margen Continental Ibérico Atántico. Cádiz, Spain.

Motos L., A. Uriarte, and V. Valencia. 1996. The spawning environment of the Bay of Biscay anchovy (Engraulis encrasicolus L.). Scientia Marina, 60 (2): 117-140.

Nixon, S.W. 1988. Physical energy inputs and the comparative ecology of lake and marine ecosystems. Limnology and Oceanography, 33(4, part 2): 1005-1025

Olson, D.O. and R.R. Hood. 1994. Modelling pelagic biogeography. Progress in Oceanography, 34: 161-205.

Ortea, J. 1986. The malacology of La Riera Cave. Arizona State University Anthropological Research Papers, 36: 289-298.

Ortiz de Zárate, V., A. Lavín, and X. Moreno–Ventas. 1997. Is there a relationship between environmental variables and the surface catch of albacore Thunnus alalunga (Bonnaterre, 1788) in the North Atlantic? SCRS/97/54.

OSPAR Commission. 2000. Quality Status Report 2000: Region IV: Bay of Biscay and Iberian Coast. OSPAR Commission, London. 134 + xiii p.

Parra, S. and E. López-Jamar. 1997. Cambios en el ciclo temporal de algunas especies infaunales como consecuencia del vertido del petrolero "Aegean Sea". Publicaciones Especiales Instituto Español de Oceanografía, 23: 71-82.

Pingree, R.D. 1979. Baroclinic eddies bordering the Celtic Sea in the late summer. Journal of the Marine Biological Association of the United Kingdom, 59: 689- 698.

Pingree, R.D. 1993. Flow of surface waters to the west of the British Isles and in the Bay of Biscay. Deep-Sea Research, 40(1/2): 369-388.

Pingree R.D. 1994. Winter warming in the Southern Bay of Biscay and Lagrangian eddy kinematics from a deep-drogued Argos Buoy. Journal of the Marine Biological Association of the United Kingdom, 74: 107-128.

Pingree, R.D. and B. Le Cann. 1990. Structure, strength and seasonality of the slope currents in the Bay of Biscay region. Journal of the Marine Biological Association of the United Kingdom, 70: 857-885.

Pingree, R.D. and B. Le Cann. 1992a. Three anticyclonic Slope Water Oceanic eDDIES (SWODDIES) in the southern Bay of Biscay in 1990. Deep-Sea Research, 39(7/8): 1147- 1175.

Pingree, R.D. and B. Le Cann. 1992b. Anticyclonic eddy X91 in the southern Bay of Biscay, May 1991 to February 1992. Journal of Geophysical Research, 97(C9): 14353-14367.

Porteiro C., J.M. Cabanas, L. Valdés, P. Carrera, C. Franco, and A. Lavín. 1996. Hydrodynamic features and dynamics of blue whiting, mackerel and horse mackerel in the Bay of Biscay, 1994-1996. A multidisciplinary study on Sefos. ICES CM. 1996/S: 13.

Quéro, J.C. and O. Cendrero. 1996. Incidence de la pêche sur la biodiversité ichthyologique marine: le bassin d'Arcachon et le plateau continental sud Gascogne. Cybium, 20(4): 323-356.

Quéro, J.C., J. Dardignac, and J.J. Vayne. 1989. Les poissons du golfe de Gascogne. IFREMER/Secrétariat de la Faune et de la Flore, 229 p.

Quéro, J.C., M.H. Du Buit, and J.J. Vayne. 1997. Les captures de poissons à affinités tropicales le long des côtes atlantiques européennes. Annales Societe des Sciences Naturelles de la Charente-Maritime, 8(6): 651-673.

Quéro, J.C., M.H. Du Buit, and J.J. Vayne. 1998. Les observations de poissons tropicaux et le réchauffement des eaux dans l'Atlantique européen, Oceanologica Acta, 21(2): 345-351.

Ríos, A., F.F. Pérez, and F. Fraga. 1992. Water masses in the upper and middle North Atlantic Ocean east of the Azores. Deep-Sea Research, 39: 645-658.

Robles, R., C. Porteiro, and J.M. Cabanas. 1992. The stock of Atlanto-Iberian sardine, possible causes of variability. ICES Marine Science Symposium, 195: 418-423.

Roemmich, D. and J. McGowan, J. 1995. Climatic warming and the decline of zooplankton in the California current. Science, 267: 1324-1326.

Roy, C. 1993. The optimal environmental window hypothesis: A non linear environmental process affecting recruitment success. ICES CM 1993/L: 76.

Sánchez, F. and J. Gil. 2000. Hydrographic mesoscale structures and Poleward Current as a determinant of hake (Merluccius merluccius) recruitment in southern Bay of Biscay. ICES Journal of Marine Science, 57:152-170.

Sánchez, F., F. de la Gándara, and R. Gancedo. 1995. Atlas de los peces demersales de Galicia y el Cantábrico. Otoño 1991-1993. Publ. Esp. Inst. Esp. de Oceanogr. 20: 99 p.

Sánchez, F., I. Olaso, and R. Goñi. 1997. Changes in bottom trawl survey catch rates in dogfish (Scyliorhinus canicula) and thornback ray (Raja clavata) in the Cantabrian Sea between 1983–1997. Working document for the 1997 WGECO meeting.

Santiago, J. 1997. The North Atlantic Oscillation and recruitment of temperate tunas. SCRS/97/40.

Saunders, P.M. 1982. Circulation in the Eastern North Atlantic. Journal of Marine Research, 40 (Supl.): 641- 657.

Sherman, K. and A.M. Duda. 1999. An ecosystem approach to global assessment and management of coastal waters. Marine Ecology Progress Series, 190:271-287.

Sournia, A., J.M. Brylinski, S. Dallot, P. Le Corre, M. Leveau, L. Prieur, and C. Froget. 1990. Fronts hydrologiques au large des côtes françaises: Les sites-ateliers du programme frontal. Oceanologica Acta, 13(4): 413-438.

Southward, A.J. and G.T. Boalch. 1994. The effect of changing climate on marine life: Past events and future predictions. Exeter Maritime Studies, 9: 101-143.

Southward, A.J., S.H. Hawkins, and M.T. Burrows. 1995. Seventy years' observations of changes in distribution and abundance of zooplankton and intertidal organisms in the western English Channel in relation to rising sea temperature. Journal of Thermal Biology, 20(1/2): 127-155.

Stocker, T.F. 1994. The variable ocean. Nature, 367: 221-222

Stouffer, R.J., S. Manabe, and K. Ya Vinnikov. 1994. Model assessment of the role of natural variability in recent global warming. Nature, 367: 634-636.

Taylor, A.H. 1996. North-South shifts of the Gulf Stream: Ocean-Atmosphere interactions in the North Atlantic. International Journal of Climatology, 16: 559-583.

Taylor, A.H. and J.A. Stephens. 1997. The north Atlantic Oscillation and the latitude of the Gulf Stream. Tellus, 48 A.

Turrell, W.R., R.R. Dickson, S. Narayanan, and A. Lavín. 1997. Ocean climate conditions in the ICES North Atlantic region 1996/1997. ICES CM 1997/EE: 9.

Uriarte, A., P. Alvarez, S. Iversen, D. Kennedy, M.M. Martins, J.M. Massó, J. Molloy, E. Mullins, S. Myklevoll, and B. Villamor. 1998. Spatial Pattern of Migration and Recruitment of Northeast Atlantic Mackerel. Report to the European Commission of Project 96-035, March 1998.

Valdés, L. and M. Moral. 1998. Time series analysis of copepod diversity and species richness in the southern Bay of Biscay (Santander, Spain) and their relationships with environmental conditions. ICES Journal of Marine Science, 55: 783-792.

Valencia, V. 1993. Estudio de la variación temporal de la hidrología y el plancton en la zona nerítica frente a San Sebastián entre 1988-1990. Informe Técnico Gobierno Vasco, 52, 105 p.

Varela, M., S. Barquero, B. Casas, M.T. Alvarez-Ossorio, and A. Bode. 1995. Are variations in salinity of north Atlantic central waters affecting plankton biomass and production in NW Spain? First JGOFS International Scientific Symposium. Villefranche-sur-Mer, France. Mayo, 8-12, 1995.

Varela, M., A. Bode, and N. González. 1998. Intrusiones de agua Nordatlántica de origen subtropical y fitoplancton durante la primavera en el Golfo de Vizcaya. 6° coloquio Internacional de Oceanografía del Golfo de Vizcaya. San Sebastián, Spain.

Villate, F., M. Moral, and V. Valencia. 1997. Mesozooplankton community indicates climate changes in a shelf area of the inner Bay of Biscay throughout 1988 to 1990. Journal of Plankton Research, 19(11): 1617-1636.

Whitehead, P.J.P., M.L. Bouchot, J.C. Hureau, J. Nielsen, and E. Tortonese. 1984. Fishes of the North-eastern Atlantic and the Mediterranean. Vols I, II, III. UNESCO, Paris. 1473 p.

Large Marine Ecosystems of the North Atlantic
K. Sherman and H.R. Skjoldal (Editors)
© 2002 Published by Elsevier Science B.V.

11

Iberian Sardine Fisheries: Trends and Crises

Tim Wyatt and Carmela Porteiro

FISHERIES GEOGRAPHY OF THE IBERIAN LME

The Iberian coastal ecosystem (LME 24) is relatively small in comparison with other such designated systems. It blends to the north with the Celtic-Biscay shelf (LME 23) across the Cap Breton Canyon, to the south with the Canary Current (LME 27) off northwest Africa, and with the Mediterranean Sea (LME 25) through the Straits of Gibraltar. The main geographical features are shown in Figure 11-1. They include: i) the north-facing Cantabrian shelf, which is quite narrow, mostly about 20 km wide but broader off Cabo de Peñas and Cabo Ortegal where it reaches a width of nearly 60 km; ii) the mostly west-facing shelf between Cabo Finisterre and Cabo de San Vicente, 30 to 50 km wide and incised by deep canyons; this coast trends mainly north to south but turns sharply eastwards at the important capes, Cabo de Roca and C. de Espichel; iii) the shelf of the Gulf of Cádiz which trends east from C. de San Vicente and gradually turns southeast towards the Straits of Gibraltar. Depths in the strait exceed 200 m so that the Spanish and Moroccan continental shelves are unconnected.

Meteorologically, this region is dominated between spring and autumn by the Azores High, which leads to a predominantly northerly airflow. The upper waters consist mainly of subpolar and subtropical branches of the Eastern North Atlantic Central Water (ENACW) which converge at the Galicia Front running NNW from Cape Finisterre (Fraga *et al.* 1982). Using sea surface temperature observations from merchant ships, Wooster *et al.* (1976) described the mean seasonal pattern of cold water anomalies from Cape Finisterre southwards, and noted that there was a good fit between this pattern and computed offshore Ekman transport values. They concluded that the temperature anomalies were due to upwelling driven by longshore wind stress, but noted that there was an unexplained phase difference between the wind stress and the temperature deficit, and that the cold anomalies extend farther offshore than expected on the basis of Ekman pumping. The cold plumes probably result from the upwelling associated barotropic instabilities, and there may be some secondary topographic trapping, especially off C. Roca and C. San Vicente where the plumes penetrate the ocean for 100 to 200 km (Coste *et al.* 1986). The general flow of the recently upwelled water is southwards near the shelf break as a summer coastal current. This flow reverses between late autumn and spring to form a poleward current (Frouin *et al.* 1990); these flow patterns have also been identified in the Cantabrian Sea and in the Gulf of Cádiz.

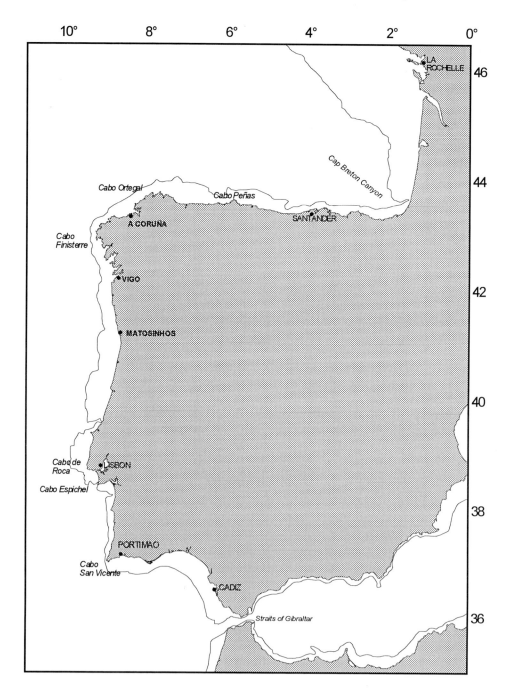

Figure 11-1. Main geographical features of the Iberian coastal ecosystem.

Seasonal patterns of phytoplankton pigment concentrations, based on Nimbus-7 satellite data, are described by Peliz and Fiúza (1999). Moderate to high concentrations occur in all coastal waters of the Iberian LME in winter, and extend well offshore beyond the shelf break; the highest offshore concentrations lie along the Galicia front, and these are present almost year round. During the upwelling season, pigment concentrations are lower than in winter and are confined to the inner shelf, except in the Gulf of Cádiz which is pigment-rich all year. The poleward current is very poor in pigments. These pigment patterns recur from year to year and appear to be strongly tied to the hydrodynamics (Peliz and Fiúza 1999).

IBERIAN SARDINE FISHERIES

Fish stocks collapse for two reasons, either because exploitation rates are too high, or because recruitment levels are inadequate to maintain the fishable stock. It is also recognized that overfishing can cause recruitment failure (Murphy 1966; Cushing 1971). To distinguish the relative importance of the natural and man-made components of change is a major challenge, especially now that many stocks are fully or over-exploited. For many fishery resources, overfishing is undoubtedly the major constraint at present on efficient economic exploitation, but if and when this problem is solved, the recruitment problem will remain, and, as several generations of biologists have emphasized, understanding recruitment processes must lie at the core of rational fisheries management.

It is generally agreed that the most significant event leading to variations in recruitment must be mortality during the early life of fish, during what is loosely called the larval phase, and that this mortality must be density dependent since fecundity is the result of natural selection. Density dependent survival depends on adequate feeding conditions, both to avoid starvation, and to escape size dependent predation. Sardine larvae found in the sea are therefore nearly always fit and healthy since those that fail to be so die of starvation or are eaten (Chícharo 1997, 1998). Climate *per se* is not the proximate factor which regulates density, but if changes in widely separated sardine stocks are synchronous, as Kawasaki and Omori (1988) and others have maintained, or are (in other species) precisely 180 degrees out of phase (Lluch-Belda *et al.* 1989) on these decadal time scales, then global-scale external forcing must be involved. Climate and tidal forcing are then the major candidates. Fecundity is high in many fish, but there are large variations in how high, and sardines are very much less fecund than cod for example, which may suggest that the former have better control over the density of their offspring than the latter.

Landings of pelagic fish like sardines undergo particularly high amplitude fluctuations in abundance, and there is evidence for large fluctuations in the abundance of some stocks even in the absence of exploitation (e.g., Soutar and Isaacs 1974; Holmgren-Urba and Baumgartner 1993). It has been established too that these fluctuations are accompanied by changes in other components of the ecosystems (Cushing 1984).

Generally, however, perceptions of the occurrence of fluctuations are based on fisheries records as distinct from real abundance estimates, and they may in principle therefore also be attributed to accessibility (as well as market forces). Without independent information, it is not possible to distinguish between the alternatives. Thus when we talk of the *abundance* of a stock, we should keep in mind that what may be meant is a combination of the abundance of the part of the stock available to the fishery, and a measure too of its economic value. In the case of the sardine stock to be examined here, there are no independent estimates of abundance prior to 1985, so that the perceived crisis of the 1940s and 1950s is based entirely on the market records.

Here we examine two sardine crises, particularly as recorded in the records of the Vigo fish market, and compare them with changes in sardine landings at other ports in the region. In earlier papers (Wyatt and Perez-Gándaras 1988, 1989), some of these records were compared with upwelling indices, and it was concluded that several distinct stocks contribute to the sardine fisheries of the region, and that each might respond independently to Ekman forcing. It was also suggested that changes in the geographical distribution of the different groups could partly account for the fluctuating fortunes of the fisheries in different parts of the region. A major redistribution of sardines (based on age classes) between the Cantabrian and western parts of the region is central to the hypothesis of Robles *et al.* (1992). Spring and winter spawning populations or races are also recognized in the central sector (Asturias) of the Cantabrian Sea (Villegas 1987), and Villegas and López-Areta (1988) note that multiannual trends in captures since the 1950s seem to occur independently in Asturias, Cantabria, and the Basque region.

NATURAL HISTORY OF SARDINES

Plankton surveys in European coastal waters have revealed that sardine eggs occur in high concentrations at or just beyond the edge of the continental shelf in the winter-spring months, and in lower numbers in summer-autumn at inshore locations. Such a pattern is described for the Bay of Biscay (Arbault and Lacroix 1977), and for the western Channel (Southward 1963, 1974; Demir and Southward 1974), where it is quite clear due to the breadth of the shelf. There are two peaks of sardine eggs farther south too, in March and November, off Matosinhos in Portugal (Ré 1981; Ré *et al.* 1990), in Ría de Vigo (Ferreiro and Labarta 1984), and off Santander (Oliver and Navarro 1952). But at these locations the shelf is narrow and it is less clear that the egg patches are produced at different distances from the coast. Sampling in the Cantabrian Sea (unpublished) between 1990 and 1995 has revealed that the autumn spawning peak in those years was weak in 1990 and absent for the next five years. It has been claimed that the autumn spawning fish are larger than those that spawn in spring (e.g., Oliver and Navarro 1952).

There may be two races of European sardine in much of its range in the Atlantic, with spring and autumn spawning seasons. In the Galician rías, juvenile sardines (*parrocha*) become the target of a fishery when about six months old. This fishery has two peaks in the year, in spring and autumn (Andreu 1969). The spring caught parrocha might then be equated with the autumn spawning and the autumn parrocha with the spring spawning fish. Southward (1963) came to the same conclusion for sardines in the Western Channel. The view that there are two (or more) kinds of sardine in Galician waters has wide currency among fishermen, who are recorded as believing that one group spawns in the open coastal waters and the other within the rías (Bernárdez 1926). Fishermen along the French Atlantic coast traditionally distinguished *rogue* sardine which approached from the west in spring or summer, successively later at more northerly fishing stations, and never found in spawning condition, from *derive* sardines which could be fished in all months of the year, and which were found to be ripe at a certain season (Wyatt 1983). Thus spring and autumn spawning fish are spatially segregated during their reproductive seasons. It is probable that the larvae that result from the inshore and offshore spawning areas remain segregated during the drift phase, as has been demonstrated for herring larvae in the North Sea (Heath 1990). The nursery grounds of sardines have not been discovered on the Atlantic coasts of either Portugal or Spain, but in the Mediterranean are close inshore near sandy beaches, where juvenile sardine were commonly caught in beach seine fisheries.

The existence of sympatric (or parapatric) spring and autumn spawning sardines in the western Channel, Galicia, and elsewhere is additional to the classical racial separations of sardines based on morphological criteria (see e.g. Furnestin 1945; Larrañeta 1968). Both sympatric and allopatric evolution of the European sardine may have resulted from the combination of temperature changes and rising sea levels that followed the end of the last Ice Age (Wyatt *et al.* 1991). As in herring, the variability of morphological and genetic patterns may be uncoupled. Genetic differentiation between inshore and offshore marine populations in the Atlantic has been described too in bottlenose dolphins (Hoelzel *et al.* 1988) and in petrels as well as in fish.

Long-term records of sardine egg abundance are only available for the Plymouth area, but may be taken to indicate the kinds of change that can be expected throughout the range of the species. Records for the Plymouth area begin in 1924. The general trends are summarized by Southward (1974). From 1924 to 1934, spring-summer spawned eggs (April to July) were rather scarce, but from 1935 to 1961 they became very abundant. Beginning about 1950, a second peak in spawning intensity was discernible from September to November. The quantities of eggs produced in the two spawning seasons appear to vary independently.

SARDINE FLUCTUATIONS IN IBERIAN WATERS

Fluctuations in the abundance of sardines have provoked speculations concerning their causes for at least two hundred years in Galicia, just as in other coastal regions of Europe where this fish has formed an essential component of the economy and culture of the local inhabitants. The same is true for other species such as the herring and the cod in more northerly regions. Much of this effort may have been barren, but without it and the publications to which it has given rise, we would know less about the long-term history of these fluctuations. This history is not only of interest in itself, but must eventually comprise an essential component of any durable understanding of the causes of these changes in abundance. Fishing communities have always reckoned with changes in catches from day to day and from year to year, but the changes are sometimes so dramatic that in the cases when there are catastrophic declines in catch over very short periods of time, one or a few years and accompanied by severe economic disruption, the word "crisis" or some equivalent is employed to describe them and to distinguish them from the lower amplitude fluctuations which the experiences of fishermen classify as normal.

Excluding the present crisis, five periods in which sardines catches were *perceived* to be notably poor in Galician waters have been recorded in the last 150 years, although the earlier ones are poorly documented. These were from 1876 to 1895, 1924 (or perhaps earlier) to 1925, 1941 to 1942, 1946 to 1957, and since the late 1980s. We say *perceived* because one could argue that, with the exception of a few years, the Galician sardine fishery has been in an almost permanent state of crisis since the early 1940s. There were of course variations in catches at other times, but in the years listed here sardines were taken in such small quantities as to generate widespread concern and economic hardship. These periods range from two to nearly twenty years in duration, which suggests that a single generic cause may not be adequate to account for all of them. A puzzling feature of the crisis which began in the 1940s was that catches were low for only two years at Matosinhos in Portugal, about 100 km to the south of Vigo (Figure 11-2).

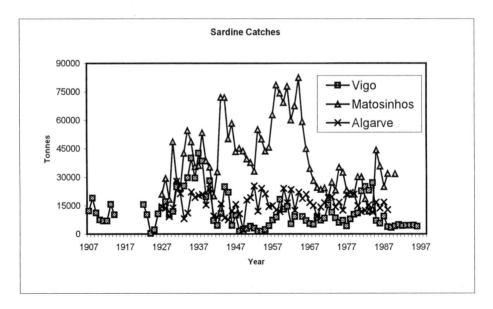

Figure 11-2. Sardine catches, 1907-1997.

Thus the spatial scale on which we need to examine these trends to understand the underlying ecology are more fine-grained than those which are used for management purposes.

The sardine crisis of 1942 to 1957

Sardine landings in Vigo and neighbouring Galician ports reached their highest levels of this century in the 1930s. The bimonthly trade journal *Industrias Pesqueras* provides a detailed view of the fortunes of this fishery. In the 1930s, editorials routinely described Vigo as the premier sardine port of the world (although catches were generally higher in Matosinhos!), referred to the fish as "our bread, our love, our prosperity" (15th January, 1938). As the fishery failed, and year succeeded year without any signs that the sardine would return, the phrase "another year without sardine" was repeated like a refrain.

We should bear in mind that while the sardine has probably changed its habits very little since the 1930s, fishing practices have evolved rapidly. In the 1940s and 1950s, the Portuguese fleet was technologically more advanced than the Galician, and had more powerful vessels equipped with echosounders and radiotelephones. After three years' experience with echolocation, the Portuguese abandoned night fishing and dispensed with the traditional reliance on surface manifestations of the shoals during the darks of the moon (*ardora*). The number of so-called American seines employed, which had been the main nets used in the 1930s, gradually fell, and they had disappeared by the end of the 1950s. This trend was compensated by a rapid increase in the number of *traineros* until 1955, and these were maintained until 1970.

Previously, it had been traditional to pay out the nets as Venus was setting. At this moment the sardines (and anchovies) were thought to be provoked to change the direction of their movements (*motivo*). The shoals were reckoned to enter the rías *en masse* after San Xoan, and especially in early September (*Golpes de Santos*) and in late December (*Golpes de Nadal*). Most of these signs had their equivalents among herring fishermen in the North Sea and elsewhere. Much of this tradition is now forgotten, and indeed has never been transmitted to the younger generations of fishermen who rely much more heavily on technological developments. Thus in the 1950s in Portugal, and somewhat later in Galicia, a fishing culture which may have had its roots in the Palaeolithic started to die.

HYPOTHESES TO EXPLAIN SARDINE CRISES IN GALICIA

Overfishing

Overexploitation was widely agreed by industrialists at the time to be a major reason for the collapse of the Galician sardine fishery in 1947 (various articles in *Industrias Pesqueras*). The scientific community did not agree. Sardine fishing at that time was almost entirely limited to the interiors of the rías, where the presence of sardines was regarded as casual. The main concentrations of fish were known to remain outside the rías. Andreu (1956) argued that fishing methods were quite primitive, and that the reduction of the fleet and the regulations against winter fishing imposed after 1953 did not lead to any increases in yields. This view was supported by Anadón (1954), who failed to find any relation between the abundance of juveniles and adults, but did establish that there had been a real reduction in abundance. The idea that the crisis could be overcome by protecting the sardine during the spawning season was also discounted, since spawning was thought to continue for eight or nine months of the year (Andreu 1955), while landings had in any case always been very low during the time of maximum sexual activity. The spawning areas had not then been identified, but spawning was known not to take place on the traditional fishing grounds. These arguments were summarized by Margalef and Andreu (1958), who commented that the precocity and high fecundity of the sardine were sufficient guarantees for adequate reproduction.

Changes in distribution

The very marked contrast between the Galician crisis and the continued abundance of sardines in the Matosinhos fishery little more than 100 km to the south suggested that there had been a shift in the distribution of the main concentrations of fish. Meristic studies, however, indicated that the Spanish and Portuguese fish were probably racially distinct (Andreu 1955), and the fact that landings peaked at the same time in both Matosinhos and Vigo was also used to argue against a displacement over this distance. Paradoxically, Andreu seems to have been unaware that sardine spawned in Galician waters, and followed de Buen (1928) in supposing that all sardines in the sector between the Miño and the French border derived from spawning areas in the Cantabrian Sea or off Setúbal in central Portugal.

Changes in reproductive success

Data for the years 1925 to 1952 from Marín (Ría de Pontevedra) analysed by Anadón (1954) revealed decadal scale changes in the relative contributions to the catches of spring and autumn progeny, principally due to a long-term decline in the abundance of juveniles derived from the spring spawning. Anadón's arguments were extremely astute, and although by and large they have not been dismissed to any extent, neither have they been considered in detail in more recent efforts to interpret sardine fluctuations. Anadón wished to calculate mortality rates of the Galician sardine from the data available, and to do so it was necessary to make several assumptions. These were a) that the population was either stationary, or that its movements were local and regular; b) that fish of different ages were not spatially segregated in the target stock; c) that fishing effort impinged equally on all age classes, and d) that fishing did not significantly affect mortality. He discussed each point in turn.

At the time Anadón wrote, it was fashionable to estimate mean vertebral numbers in fish to define racial distinctions between stocks from different areas. It was already recognized that the mean vertebral count of a fish contained signals of both the genetic origin of the individual and of the environmental conditions, mainly temperature, during a critical phase during early morphogenesis. Ruivo (1950) had implied that the mean vertebral number of a group of fish could be used to determine the geographical origin of the sample, and Anadón made this hypothesis explicit. He argued that if two or more generations of sardine appeared successively in the same region, it would be very likely that each group would have a distinct mean vertebral count, and that they would therefore appear to be mixtures of fish of distinct origin. Literature data indicated that this might be so for various local sardine fisheries on the Atlantic coasts of France and Spain and in the Mediterranean.

He then suggested that if mean vertebral numbers remained constant at a given locality over periods of several years, and were different from the means at adjacent localities, the populations must be relatively stationary. He cited data that indicated that the sardines in northern Portugal were distinct from those of Vigo, and that those of Vigo, Ferrol, and Gijón were all different from each other. Anadón therefore concluded that the sardine has different local spawning areas in each coastal zone, that the annual migrations must be short, and that the routes of these migrations could not be parallel to the coasts of Portugal and Galicia.

The second and third assumptions, that age classes are neither spatially separated nor differentially captured, were based on the fishermen's experience a) that sardines are never taken in depths beyond the 100 m isobath (and in Galicia this means never further from the coast than 2 to 6.5 miles), even though the same nets were used in greater depths to capture anchovy, b) the unselective nature of the gear (*cerco*, which reached depths of 85-100 m, and hence could reach all age classes even if they were vertically segregated), and c) the absence of sardines from the stomachs of albacore and bonito which reach the outer shelf but are not caught inshore.

The landing data at Marín available to Anadón (1954) were recorded as either juveniles (*parrocha*) or adults. The exact size of *parrocha* is a little uncertain, but usually about 9 to 15 cm in Galicia, so includes fish from 6 to 12 months old. The numbers of these young fish reach maximum abundance (*arribazones*) in the landings twice each year, in spring and autumn. Knowledge of the growth rate of the Galician sardine and the time of spawning allowed Anadón to relate these peaks to eggs produced in autumn and spring respectively. It was believed then that spawning occurred continuously from spring to autumn, and Anadón therefore presumed that the two peaks were due to the reduced capture rates of the winter months (as a result of the fish being more dispersed), and not to the possibility that there were two peaks in spawning.

A closer examination of the Marín data (his Table II and Figure 4) may suggest that the two peaks identified by Anadón (1954) were identified by intuition rather than rigour. The autumn parrocha peak is a strong feature, but the spring peak is quite ephemeral and only clearly recognizable in about seven years of the 28-year series; also in some years the two peaks cannot easily be distinguished (e.g. 1931, 1932). Furthermore, the timing of parrocha maxima is quite variable. Much of this variability can be attributed to the fact that parrocha varies in length from 9 to 15 cm and thus in age over several months.

We can identify the following general trends in the Marín time series. From 1924 to 1928 parrocha landings were very small, perhaps indicating that recruitment failed almost completely in those years. From 1929 to 1947, nineteen years altogether, the autumn parrocha fishery was generally good in most years (but poor in 1935, 1938, 1940, and 1946), and in three years exceptionally good (1944, 1945, 1947); it is during this period that the spring peak can be identified. Finally, from 1948 to 1952 there was a return to the very poor parrocha fishery of the 1920s.

Changes in oceanographic conditions

A climatic hypothesis was favoured by Margalef (1956), who associated the sardine crisis with changes in the phytoplankton composition and with changes in patterns of water exchange between the rías and the open sea. He linked good sardine fishing with presence in the plankton of the diatom *Melosira (Paralia) sulcata*, and poor fishing with *Thalassiosira rotula* "invasions". *M. sulcata* is predominantly benthic, but is found in the water column either during relatively intense upwelling or during strong vertical mixing in winter, while *T. rotula* is a planktonic species associated with upwelling pulses in spring. Margalef proposed that during years favorable to the fishery, i.e., when the sardine shoals were abundant inside the rías, the surface water flow into the rías was frequent and provided them easy access, especially in autumn. These conditions he thought would favour the entry of the fish and hence their accessibility to the fishermen. During unfavorable years, the flow of surface water into the rías would be reduced, and the entry of sardines discouraged. These good and bad years were associated respectively with high and low rainfall, particularly in autumn and winter. Thus Margalef did not at that time explicitly invoke reduced productivity as the ultimate cause of the sardine crisis, nor suggest that recruitment variability was critical.

THE PRESENT SARDINE CRISIS

Stock evaluations since 1976 have shown that there has been a reduction in spawning stock biomass since the mid 1980s. Also, surveys in the Cantabrian region indicate that spawning there has been very low. During the 1980s, the fishery was mainly sustained by the '83 and '87 year classes. The moderate '91 year class was the last and the fishery is now very poor.

Data from the present crisis has provoked another explanation for the changing yields based on analogies with the Norwegian spring spawning herring and the Pacific sardine.

In this view, reductions in the size of the stock lead to behavioral changes; some parts of the traditional feeding and spawning areas are abandoned, and the fish then confine themselves to reduced areas closer to the coast, in what can be called core areas or refuges. For the Iberian sardine, the main core area is the northern Portuguese shelf, and when the stock is concentrated there, the spawning areas in the Cantabrian Sea are not visited. Changes in the upwelling regime drive this model as in the hypotheses already mentioned. An increase in the strength and frequency of the northerly wind component during summer is thought to favour larval dispersal, which transports them away from the recruitment areas and increases mortality. It is not possible with data currently available to distinguish between this model and one based on the varying success of independent stocklets, but a crucial component of this view is the extent to which the shoals migrate.

OTHER CHANGES IN THE IBERIAN LME

Two other notable changes have taken place in Iberian seas over the decades considered here. Neither of them is linked in any obvious way with fluctuations in the sardine fishery. The first is a rather marked change in the abundance of certain dinoflagellate species, and the second the expansion from small beginnings in the 1950s of a blue mussel farming industry. From about the beginning of this century until the 1950s, red tides were a more or less annual occurrence in the Rías Bajas; these were almost always due to the dinoflagellate *Gonyaulax*, and sometimes the ciliate *Mesodinium*. But since the 1970s, *Gonyaulax* has not been reported to form blooms in the Rías Bajas, and instead there have been occasional blooms of toxic dinoflagellates, *Alexandrium tamarense* and *A. minutum*, and *Gymnodinium catenatum*. It is not clear whether a similar trend has taken place in Portuguese waters. These changes in the phytoplankton of the region have been seen as part of a worldwide increase in the frequency and intensity of harmful algal blooms, called the "global spreading hypothesis" (Wyatt 1995), and attributed to various causes including eutrophication and ballast water transport.

Species of *Alexandrium* are almost certainly autochthonous in these waters, and rare toxic events (PSP) which are probably attributable to them are known since at least 1946 in Portugal. The vectors to humans are shellfish. We are not aware that any systematic search of historical medical records for PSP has been made in either Portugal or Spain, but by 1976 when the first major toxic event in Spain took place, mussel cultivation on rafts had already been practiced for about twenty years. It is therefore possible that there has been a genuine increase in the frequency or magnitude of toxic *Alexandrium* outbreaks. It is possible too that *Gymnodinium catenatum* has been introduced to this region, but it too may have been a cryptic species prior to its becoming a pest. Suggestions for source range from Morocco to Argentina (Wyatt 1992), but the taxonomic status is not yet fully resolved. There may have been other changes in the phytoplankton composition which have escaped attention since they are not of significance to human health. Unfortunately, there have been no follow-up studies in the region of the diatoms which Margalef used as indicator species.

The second major change in the Rías Bajas is the introduction of raft cultivation of blue mussels, as just mentioned. This industry started slowly in the 1950s, and now produces about 200,000 tons annually. It is thought that in the main raft area, Ría de Arosa, the standing stock of mussels is near or above the ideal carrying capacity, and that mussels alone could consume the entire annual production of phytoplankton. Cabanas *et al.* (1979) provided budgetary estimates for a mussel raft in Ría de Arosa over a 20-day period. Samples upstream and downstream of the raft indicated that about 60% of the chlorophyll in the water column was removed during passage through the mussel ropes. The daily "fallout" of organic matter from the raft amounted to 190 kg of dry sediment daily. This last number, multiplied by the number of rafts in Arosa, amounts to around 175,000 tons annually and has led to speculation that it may lead to changes in the species diversity of the ría. Whether or not the growth of mussel farming in the rías is linked to the changes in the phytoplankton already mentioned remains an open question.

DISCUSSION

It is tempting to suppose that fluctuations in sardine abundance are ecologically equivalent to those of other pelagic species such as herring, but information on the age structure of the two species indicates that parallels are not exact. For example, in the case of the Atlanto-Scandian herring, it is well known that the success of the fishery over many decades has depended on a very small number of large year classes. This possibility is due of course to the longevity of this herring. The shorter-lived European sardine needs a higher frequency of years with good recruitment if it is to support a successful fishery. In the northern part of the sardine's range, in Cornwall and Ireland, the fishery has always been seasonal and intermittent, and declines in catches have nearly always been attributed to movements of the shoals beyond the range of the vessels. But in Spain and Portugal, the sardine fishery is prosecuted throughout the year, weather permitting. Nor is it clear to what extent the Pacific sardine provides a model for the European sardine, since those stocks seem to have distinct patterns of genetic differentiation from the European sardine (Hedgecock 1991).

Current views on the regulation of fish recruitment in eastern boundary current ecosystems are based on the studies of Lasker and his colleagues (e.g. Lasker 1975, 1978), Bakun (1985), Cury and Roy (1989), and others. Bakun (1996) provides a synthesis of these ideas, and builds on them a model that he calls *ocean triads*. The essential elements of the triad are enrichment due to upwelling, concentration of the larval food brought about by convergences, and mesoscale circulation patterns that help to maintain larval retention. When the temporal sequence of these sometimes conflicting oceanographic processes is optimal, larval survival and subsequent recruitment are maximized. It has generally been taken for granted that within limits, increased upwelling rates raise productivity, and the extra planktonic food becomes available to the feeding larvae. But in Lasker's hypothesis and in the ocean triad model, this only happens if the plankton is subsequently concentrated at fronts or other convergences. Larval feeding success can therefore vary even though productivity does not. But fluctuations in upwelling rates do lead to changes in primary production, and to changes in species composition and phytoplankton size distribution (Wyatt 1980), so that there are four extreme possibilities determined by the oceanography, as follows:

productivity	convergence	recruitment
high	effective	good
high	ineffective	bad
low	effective	moderate
low	ineffective	bad

Within upwelling systems, suitable spawning sites are near the areas where upwelling is strongest but not coincident with them; this leads to a decrease in dispersive loss rates (Parrish *et al.* 1981). Shelter from the more dynamic divergence of core upwelling areas can be provided, for example, by capes (Roy 1997) or topographically trapped eddies (Navarro-Perez and Barton 1997). The distribution of sardine eggs and larvae in Galician waters seems to indicate that capes are especially important, and we can assume that this is generally true (Bakun 1996). We can surmise that topographic features and circulation patterns constrain the natural history of each putative stocklet, and thus the stock structure of the region as a whole. The same principle has been established for the herring in the North Sea (Heath 1990).

Fish are normally faithful to their spawning sites, and thus in the absence of vagrancy (i.e., failure to return to the natal site at the spawning season) must be reproductively isolated from fish at other spawning sites. Even if vagrancy occurs, genetic exchange may be reduced or absent due to variations in the times at which the different groups reach maturity. Existence of this pattern of local demes means that the variability of a stock as a whole is damped. Small climate shifts that lead to negative trends in one deme can be balanced by positive trends in another deme. Then crises generated by the influence of larger climatic shifts might cause lowered recruitment to all or most demes simultaneously. The French sardine crises of the late 19th and early 20th centuries provide an example of the latter. Modellers search for simplicity by looking for transcendent principals, and strip away the messy complexity that obscures them, perhaps stigmatizing it as noise. In some fisheries management, this process of simplification has perhaps been pursued too far. If the smaller scales, which may be the correct ones within which to interpret sardine population dynamics, are not resolved, then the larger scale "maps" which we use for management will contain large errors.

Sherman (1998) groups the sources of variability in LMEs under three categories, climate shifts, exploitation, and pollution. The third is not of major importance in the context of the Iberian LME, except in a few local regions. Climate, and in this case we think mainly in terms of upwelling, has certainly played the primary role in the regulation of productivity and resource abundance historically. But the impact of exploitation is now challenging climate as a significant source of variability in this region as in so many other parts of the ocean, and some stocks are depressed below the levels at which yields can be optimized.

It is well known that major changes in fish abundance are accompanied by equally radical changes in other trophic levels. We mentioned above that Margalef identified marked changes in the phytoplankton composition in Galician waters during the 1950s sardine crisis there, and that he associated *Thalassiosira rotula* abundance with poor yields. There are no reports to indicate whether this diatom became less abundant when the fishery recovered after 1957, but there have been significant changes in phytoplankton composition in recent decades.

As a test-bed for exploring and developing the ideas of *ecosystem management* (Christensen *et al.* 1996), or for the more ambitious project of *ecosystem rebuilding* (Pitcher and Pauly 1998), the Iberian LME offers several advantages. It is one of the smallest LMEs of the fifty or so that have so far been designated, sufficiently small that the entire system can be surveyed with the national resources already available on an annual or better frequency. It is bordered by only two nations, Spain and Portugal, which already collaborate effectively in various fisheries contexts. Both are members of the European Union, and can therefore lay claim to substantial resources in addition to those available nationally. Thus the geographical, political and financial barriers to acquiring the background information necessary to convert ecosystem management from a paradigm to a practical affair are less severe here than in many other LMEs. Spain, which has the largest distant water fleet of any European country, is also gradually being excluded from several of its traditional extraterritorial fisheries, and needs to turn more attention to exploiting local resources in a more rational way, with long-term sustainability as a priority goal, in the spirit of Agenda 21 (Earth Summit).

REFERENCES

Anadón, E. 1954. Estudios sobre la sardina del noroeste español. Publicaciones del Instituto de Biología Aplicada, Barcelona, 18: 43-106.

Andreu, B. 1955. Consideraciones sobre la sardina gallega. II Reunion de Productos y Pesquerias, Barcelona. 65-69.

Andreu, B. 1956. La crisis de sardina en Galicia y sus perspectivas. Industrias Conserveras 20: 40-44.

Andreu, B. 1969. Las branchispinas en la caracterización de las poblaciones de sardina, *Sardina pilchardus* (Walb.). Investigación Pesquera 33: 425-607.

Arbault, S. and N. Lacroix. 1977. Oeufs et larves de Clupeides et Engraulides dans le Golfe de Gascogne (1969-1973). Distribution des frayeres. Revue de Travaux de l'Institute de Pêches Maritimes 41: 227-254.

Bakun, A. 1985. Comparative studies and the recruitment problem: searching for generalizations. California Cooperative Fisheries Investigation Report 26:30-40.

Bakun, A. 1996. Patterns in the Ocean. Ocean Processes and Marine Population Dynamics. University of California Sea Grant Program (La Paz, Mexico). 323 p.

Bernárdez, A. 1926. La Pesca en Galicia. In: F. Carreras y Candi (de.). Geografía General del Reino de Galicia (Albert Martín, Barcelona). 533-550.

Buen, F. de. 1928. Fluctuaciones en la sardina, *Sardina pilchardus* (Walb.). Instituto Español de Oceanografía, Notas y Resumenes 2: 80 p.

Cabanas, J.M., J.J. Gonzalez, J. Mariño, A. Perez, and G. Roman. 1979. Estudio del mejillón y de su epifauna. III. Observaciones previas sobre la retención de partículas y la biodeposición de una batea. Boletin del Instituto Español de Oceanografía 5: 43-50.

Chícharo, M.A. 1997. Variation of field-caught *Sardina pilchardus* larvae starvation with length. Scientia Marina 61: 137-150.

Chícharo, M.A. 1998. Nutritional condition and starvation in *Sardina pilchardus* larvae off southern Portugal compared with some environmental factors. Journal of Experimental Marine Biology and Ecology 225: 123-137.

Christensen, N.L. and 12 co-authors. 1996. The Report of the Ecological Society of America Committee on the scientific basis for ecosystem management. Ecological Applications 6: 665-691.

Coste, B., A.F.G. Fiúza, and H.J. Minas. 1986. Conditions hydrologiques et chimiques associées à l'upwelling côtier du Portugal en fin d'été. Oceanologica Acta 9: 149-158.

Cury, P. and C. Roy. 1989. Optimal environmental window and pelagic fish recruitment success in upwelling areas. Canadian Journal of Fisheries and Aquatic Sciences 46: 670-680.

Cushing, D.H. 1971. The dependence of recruitment on parent stock in different groups of fish. Journal du Conseil 33: 340-362.

Cushing, D.H. 1984. Climate and Fisheries. Academic Press. 373 p.

Demir, N. and A.J. Southward. 1974. The abundance and distribution of eggs and larvae of teleost fishes off Plymouth in 1969 and 1970. 3. eggs of pilchard (*Sardina pilchardus* Walbaum) and sprat (*Sprattus sprattus* (L.)). Journal of the Marine Biological Association of the United Kingdom 54: 333-353.

Ferreiro, M.J. and U. Labarta. 1984. Spawning areas and seasons of three clupeid species (*Sardina pilchardus, Sprattus sprattus* and *Engraulis encrasicholus*) in the ría de Vigo. Cybium 8: 79-96.

Fraga, F., C. Mouriño, and M. Manríquez. 1982. Las masas de água en la costa de Galicia, junio-octubre. Resultados Expediciones Científicos 10: 51-77.

Frouin, R., A.F.G. Fiúza, I. Ambar, and T.J. Boyd. 1990. Observations of a poleward surface current off the coasts of Portugal and Spain during winter. Journal of Geophysical Research 95: 679-691.

Furnestin, J. 1945. Contribution à l'étude biologique de la sardine (*Sardina pilchardus* Walbaum). Office Scientifique et Technique, Pêches Maritimes 13: 221-386.

Heath, M. 1990. Segregation of herring larvae from inshore and offshore spawning grounds in the north-western North Sea-implications for stock structure. Netherlands Journal of Sea Research 25: 267-278.

Hedgecock, D. 1991. Contrasting population genetic structures of pelagic clupeoids in the California Current. In: T. Kawasaki, S. Tanaka, Y. Toba, and A. Taniguchi (eds.). Long-term Variability of Pelagic Fish Populations and their Environment. Pergamon. 199-207.

Hoelzel, A.R., C.W. Potter, and P.B. Best. 1988. Genetic differentiation between parapatric 'nearshore' and 'offshore' populations of the bottlenose dolphin. Proceedings of the Royal Society of London B, 265: 1177-1183.

Holmgren-Urba, D. and T.R. Baumgartner. 1993. A 250-year history of pelagic fish abundances from the anaerobic sediments of the central Gulf of California. CalCOFI Report 34: 60-68.

Kawasaki, T. and M. Omori. 1988. Fluctuations in the three major sardine stocks in the Pacific and the global trend in mean temperature. In: T. Wyatt and M.G. Larrañeta (eds.). Long Term Changes in Marine Fish Populations. Real, Baiona, Spain. 273-290.

Larrañeta, M.G. 1968. Unités de stock de la sardine de la Méditerranée occidentale et de l'Adriatique. Études et Revues. FAO. 33. 54 p.

Lasker, R. 1975. Field criteria for survival of anchovy larvae: the relation between inshore chlorophyll maximum layers and successful first feeding. Fisheries Bulletin, U.S. 73: 453-462.

Lasker, R. 1978. The relation between oceanographic conditions and larval anchovy food in the California Current: identification of the factors leading to recruitment failure. Rapports et Procès-Verbaux de Réunions, Consil International pour l'Exploration de la Mer 173: 212-230.

Lluch-Belda, D., R.J.M. Crawford, T. Kawasaki, A.D. MacCall, R.H. Parrish, R.A. Schwartzlose, and P.E. Smith. 1989. World-wide fluctuations of sardine and anchovy stocks: the regime problem. South African Journal of Marine Science 8: 195-205.

Margalef, R. 1956. Paleoecología postglaciar de la ría de Vigo. Investigaciones Pesqueras 5: 89-112.

Margalef, R. and B. Andreu. 1958. Componente vertical de los movimientos del agua en la ría de Vigo y su posible relación con la entrada de sardina. Investigaciones Pesqueras 11: 105-126.

Murphy, G.L. 1966. Population biology of the Pacific sardine (*Sardinops* caerulea). Proceedings of the California Academy of Sciences 34: 1-84.

Navarro-Perez, E. and D. Barton. 1997. The physical structure of an upwelling filament off the north-west African coast. South African Journal of Marine Science 19: 61-73.

Oliver, M. and F. de P. Navarro. 1952. Nuevos datos sobre la sardina de Vigo (febrero de 1950 a marzo de 1952). Boletin del Instituto Español de Oceanografia 56: 26-39.

Parrish, R.H., C.S. Nelson, and A. Bakun. 1981. Transport mechanisms and reproductive success of fishes in the California Current. Biological Oceanography 1: 175-203.

Peliz, A.J. and A.F.G. Fiúza. 1999. Temporal and spatial variability of CZCS-derived phytoplankton pigment concentrations off the western Iberian peninsula. International Journal of Remote Sensing 20: 1363-1403.

Pitcher, T.J. and D. Pauly. 1998. Rebuilding ecosystems, not sustainability, as the proper goal of fishery management. In: T.J. Pitcher, P.J.B. Hart, and D. Pauly (eds.). Reinventing Fisheries Management. Kluwer, London. 311-329.

Ré, P. 1981. Seasonal occurrence, mortality and dimensions of sardine eggs (*Sardina pichardus* Walbaum) off Portugal. Cybium 3, Séries 5 (4): 41-48.

Ré, P., R. Cabral e Silva, E. Cunha, A. Farinha, I. Meneses, and T. Moita. 1990. Sardine spawning off Portugal. Boletin, Instituto Nacional de Investigação das Pescas (Lisboa) 15: 31-44.

Robles, R., C. Porteiro, and J.M. Cabanas. 1992. The stock of Atlanto-Iberian sardine: possible causes of variability. ICES Marine Science Symposia 195: 418-423.

Roy, C. 1997. Upwelling-induced retention areas: a mechanism to link upwelling and retention processes. South African Journal of Marine Science 19: 89-98.

Ruivo, M. 1950. Sobre as populaçoes da sardinha (*Clupea pilchardus* Walb.) da costa portuguesa. Bol. Soc. Port. de C. Nat., III, 18: 80-121.

Sherman, K. 1998. Large marine ecosystems: assessment and management from drainage basin to ocean. With Rivers to the Sea: interaction of land activities, freshwaters and enclosed coastal seas. Proceedings of Joint Stockholm Water Symposium/EMCS. Arkpressen, Stockholm. 35 p.

Soutar, A. and J.D. Isaacs. 1974. Abundance of pelagic fish during the 19th and 20th centuries as recorded in anaerobic sediment off the Californias. Fisheries Bulletin, U.S., 72: 257-273.

Southward, A.J. 1963. The distribution of some plankton animals in the English Channel and approaches. III. Theories about long-term biological changes, including fish. Journal of the Marine Biological Association of the United Kingdom 43: 1-29.

Southward, A.J. 1974. Long term changes in abundance of eggs of the Cornish pilchard (*Sardina pilchardus* Walbaum) off Plymouth. Journal of the Marine Biological Association of the United Kingdom 54: 641-649.

Villegas, M.L. 1987. Aspectos reproductivos de los peces en la zona costera asturiana (N. de España) con especial referencia a la especie *Sardina pilchardus* (W., 1792). Boletin del Instituto Español de Oceanografia 4: 121-132.

Villegas, M.L. and J.M. López-Areta. 1988. Cambios anuales y estacionales en las capturas de *Sardina pilchardus, Trachurus trachurus, Engraulis encrasicolus,* y *Scomber scombrus* en las costas asturianas (1952-1985). In: T. Wyatt and M.G. Larrañeta (eds.). Long Term Changes in Marine Fish Populations. Real, Baiona, Spain. 301-319.

Wooster, W.S., A. Bakun, and D.R. McLain. 1976. The seasonal upwelling cycle along the eastern boundary of the North Atlantic. Journal of Marine Research, 34: 131-141.

Wyatt, T. 1980. The growth season in the sea. Journal of Plankton Research 2: 81-96.

Wyatt, T. 1983. Fluctuations in North Atlantic fish populations. Oceanologica Acta, Volume Special: 219-222.

Wyatt, T. 1992. *Gymnodinium catenatum* in Europe. Harmful Algae News, IOC, 2: 4-5.
Wyatt, T. 1995. Global spreading, time series, models and monitoring. In: P. Lassus, G. Arzul, E. Erard-le Denn, P. Gentien, and C. Marcaillou-le Baut, (eds.). Harmful Marine Algal Blooms. Lavoisier, Paris. 755-764.
Wyatt, T. and G. Perez-Gándaras. 1988. Ekman transport and sardine yields in western Iberia. In: T. Wyatt and M.G. Larrañeta (eds.). Long Term Changes in Marine Fish Populations (Real, Baiona, Spain). 125-138.
Wyatt, T. and G. Perez-Gándaras. 1989. Biomass changes in the Iberian ecosystem. In: K. Sherman and L.M. Alexander (eds.). Biomass Yields and Geography of Large Marine Ecosystems (AAAS Selected Symposium Series, 111). 221-239.
Wyatt, T., D.H. Cushing, and S. Junquera. 1991. Stock distinctions and evolution of European sardine. In: T. Kawasaki, S. Tanaka, Y. Toba, and A. Taniguchi (eds.). Long-term Variability of Pelagic Fish Populations and their Environment. Pergamon. 229-238.

Large Marine Ecosystems of the North Atlantic
K. Sherman and H.R. Skjoldal (Editors)
© 2002 Published by Elsevier Science B.V.

12

The North Sea Large Marine Ecosystem

Jacqueline M. McGlade

INTRODUCTION

The North Sea Large Marine Ecosystem (LME) is situated on the continental shelf of northwestern Europe, bounded by the coastlines of England, Scotland, Norway, Sweden, Denmark, Germany, the Netherlands, Belgium, and France (Figure 12-1). To the south it is delimited by the English Channel (48°30'N 5°W), to the north by a line between Scotland and Norway (62°N), and to the east by the Kattegat between Sweden and Denmark. It is a large semi-enclosed continental sea with a surface area of 750,000km², a volume of 94,000 km³ and a mean depth of 90m. Although a number of classifications for the North Sea have been developed, the dominant physical division is between the north and south. The northern part is comparatively deep (50-200m, with the Norwegian Trench dropping off to 700m) and is subject to strong oceanic influences, significant atmospheric deposition, and modest inputs of land-based waste. The southern part is shallower (20-50m) and is subject to oceanic inputs via the English Channel; it has strong tidal currents, high levels of sediment load and notable land-based waste inputs.

The North Sea has a long geological history. Around 350 million years ago (mya), its land mass was situated approximately 2°S, where the present day Brazilian rain forest exists. Deposits of mud and vegetation accumulated and gradually changed into what are now rich coal seams. In the Permian era, 240mya, this swamp became a sea and the rock beneath it began to subside. This allowed the area to become flooded, except for the mountainous regions. The land mass continued to migrate northwards to 20°N, until 168 mya, when it rose again creating a series of deltas. Marine deposits were formed at this time and, compressed under the weight of sediments, the hydrocarbons in the deposits became sealed, forming the crude oil and gas reserves that are currently being exploited.

By 100 mya, the region was 40°N and isolated except for an opening to the north; the water levels rose gradually until most of northwestern Europe lay under a chalky sea. Chalk stopped being formed at approximately the same time as the dinosaur's extinction. Then, until the ice ages began 2 mya, water levels rose and fell and the North Sea basin continued to drift into its current position. During the Holocene, the most recent ice age, 20kya, covered the region northwards from the Wash in England to

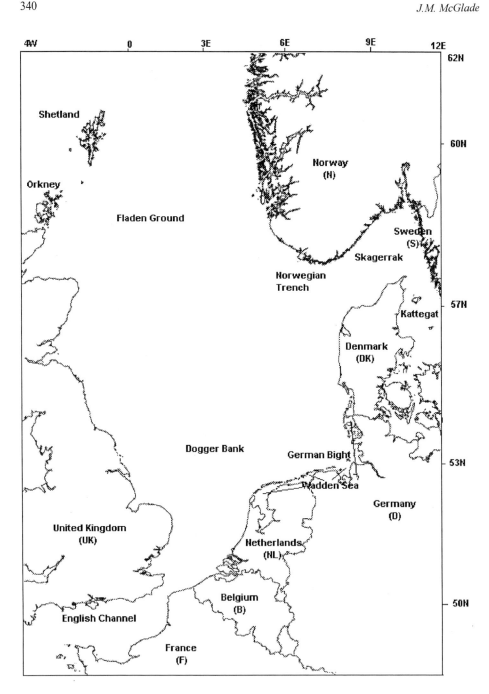

Figure 12-1. Map of the North Sea Large Marine Ecosystem.

Jutland with ice that was 2km thick and that carved out the deep river valleys and the English Channel of today. When the ice retreated, the land mass began to undergo isostatic rebound and the valleys and channels filled with sediments, sands and gravels that form the basis for today's marine aggregate industry.

Today eustatic movements are still going on, with negative trends in sea-level in Scotland and Sweden, and positive trends (i.e. sinking landmass) in southern England, the Netherlands, and Germany due to isostatic rebound. Woodward, Ekman and others have been able to create the world's oldest sea-level related records beginning in 1682, through careful reconstruction, using high water tidal information from different ports around Europe (Liverpool, Amsterdam, Brest, and Sheerness) (Ekman 1988; Woodward 1999). After compensating for changes in the amplitude of the ocean tide, they have shown that the annual mean high water values have increased at an average rate of 0.83+/- 0.06 mm/yr over the past three centuries, with an acceleration of 0.33 +/- 0.10 mm/yr/century. Comparisons with results from geological studies indicate that the apparent rise of 1mm/yr observed in the twentieth century in the regional Mean Sea Level (Shennan and Woodworth 1992) took place primarily as a consequence of an acceleration during the second part of the nineteenth century. The current overall estimate of sea-level rise for the next century is c. 50 cm for the North Sea (Watson *et al.* 1997).

The North Sea is one of the most varied coastal regions in the world, containing rocky, fjordic and mountainous shores, chalk cliffs, shingle banks, sandy beaches with dunes, intertidal mudflats, salt marshes, and a range of subtidal rocks and banks. A high proportion of the biotopes of major conservation importance reported in the European Community's Coordination of Information on the Environment (CORINE: http://etc.satellus.se/the_data/overview.htm) occur within and around the North Sea, and several coastal areas of exceptional scientific and environmental interest have been designated as sites of international and national importance (see MarLIN: www.marlin.ac.uk for access to biotope data on more than 30,000 marine locations). One of these is the Wadden Sea, lying to the north of the Netherlands and abutting the German and Danish coasts. It is renowned for its biological productivity, providing habitats for millions of wading birds, ducks and geese (Zweip 1990), 20% of the North Sea's salt marsh and nursery areas for 80% of all plaice (*Pleuronectes spp.*), and 50 % of all sole (*Solea spp.*) caught in the North Sea (van Dijk 1994).

Although the North Sea represents less than 1% of all the oceans, it plays a key role in one of the world's major economic regions. There are approximately 164 million people in the North Sea catchment who use the coastline and marine environment in a multitude of ways. The North Sea itself supports highly productive extractive industries for fisheries, hydrocarbons and aggregates, and the Straits of Dover and the North Sea proper are among the most heavily utilised sea routes in the world, serviced by large commercial harbours and ports. Before the last century, human activity had little impact on the North Sea, but with increased growth in populations and industrial activity, it has become very clear that many of its resources are close to over-exploitation, with conflicts over access increasing. Thus the future of this LME and its management as an integrated water body is of critical interest not only regionally, but also globally.

An assessment of the health of the North Sea was initiated in 1987 as part of the international ministerial activities to address concerns over the impact of human activities and climate change on the marine ecosystem (International Ministerial Conference on the Protection of the North Sea 1984, 1987, 1990, 1995; Quality Status Report 1987; North Sea Task Force 1993; Intermediate Ministerial Meeting 1993, 1997; Tromp and Wieriks 1994). The outcome and experiences gained from the North Sea Task Force's Quality Status Report (1993) have been used to build a more integrated environmental approach to management of the North Sea, together with the 1992 Convention on Biological Diversity (CBD) and the Oslo and Paris Conventions (OSPARCOM) which contain a number of supporting legislative and policy instruments. In particular, the CBD proposes that there should be a focus on the functions of ecosystems; promotion of fair and equitable sharing of the benefits derived from these functions; use of adaptive management practices; inter-sectoral co-operation; and management actions at a scale appropriate for the issues being addressed. To look at some of these and other issues facing the North Sea, the following review has been divided into five areas, as suggested by Sherman (AAAS 1986; Alexander 1986; Sherman 1994): i.e. productivity, fisheries, pollution and ecosystem health, socio-economics, and governance.

PRODUCTIVITY

Climatology and physical oceanography

The climatology of the North Sea is largely driven by the mixing of oceanic water from the North Atlantic and freshwater river run-off, and the prevailing westerly winds. In general, warmer oceanic water enters through the English Channel, moving to the Belgian/Netherlands coast, and colder oceanic water enters from the north and moves along the western coast of the North Sea. There is an annual temperature cycle ranging from $7°$ to $15°C$ in the north and from $6°$ to $18°C$ in the English Channel. Freshwater input from river run-off and melt water from Scandinavia totals approximately 300-500km^3/yr; in the southern North Sea run-off is about 190 km^3/yr and is dominated by the Rhine. Salinity variations are small, but in combination with temperature, it is possible to characterise the main water masses in the North Sea (Table 12-1). Winds increase away from the coast and are seasonal, with high speeds in springtime. Where there is no wind influence, i.e. below 70m, depressed oxygen levels can occur. The tidal regime is semi-diurnal and the dominant solar and lunar tides are sinusoidal (Hardisty 1990): tides enter mainly from the northwestern edge where three amphidromic systems develop off the Netherlands, Denmark and Norway. Tidal currents, which dominate the kinetic energy spectrum, vary widely: for example, the mean spring amplitude which increases southwards, exceeds 1.2 ms^{-1} in some places whilst in others it barely reaches 0.2 ms^{-1} (Howarth *et al.* 1994). Estimates of the flushing time for the whole North

Table 12-1. Average salinity (ppt) and temperature (°C) values for water masses in the North Sea (source: North Sea Task Force 1993).

Water mass	Salinity (ppt)	Temp (°C)
Atlantic	>35	7-15
Atlantic (deep)	>35	5.5-7.5
Channel	>35	6-18
Baltic	8.5-10	0-20
Northern North Sea	34.9-35.0	6-16
Central North Sea	34.75-35.0	5-10
Southern North Sea	34-34.75	4-14
Scottish coastal	33-34.5	5-15
Continental coastal	31-34	0-20
Norwegian coastal	32-34.5	3-18
Skagerrak	32-35	3-17
Skagerrak coastal	25-32	0-20
Kattegat surface	15-25	0-20
Kattegat deep	32-35	4-15

Sea, obtained from models and tracer-experiments of the caesium-137 discharges from Sellafield and atmospheric inputs from Chernobyl, vary from 1 year to 500 days (North Sea Task Force 1993).

In the north, metocean characteristics are influenced by a number of large-scale oceanic forces. The seasonal cycle of water masses depends upon a balance between the inflowing North Atlantic Current and the shelf-edge Slope Current (Davies and Xing 1997). Transects across the East Shetland Basin (between 60°-61°N and 10°W-10°E) and off the Faeroes, i.e. the northern mouth of the North Sea, indicate that well-mixed waters exist up until April, whereupon thermal stratification commences offshore, with a deep thermocline intensifying to 30-40m, but with well-mixed water inshore from the front (Turrell *et al.* 1996). This situation remains until October when strong winds result in the weakening of the thermocline. In this way warm, low salinity water (8.5°C, 35.1-35.3ppt) occurs along the eastern boundary of the Basin and cool saline water (6.5°C, 35.2-35.3ppt) in the central portion - a remnant of the previous year's winter in the northern North Sea, which has been trapped beneath the thermocline during the summer. Between these is the water mass known as Atlantic Water (7.5°C, 35.3ppt), which enters from the north and lies within the well-mixed zone along the western boundary of the Basin; it results from broad-scale wind-driven and localised density-driven inflows. The Atlantic Water is made up of northeast European continental shelf-edge slope water, which has been shown to be a persistent year-round feature with little variability, and North Atlantic Water derived from the North Atlantic Current with origins in the Gulf Stream (McCartney and Talley 1982; Krauss 1986). The salinity maximum is the result of the changing proportions of Northeast and North Atlantic

Waters, which in turn affects the composition of the oceanic inflow through density effects.

In the south, the residual flow in the English Channel is largely driven by wind and tides (Pingree and Maddock 1977); on average this flow is from west to east, feeding a narrow saline core of Atlantic Water through the Strait of Dover. Studies of salinity data from stations in the English Channel (E1 off Plymouth 1930-1985) and off the Dutch coast (1954-1985) show no regular pattern, although relatively long periods of up to five years with salinities above or below the average have been observed (Laane *et al.* 1996; Prandle *et al.* 1997). Overall, salinity values are highest at E1, reflecting the greater influence of Atlantic Water in the English Channel. Because the transit-times are long (a year or more), the signal from this water is gradually modified by coastal processes as it passes up into the southern North Sea.

The general circulation patterns within the southern North Sea are strongly influenced by winds and bottom topography (especially during periods of weak winds) (Pingree and Griffiths 1980). Tidally generated internal waves occur from interactions of the tide with the Dogger Bank. The large areas less than 40m deep (the Southern and German Bights), the two deeper regions to the east and west of the Dogger Bank where depths can exceed 90m, and the extensive sandbanks off the Norfolk coast all help to create an overall cyclonic circulation pattern.

The mean transport resulting from these topographic features, nonlinear tides, wind, storms, and horizontal density distribution is northeastwards, with most of the water flowing through the Skagerrak before exiting via the Norwegian Coastal Current (Eisma 1987; Otto *et al.* 1990). Water has a long residence time in the centre of the Skagerrak, so there is significant sedimentation here; inputs of Baltic water and freshwater from Scandinavian rivers are important in maintaining the Norwegian Coastal Current and creating a stable surface layer in the Kattegat and Skagerrak and large parts of the northern North Sea (North Sea Task Force 1993).

The combination of tidal currents and depths means that heat inputs at the surface give rise to a complex pattern of mixed and stratified waters, separated by tidal mixing and thermal fronts, gyres and eddies. Following the pioneering studies on tidal mixing fronts on the European shelf (Pingree and Griffiths 1978; Simpson 1981), they have become the focus of considerable oceanographic and biological interest as sites of enhanced primary and secondary production (Le Fèvre 1986; Hill *et al.* 1994). For example, Proctor and James (1996) have shown that there is a link between the formation of the tidal fronts in the southern North Sea and the onset of the spring thermocline. Tidally mixed frontal systems occur off Flamborough Head and the Frisian coast, and in the Skagerrak and German Bight, as a result of strong upwelling events; freshwater plumes and fronts are also evident in the estuaries of the Rhine, Thames and Humber. The location of these fronts affords subdivision of the open sea shelf into three ecologically relevant areas. Thus, in summer there is a central area of vertically mixed water, a stratified area occupying the whole of the northern area above a line from Denmark to the Humber estuary on the east coast of England, and a transitional zone across which the shelf sea fronts migrate in the western English Channel up to an arc across the southern North Sea, from the east coast of England to the Netherlands coast and up to Denmark.

Several 1-, 2- and 3-D models for the North Sea exist, which give good agreement with observations of hydrographic and biogeochmical distributions (Howarth *et al.* 1994; Prandle *et al.* 1994). The circulation patterns can be seen using an animation-programme based on the hydrographic flow fields derived from a three-dimensional baroclinic primitive equation North Sea model (Pohlmann 1996; Luff *et al.* 1996 http://www.agu.org/eos_elec/96197e.html). The driving forces for the model are weekly sea surface temperatures (SSTs), climatological salinity distributions, and surface wind stress and air pressure fields measured every 3 hours. The model has a meridional spacing of 12 minutes, a zonal spacing of 20 minutes, and the vertical resolution in 19 layers is 5m per layer in the upper 50m to give a better description of the thermocline dynamics. The flow fields are used in a dispersion model, based on the three-dimensional Eulerian transport model, to estimate the influence of the different water masses into the North Sea (Luff and Pohlman 1995), and a primary production model that includes advection-diffusion-reaction processes (Moll 1997).

Within the past three decades, a number of extreme events and trends have been observed that have had a direct impact on oceanographic conditions and productivity in the North Sea. The coldest years in the twentieth century occurred in 1942, 1962, 1977, and 1979; winds were weaker in the 1960s than subsequently (Coeling *et al.* 1996). During the period 1985 to 1994, conditions were exceptional, with extreme cold winters followed by extreme mild winters. There were also multi-year cycles of river discharges, with exceptional dryness in southern England and on the continent in 1989-1990; the manifestation was low salinities in the 1970-1980s and higher values in 1989-1995.

Time-series of Atlantic Water salinity show two other principal climatic events that have occurred since 1970: the 1975-1990 "Great Salinity Anomaly" (GSA) (Dooley *et al.* 1984; Dickson *et al.* 1988b) and the high salinity event of 1989-1990 (Ellett and Turrell *et al.* 1992). The GSA, an outcome of a decline in salinity in the Atlantic Water (0.1ppt) and the Fair Isle Current (0.5ppt) to the west of the northern entrance of the North Sea, appeared in 1976 above the Rockall Channel, and then propagated into the North Sea in 1977-1979, reducing salinity in the central regions (Becker and Pauly 1996). The implication is that during this period less oceanic water crossed onto the continental shelf, a suggestion supported by evidence from biological and fisheries data, nutrient and production data, and indices of physical flows and local wind forcing (Martin *et al.* 1984; Aebischer *et al.* 1990; Corten 1990; Turrell *et al.* 1992; Lindeboom *et al.* 1995; Witbaard 1996). Several causes for the GSA have been suggested; Dickson *et al.* (1988b) thought it was the result of a "bolus" of freshwater created by atmospheric conditions east of Greenland in the late 1960s, which circulated around the sub-polar gyre arriving at the Faroe-Shetland Channel two decades later. Others have suggested that it was due to changes in the evaporation/precipitation ratio (Pollard and Pu 1985) or changes in the strength of the transport of the Gulf Stream (Taylor 1996) combined with the East Icelandic Current (Turrell *et al.* 1996).

During the GSA period, the North Sea annual mean sea surface temperature anomaly was noticeably negative. Becker and Pauly (1996) subsequently studied the weekly sea-surface temperatures in eight sub-regions, and observed that the eastern and western sides of the North Sea had exhibited strikingly different behaviours over the past 25 years: the east was more heavily influenced by continental processes and the west by oceanic trends. Spectral analysis of the data indicated a climatic fluctuation peak across the North Sea with a period of 8 years, similar to that in the North Atlantic Oscillation index. However, they did not detect any significant long-term trend.

The long-term climatology of the Skagerrak is also relevant, because about 70% of the water entering the North Sea is assumed to pass through this passage before leaving (Svansson 1975; Danielssen *et al.* 1991). Danielssen *et al.* (1996) analysed data from a section across the Skagerrak and found that: i) the years 1989-1993 were characterised by an exceptionally warm, multi-year event associated with a large inflow of Atlantic Water during the winter months; ii) there were clear long-term variations with a decadal period in the deeper parts of the Skagerrak; iii) the lowest temperatures and salinities were observed during the GSA of 1977-1981 and the highest temperatures and salinities during 1989-1993; iv) the intermediate and deep water below the seasonal pycnocline was characterised by higher temperatures and lower salinities during winter; and v) the Central and Northern North Sea Water contributed less, and the Atlantic Water more, to the volume of water in the Skagerrak during the summer than during the winter.

The wave climate of the North Sea has also been extensively studied, as it is crucial to oil and gas operations, shipping and navigation (Kushnir 1994; Kushnir *et al.* 1997; http://www.geos.com); in particular, extreme wave height, which in the northern North Sea has been observed at 30m, is part of the design criteria for offshore installations. Analysis of wave height observations from Seven Stones Light Vessel (1962-1986) and Ocean Station Lima (1975-1988) suggests that there has been a positive trend in mean wave height in the eastern North Atlantic of about 2% over the year between 1960 and 1988 (Bacon and Carter 1991) However, analysis of observed wind speeds do not show a comparable trend (Hogben 1994), suggesting that the increase is mainly in the swell component of the sea state. The Dutch Meteorological Institute (KNMI) has recently published a study of the historical wave height that tends to support the wave height increase but at a lower rate (less than 1%/yr) during the period 1961-1987 (Bouws *et al.* 1996). However, because weather observations from transient vessels and even weather ships are subject to errors arising from changes to platforms and instrument exposure, air pressure data from land stations and water level measurements from tide gauges at coastal stations have been used to create hindcasts (Storch 1996). These analyses indicate a considerable amount of inter-decadal variability and strong increases in severity since 1960, especially for the United Kingdom, North Sea and Norwegian Sea. Results from the Norwegian Meteorological Institute (DNMI) project on Waves and

Storms in North Atlantic (WASA), which used hindcast data for 1955-1994, indicate that the wave climate of the North Sea has undergone significant inter-decadal variability resembling in part the North Atlantic Oscillation (Gunther *et al.* 1998). At present, computer models in conjunction with data assimilation are used for operational wave forecasting (Flather *et al.* 1991; Voorrips *et al.* 1997; United Kingdom Meteorological Office http://www.met-office.gov.uk).

Changes in the storm climate have also been assessed in terms of the frequency and intensity of depressions depicted on archived surface pressure charts (Stein and Hense 1994). Because improvements in the charts have led to a creeping inhomogeneity in the extracted time-series that has tended to undermine the idea of an apparent worsening storm trend, several re-analyses are now underway at the European Centre for Medium-Range Weather Forecasting (ECWMF) and DNMI. Analyses of storm surges are also important: few in Europe will forget the storm surge disaster of 1 February 1953 when the dykes in the southwestern part of the Netherlands broke and 136,000 ha of land were inundated and 200 people drowned. The goal of current storm surge models is to provide accurate predictions for areas around the North Sea coast, especially in the Netherlands, where they are used to indicate whether dykes are at risk, and whether the storm surge barrier in the Eastern Scheldt or the Rotterdam Waterway should be closed (Flather 1987; Verboom *et al.* 1992).

The movement of sediments in the North Sea, particularly fine sediments ($<63\mu m$ diameter), and the transport of dissolved and suspended particulate matter (SPM or seston) are major elements in the land-sea interaction and play key roles in determining biological productivity (Jago *et al.* 1994). Sediment transport takes place in shallower waters, with net sediment transport leading to deposition in the shallower areas in the southern North Sea; sinks are found all along the margins, in estuaries, the Wadden Sea, and deltas (Doody *et al.* 1993). These movements are largely driven by the anticlockwise gyre in the North Sea, tides, wind, and waves: over a tidal cycle, currents can thus erode or deposit sediments. Together, these processes have led to the creation of complex bed-forms, including migrating ripples, especially in the southern North Sea (Huntley *et al.* 1994; Guillen and Hoekstra 1997). Seston concentrations are determined by local resuspension and advection, rather than via bioturbation of the sediments themselves; the resuspended particles are comprised of "fluff" - the surficial layer of material enriched in organic carbon, and as such are strongly influenced by the seasonal presence of algal blooms (Jago *et al.* 1994). In stratified waters during the spring bloom, deposition of organic-rich detritus can cause seabed anoxia and efflux of trace metals (Fe and Mn); settling, deposition and resuspension of fluff are important controls of metal exchange across the seasonal thermocline.

Terrigenous and aerial substances entering the sea may be dissolved or attached to SPM. The mass balance of SPM is controlled by the input from rivers, atmosphere and adjacent seas, via advective and diffusive fluxes and by deposition and resuspension along the seabed. Only a few comprehensive data sets from shipboard measurements are available (e.g. Eisma and Kalf 1987a,b; Eisma and Irion 1988), but in recent years these have been augmented by satellite remote sensing. Not only do sediments convey contaminants, they can influence biological productivity by their extinction of light, thereby severely limiting photosynthesis. To take this into account, two types of waters are now generally defined with respect to SPM: Case I waters which are low in suspended sediments and other particulate matter and are biologically dominated, and Case II waters which are high in sediment load and coloured dissolved organic matter. Because of hydrographic and topographic effects, the nature of the material and the quite diverse time scales involved, terrestrial inputs are very differently distributed and the transport and sedimentation of SPM and the erosion of fine sediments are hard to distinguish. However, various numerical models of SPM transport have been developed for the North Sea (Pohlmann and Puls 1993; Sündermann 1994), to simulate SPM deposition and fine sediment erosion. The results show that SPM dynamics are dominated by inputs from the Atlantic Ocean, which produce a net annual gain of 26 million mt, compared to terrestrial and atmospheric inputs estimated at 4 million mt per year. However, the terrestrial inputs are heavily loaded with contaminants so they are significant for the health of the North Sea ecosystem. Overall, SPM is transported to the southern central North Sea and then to the Skagerrak; the Dutch-German inflows are moved northwards along the Danish coast, whilst the bulk of the released SPM is deposited in the Norwegian Trench. The accretion rate is now about 10cm per century, reflecting the history of European industrialisation (Puls and Sündermann 1990).

Chemical oceanography

The production of phytoplankton depends on nutrients and light; net growth occurs in depths of less than 10cm in turbid waters down to 30m in offshore areas. In much of the North Sea, nutrients in the upper layer are depleted due to summer growth following stratification; however, in autumn and winter when mixing is intense and light conditions poor, nutrients accumulate and reach a maximum in late winter. Silicate is typically the first nutrient to become depleted from coastal waters in spring. Stimulated production due to extra inputs of nutrients can lead to oxygen deficiencies in bottom waters, particularly if the waters become stratified and the oxygen cannot be replenished or exchanged, e.g. in fjords with narrow sills and intertidal estuaries. Oxygen deficiencies have been observed in large areas in bottom waters in the southeastern North Sea in the Wadden Sea area, particularly during the summers of 1981-83 coinciding with a large phytoplankton bloom (von Westernhagen et al. 1986; Hickel et al. 1989). Trends of decreasing oxygen concentration have also been recorded for deep water in the Kattegat and basin water in the Swedish and Norwegian fjords (Andersson and Rydberg 1988). It is highly likely that this trend is related to increased sedimentation and decomposition of organic materials caused by eutrophication and increased oxygen consumption due to an increase in suspended organic matter.

Typical background nutrient concentrations entering the North Sea from the Atlantic Ocean are 12 μmol/l nitrate, 0.8 μmol/l inorganic phosphate and 6 μmol/l silicate. Slightly higher concentrations can be found in deeper stagnant waters and much higher values in coastal waters. Offshore the values drop in winter. Riverine inputs have nitrate concentrations about 50 times higher than background values, with concentrations close to the river mouth (i.e. zero salinity) ranging from 350 to more than 600 µmol/l; dilution through mixing is the main process governing nitrate concentration in estuaries during winter. For phosphates the patterns are less clear, as flocculation and sedimentation processes affect concentrations. Overall spatial patterns of nutrients reflect the seasonality of major sources of input from rivers (highest in spring and lowest during summer) and the Atlantic, so that despite significant interannual variability, the overriding patterns remain the same. Thus, in February there are elevated nitrate concentrations along the south and east coasts of the North Sea from northern France to Denmark (c. 20 µmol/l), lower concentrations along the southeast coast of England (c. 10 µmol/l), and even lower concentrations in the central North Sea (c. 8 µmol/l). Low conditions prevail over the shallow offshore Dogger Bank because of phytoplankton growth during winter. Phosphate distributions are generally the same as nitrate, with concentrations exceeding 1 µmol/l in most coastal areas of the southern North Sea, and up to 3 µmol/l in near-shore waters from France to Denmark. From November to January, nutrient supply is thought to be dominated by regeneration rather than agrochemical inputs from rivers (Howarth *et al.* 1994).

Time-series of nutrient inputs do exist for a number of rivers, such as the Rhine and Elbe. In both, nitrate has increased steadily during recent decades, offset by a decrease in concentrations of ammonium (due to improved treatment facilities) leading to a stable level of total nitrogen. Concentrations of inorganic phosphate and total phosphorus have shown a decreasing trend during the 1980s and 90s. The only extensive data set for nutrients in the southern North Sea prior to 1960 is for phosphate during the winters of 1935 and 1936; these suggest that phosphate concentrations have increased by a factor of 3 to 4 along the coast of the southern and eastern North Sea. Other comparative data from the Belgian coast in the 1960s suggest that levels from 1975-1990 were 1.5 to 3 times higher for nitrate and 2 to 4 times higher for phosphate. Similar trends have been observed in the inner German Bight (Hickel *et al.* 1989), but none have been observed off the east coast of the United Kingdom (Dickson *et al.* 1988a). Although riverine discharges carry high levels of silicate, no increase in concentration has been observed in the North Sea in recent decades. Silicate is thus typically the first nutrient to become depleted in coastal waters, arresting the growth of diatoms in the spring bloom; the excess nitrate and phosphate is usually taken up by flagellates and other plankton such as *Phaeocystis sp.* (Billen *et al.* 1991). Abatement of phosphorus has also led to changes in the ratio of N/P in the North Sea: there is generally a band of water with a high N/P ratio stretching from the southeast coast of England to northern Denmark, with a more pronounced effect in brackish waters where ratios of up to 200 can occur. The change in N/P ratio in some of these areas has a direct effect on the plankton community.

Sediments act as a source and sink for nutrients and play a key role in determining the fluxes from estuaries to the sea. In the pore water, denitrification occurs as a result of microbial breakdown of organic material with nitrate as a substrate in the absence of oxygen. Law and Owens (1990) found that denitrification rates in samples from the North Sea were low compared to most estuaries but varied over three orders of magnitude; for example, in samples from the Norwegian Trench rates went as high as 150 µmol N/m^2/day. They estimated that 10% of the anthropogenic inputs of nitrogen to the North Sea are lost to the atmosphere via this process. For the Danish Belt Sea and the western and eastern Kattegat annual denitrification rates of 140 to 350 µmol N/m^2/day have been recorded, suggesting that denitrification is an important part of the North Sea nitrogen budget (Blackburn and Henriksen 1983; Lomstein and Blackburn 1992). Concomitantly, the atmosphere is a minor source of nitrogen species (ie. HNO_3, NO_3, NH_3, NH_4^+) when compared to the Atlantic inflow, but constitutes 25% of the terrestrial inputs; thus in stratified areas away from the coast, wet deposition may be the dominant source of nitrogen (Ottley and Harrison 1992).

Biogenic dimethyl sulphide (DMS), which is ubiquitous to the surface waters of the North Sea, is also important in determining the acidity of rain and atmospheric aerosol. A comparison of the air-sea fluxes from the North Sea with man-made emissions of sulphur from Europe to the atmosphere showed that in June the North Sea emits about 10% of the rate of anthropogenic inputs; whereas, in comparison with Scandinavia the North Sea flux represents approximately 75%, suggesting that for these areas the North Sea is a significant source of sulphur in spring and summer (Liss *et al*. 1994). DMS concentrations show considerable patchiness, underlined by the distribution of algal species, such as *Phaeocystis pouchetii* along the Wadden Sea, which ultimately lead to the production of DMS. Methyl iodide, which is thought to be formed entirely in seawater, has a net flux from sea to atmosphere in a marked seasonal cycle, with high values in summer. In contrast, monthly values of carbon tetrachloride, a man-made substance, show maximum values in the winter, in part as a result of the fact that colder water is able to dissolve more of it.

Trace metals (i.e. dissolved and particulate aluminum, manganese, iron, nickel, copper, cobalt, zinc, cadmium, lead, tin, barium) enter the North Sea from riverine, marine and atmospheric sources. Their distributions reflect particle scavenging, e.g. for iron and manganese in SPM (Dehairs *et al*. 1989), and inputs from benthic recycling (for which manganese is an indicator) (Burton *et al*. 1994). There is a south-north decrease in the emission of trace metals to the atmosphere from the land surrounding the North Sea. Although there are short-term variations in the concentrations of metals in aerosols and rainwater over the North Sea, on an integrated, long-term basis the concentrations reflect those of these original land source emissions, and it is these atmospheric fluxes that contribute substantially to the total inputs of trace metals to the North Sea (Chester *et al*. 1994). Studies of the distribution of particulate trace organics (e.g. saturated hydrocarbons, polynuclear aromatic hydrocarbons and complex materials) suggest that terrestrial sources dominate over marine sources. The major source of both trace organics and metals to the North Sea atmosphere is air that has just crossed over the United Kingdom, even though concentrations in terms of origins are higher in aerosols having a continental origin.

Biological productivity

Primary production varies considerably across the North Sea. The highest values of primary productivity occur in the coastal regions, influenced by terrestrial inputs of nutrients, and in areas such as the Dogger Bank and tidal fronts (Varela *et al.* 1995). Both spring and autumn blooms occur throughout the North Sea. Although there has been no survey to estimate primary production for the whole North Sea over an annual cycle, it is probably in the range 150-250 gCm^{-2}/yr. In coastal areas the annual production can reach 400 gCm^{-2}/yr (Cadée 1992), and for the southern North Sea production has been estimated at 150-200 gCm^{-2}/yr (Adams 1987; Reid *et al.* 1990; Heip *et al.* 1992; Joint and Pomeroy 1993; Joint 1997). Within the North-West European Shelf Programme (NOWESP, van Leussen *et al.* 1996) primary production and chlorophyll data have been compiled and analysed. Results from two stations in the Netherlands coastal zone Marsdiep and 6km offshore showed increases in chlorophyll concentrations in the early 1980s, followed by a decrease, during the end of the 1980s and 1990s, to values comparable with those of the 1970s. This increase was almost exclusively due to higher summer values. The estimate of gross annual primary production between 1976 and 1992 for the station 6km offshore was $375gCm^{-2}$ (Bot and Colijn 1996). In the Kattegat primary production rose from $67gCm^{-2}$ in the 1960s to $190gCm^{-2}$ in the period 1984-1993, an increase which Richardson and Heilmann (1995) attributed to increases in nitrogen loading.

Annual production of copepods in the North Sea is estimated at 5-20gCm^{-2}, macrobenthos production at approximately 2.4gCm^{-2}, and fish production at 1.8gCm^{-2}. Measurements of the meiofauna do not exist but annual consumption can be estimated indirectly to be 10gCm^{-2} from measurements of respiration, body weight and life history. Calculations of benthic bacterial production and benthic community respiration indicate that internal recycling is important. As much of the North Sea is deeper than 30m, the contribution of the phytobenthos (i.e. macroalgae, seagrasses and microphytonbenthos) contributes very little to overall productivity. Of the total fish production, one third is consumed by fish, one third lost to predators, disease etc. and the remainder (approximately 2.5 million mt) is taken via fisheries (Daan *et al.* 1990). Given that the total annual average landings for the northeast Atlantic are 10 million mt, the North Sea has one of highest productivity levels in the North Atlantic shelf sea area.

Plankton communities

Van den Hoek and colleagues (1979) have estimated that there are more than 500 species of phytoplankton in the Wadden Sea alone, and De Wolf and Zijlstra (1988) have estimated less than 300 zooplankton, including 5 Brachiopoda, 22 Calanoid copepods, 4 planktonic gastropods and 4 tunicate species, several chaetognaths, 3 genera of cyclopoid copepods, 1 polychaete (*Tompteris*), and 7 families of Malacostraca. Bacterioplankton in particular are now recognised as being a critical element in the dynamics of marine ecosystems (Billen *et al.* 1991); seawater contains around 10^9 bacteria per litre which live mainly on organic matter. Some 60% of the primary production may enter the microbial food web in which the main consumers are microflagellates.

In the North Sea, there is a complex interaction between phytoplankton abundance and productivity, nutrient and light availability, and the degree of mixing; generally conditions are usually most favourable for phytoplankton in spring when blooms become widespread. Exceptional blooms can also occur; these represent cause for public concern as they are often made up of species that contain toxins harmful to humans (e.g. *Alexandrium spp.* and *Dinophysis spp.*), or fish and invertebrates (e.g. *Chrysochromulina spp.* and *Gyrodinium*) or discolour the water (e.g. *Noctiluca spp.*). Over the past 3-4 years, the United Kingdom Environment Agency has begun using remotely sensed SeaWiFS data (http://seawifs.gsfc.nasa.gov), adjusted for the North Sea Case II waters, to monitor seasonal changes in chlorophyll-a concentrations; the results have enabled them to track algal blooms, such as the coccolithophore bloom (*Emiliana huxleyi*) which occurred in the Channel in 1999 and a red tide of *Alexandriun tamarense* in the Fal estuary on the southwest coast of England (Chambers *et al.* 2000).

Modelling studies of the southern North Sea support the hypothesis that plankton growth and seasonal cycles are widely controlled by tidal stirring, which determines the response in terms of stratification and turbidity to seasonally varying inputs of heat and wind (Tett *et al.* 1994). Peak biomass is controlled by eutrophication and nutrient inflows from rivers (Pingree *et al.* 1978; Brockmann *et al.* 1988; Radach and Lenhart 1995).

Long-term plankton studies (phytoplankton and zooplankton) in the North Sea have been supported by data from the Continuous Plankton Recorder (CPR) survey, which has run since 1938. CPRs are towed by merchant ships and Ocean Weather ships on regular routes (Glover 1967) and then the plankton abundance for alternate 10 nm sections determined (Colebrook 1960). The detailed maps of plankton abundance show that, in general, populations of most species of zooplankton and large phytoplankton declined between 1950 and 1980 but have increased since then (Fransz *et al.* 1991; Warner and Hays 1994). Despite some difference between CPR data and localised studies (e.g. Broekhuizen and McKenzie 1995; Greve *et al.* 1996), it is clear that the CPR data provide a coherent time-series for comparisons against other data and to test different hypotheses. For example, Backhaus *et al.* (1994) use data on the distribution and succession of *Calanus finmarchicus* from Fransz *et al.* (1991) to help explain not only the life-cycle of *Calanus*, but also to support the circulation patterns along the continental slope of the North Sea and in the southern Norwegian Sea.

Another excellent example of the importance of the CPR can be seen in the work of Taylor and co-workers (Taylor and Stephens 1980; Taylor *et al.* 1992; Hays *et al.* 1993; Taylor 1995), in which they show that there is a relationship between the abundance of zooplankton observed in the CPR and the north-south displacements of the Gulf Stream, with higher abundances occurring during years when the north wall is displaced northwards. This relationship holds both for total counts of copepods as well as individual species, and supports the hypothesis of a "teleconnection" between the two sides of the Atlantic Ocean, possibly arising out of local perturbations in the atmospheric circulation caused by displacements of the Gulf Stream. A similar relationship in seasonal dynamics has been documented for coastal zooplankton off Northumberland in the northwestern part of the North Sea (Frid *et al.* 1994; Frid and Huliselan 1996).

In the southern North Sea, variations in the meso- and macrozooplankton populations in the German Bight for the period 1974-1994 (Greve *et al.* 1996) coincided with changes in the benthos (Lindley *et al.* 1995). The causal factors are not clear, but it has been suggested that they may be related to changes in boundary conditions, such as the intensity of westerly weather (Aebischer *et al.* 1990), run-off, biomass, or regional eutrophication (Hickel *et al.* 1993).

Benthic communities

The spatial and temporal dynamics of the microphytobenthos are relatively poorly understood because they are difficult to sample; however, diatoms are often associated with particular substrates which can be thus used as a classification criterion. Macrophytobenthic species, including brown and red seaweeds, seagrasses and eelgrass (*Zostera marinus*), and tasselweeds, are found throughout the coastal North Sea. In recent years some communities have been affected by the introduction of non-native species, pollution, and disturbance; as a result there has been a general decline in marine plant biodiversity during the past century (Hiscock 1997). Several plants have, in fact, disappeared in some areas, such as red macroalgae in the tidal creeks of the Wadden Sea (Ducrotoy 1999). Seagrasses have also declined along the Dutch, Danish and French coasts as a result of disease, circulation patterns, fisheries, and interspecific competition by green algae (Den Hartog 1987, 1994; Reise *et al.* 1989). Other species have increased dramatically; for example, the green algae *Ulva spp.* has grown prolifically along the beaches of Britanny (Piriou *et al.* 1991) and in enclosed mudflat areas on the English Channel coast along with *Enteromorpha spp.* In the Wadden Sea, there have been mass developments of green algae since 1989, and in the Skagerrak dense coverings of the brown algae *Halidrys siliquosa*. It has been suggested that an increase in phytoplankton density may have reduced light penetration in favour of red algae. In other regions, however, no significant changes in macroalgal flora have been observed; for example, at Helgoland the dense *Laminaria hyperborea* and deepest coralline crusts were at the same levels in the 1990s as in the 1960s (North Sea Task Force 1993).

Overall, nematodes are the dominant meiofaunal taxon; they increase in abundance northwards to 53°30'N and then decrease again. Copepods occur in their highest numbers and diversity (approximately 1500 species) in the Southern Bight, and exist in distinct assemblages associated with sediment type and depth (Huys *et al.* 1992).

Table 12-2. Zooplankton in the North Sea (from De Wolf & Zijlstra 1988).

Area	Polychaetes	Molluscs	Crustaceans	Echinoderms
Infauna				
German Bight (1985)	90	54	52	12
German Bight (1978)	68	38	55	9
Central North Sea	-	-	-	57
Doggerbank	143	-	-	-
Oyster Ground	-	63	-	-
West of Scheveningen	62	16	49	5
Epifauna				
57°20' - 54 ° 40'N and 2 ° 20'-5 ° E	2	75	56	20

The macrobenthic fauna in the North Sea have been widely studied (e.g. De Wolf and Zijlstra 1988; Eleftheriou and Basford 1989; Heip *et al.* 1992; Künitzer *et al.* 1992; Heip and Craeymeersch 1995; Basford *et al.* 1996; Jennings *et al.* 1999b) (Table 12-2). There are approximately 700 taxa, with northern species, such as *Ophelia borealis, Exogone verugera, Spiophanes bombyx, Polycirrus sp., Minuspio cirrifera, Thyasira sp. Aricidea catherinae, Nepthys longosetosa, Nucula nitosida, Callianassa subterranea* occurring from the north down, in some cases as far south as the Dogger Bank; and southern species, such as *Nucula hanleyi, Venerupis rhomboides, Echinocardium cordatum, Ophiotrix fragilis, Aonides paucibranchiata, Phoxocephalus holbolli* and *Pisione remota,* extending up to the 100m bathymetric contour. Macrofaunal abundance and diversity increase linearly northwards; the average biomass for the whole area is 7g ash free dry weight (adwt) m^{-2}, and it decreases northwards. The reasons for these patterns are complex, and reflect not only latitude, sediment type, water depth, erosion, and deposition patterns, but also ecological interactions (Zuhlke and Reise 1994; Heip and Craeymeersch 1995). For example, analyses of the macrobenthic fauna off Northumberland suggest that abundance is regulated in part by density-dependent processes (Frid *et al.* 1996). Witbaard (1996) concluded that the variations in density and growth in the mollusc *Arctica islandica* from the Fladen Ground, 150 km off the east coast of Scotland, resulted from the import and accumulation of organic matter.

Long-term analyses of changes in the benthic communities over the past 30 years suggest that a number of processes are involved. Climatic and meteorological changes, such as the position of the Gulf Stream, can exert a strong impact on intertidal communities and benthic biomass (Wieking and Kröncke 1999); for example, mortality in overwintering populations in the Dutch Wadden Sea increased (Beukema 1992a,b), whereas after eight successive mild winters off Northumberland, mortality decreased in the soft-sediment community at a depth of 55 m, and biomass stabilised (Buchanan and Moore 1986a,b; Beukema 1992b). Off Weymouth and Poole Bay in the English Channel fluctuations were observed after the severe winter of 1962-1963 (Holme 1983). Storms and fluctuations in swell can also affect benthic communities.

Benthic-pelagic coupling can also create long-term changes in benthic communities. Benthic infauna associated with soft-sediments consume organic matter that descends from the pelagic zone; most of this matter is comprised of phytoplankton, changes of which are linked to large-scale climatic and meteorological variability (see above; Frid and Clark 2000). This type of coupling has been observed and/or postulated for many sites in the North Sea (Austen *et al.* 1991; Buchanan 1993; Lindeboom *et al.* 1995); Josefson *et al.* (1993) and Beukema (1986) have attributed the doubling in biomass and abundance in the benthic community in the Skagerrak-Kattegat area and Wadden Sea to eutrophication.

Frid and co-workers also noted shifts in the benthic communities in the southern and central North Sea between the early 1920s and the late 1980s (the lack of change in the Dogger Bank was thought to be because the area had already attained an alternative stable state as a result of fishing activities prior to 1920) (Frid and Clark 2000). However, Kröncke and her colleagues resampled the data from the Dogger Bank in the 1950s onwards and found changes in faunal abundance, including an increased abundance of small opportunistic polychaete worms and the disappearance of *Spisula spp.*, leading to a 30% decline in total biomass over the period from the 1950s to the 1980s (Kröncke 1990, 1992; Kröncke and Rachor 1992; Kröncke and Knust 1995). Kröncke and colleagues interpreted these shifts as responses to food and organic carbon availability arising from eutrophication and/or climatic fluctuations. Similarly, shifts in the benthic community in the German Bight have been associated with sediment changes, nutrient flux, predation pressure, and fishing (Schroeder and Kunst 1999).

Fishing activities, especially those of the demersal fleet, have changed the benthic community structure in many areas from ones dominated by long-lived species to ones with short-lived opportunistic and scavenging species; thus populations of certain long-lived buried bivalve species, such as quahog (*Arctica islandica*) and horse mussel (*Modiolus modiolus*) have almost certainly decreased. Fonds and Groenewold (2000) have calculated that on average, beam trawling for sole and plaice in the southern North Sea generates c. 180g (ash free dry weight)/100m^2/yr from damaged benthos and 15-38g (afdw)/100m^2/yr of discard fish compared with a potential annual food demand of c. 2450g (afdw)/100m^2/yr for benthic invertebrate carnivores and 550g (afdw)/100m^2/yr for demersal fish, i.e. 7% of the maximum annual food demand of all common benthic predators.

Other controlling factors, such as climate, physical disturbance, emigration, immigration, and benthic predators, have also affected the structure of the benthic community. The disappearance of the oyster (*Ostrea edulis*) from the North Sea during the first half of the century was not only due to a number of severe winters and disease but also partly to the fishery. In the Wadden Sea, the development of immature mussels on intertidal beds in the sheltered zone behind the barrier islands is controlled by winters with heavy ice coverage, when most of the mussels are removed by the mechanical force of the ice, giving rise to strong recruitment the following summer. However, between 1987 and 1996 the beds have shown a continuous decline, due to a combination of low recruitment, predation by seabirds, heavy gales, and the emergence of a fishery for seed mussels. The heaviest decline has been in the lower Saxony subregion (Germany), even in areas without a fishery. Here the lifetime of subtidal beds of mussels is often restricted to one or two years following formation, due to predation by eider ducks (*Somateria spp.*), starfish (*Asteroidea*) and shore crabs, and the effects of gales.

Fish predation also plays an important role in benthic community evolution as many benthic feeding fish show prey selectivity. The removal of certain species can cause small-scale disturbances that in turn contribute to the spatial heterogeneity of the benthos (Wilson 1990; Hall 1994). Results from analysis of fish diets in the North Sea from 1970-1993 by Frid and co-workers (Frid and Clark 2000) indicate that the consumption of North Sea benthos may have changed as stock sizes of commercially important fish species have changed; they have estimated that the level of predation on the benthos has increased from approximately 23 million mt yr^{-1} in 1970 to 29 million mt yr^{-1} in 1993. The intensity of fish predation, i.e. 20-45% of the production consumed by fish, is similar to the 39% macrobenthic consumption by fish estimated by Greenstreet *et al.* (1997).

Fish and shellfish communities

A total of 224 species of fish and crustacea have been recorded from the North Sea (Wheeler 1978; Knijn *et al.* 1993). A high proportion of the total biomass of fish (approximately 95%), however, is comprised of a small number of species, most of which are commercially exploited and hence managed (34 species of fish and 14 shellfish and invertebrate species, details of which are given in the next module). There are essentially three fish assemblages: one along the slope edge and northern area, one in the central area and one in the southeast. In the northern North Sea, the fish community is dominated by the gadoid species of saithe (*Pollachius virens*) and haddock (*Melanogrammus aeglefinus*), with significant levels of Norway pout (*Trisopterus esmarkii*), whiting (*Merlangius merlangus*), blue whiting (*Micromesistius poutassou*) and cod (*Gadus morhua*). In the central North Sea, half of the biomass is made up of haddock, whiting and cod occur in depths of 50-200m, and whiting and the common dab (*Limanda limanda*) in shallower areas. In the southern area, anchovy (*Engraulis encrasicolus*) and sardine (*Sardina pilchardus*) occur temporarily. Some key southern species, i.e. those in the southern North Sea and English Channel, are poor cod and bib (*Trisopterus minutes* and *T. luscus*), weever (*Echiichthys vipera*), horse mackerel (*Trachurus trachurus*), and mackerel (*Scomber scombrus*). Other species which contribute significantly to the biomass of the North Sea, either *in situ* or as they migrate through, include plaice (*Pleuronectes platessa*), turbot (*Psetta maxima*), brill (*Scophthalmus rhombus*), witch (*Glyptocephalus cynoglossus*), flounder (*Platichthys flesus*), grey gurnard (*Eutrigla gurnardus*), sandeels (*Ammodytes marinus* and *Hyperoplus lanceolatus*), argentine (*Argentina spp.*), sprat (*Sprattus sprattus*), and herring (*Clupea harengus*). Although there are high fishing mortalities amongst the most important commercial species, a significant component of mortality derives from inter- and intra-specific predation.

In the 1960s there was a large and sudden increase in the abundance of several gadoid species (whiting, haddock, cod and Norway pout) - the "gadoid outburst" as it is commonly called (Cushing 1984); but the causes of this phenomenon are still unclear (Hislop 1996). The four species have many characteristics in common (i.e. high fecundity winter/spring spawners with pelagic eggs and larvae), but their relative year-class strengths did not coincide during this period. Hislop (1996) concluded that the phenomenon could have three possible explanations: i) the changes in the gadoids coincided with a reverse pattern in prey species, prompting Cushing, for the case of cod and *Calanus,* to suggest that it was the "match-mismatch" between predator and prey abundance that was the cause; ii) environmental effects, such as the GSA and winds could have caused changes to circulation patterns during the crucial larval period; and iii) the high recruitment in gadoids coincided with a reverse pattern in herring; such changes in overall abundance patterns suggest that species interactions could also be the cause. But whatever the causes were, the gadoid outburst had a significant impact on the whole North Sea ecosystem, and has led to a somewhat distorted view of what should be considered "normal" for the system. Overall, long-term changes in the northern North Sea are rather subtle, despite a century of intensive fishing. Comparisons of fish communities in the southern North Sea suggest that the abundance of demersal species has declined since the beginning of the century, noticeably haddock, and that there has been a change in the size composition downwards (Pope and Macer 1996).

Other fluctuations and trends in the biological characteristics of fish species indicate that possible causes of change include eutrophication, fisheries, alterations in wind components, transport rates of larvae, and quality of nursery areas (e.g. Boddeke and Hagel 1995; Corten and van de Kamp 1996; Heessen 1996; Heessen and Daan 1996; Millner and Whiting 1996; Philippart *et al.* 1996; Rijnsdorp and van Leeuwen 1996; Rijnsdorp *et al.* 1996; Serchuk *et al.* 1996). In the case of a number of southern species in the southern North Sea, Corten and van de Kamp (1996) suggest that it was temporary changes in southerly winds over the southern North Sea that resulted in increased transport of southern fish species, such as *Sardina pilchardus*, through the Strait of Dover. Changes in physical environmental conditions have also led to infrequent occurrences of species from adjacent biogeographic zones. For example, Rogers and Millner (1996) found that the occurrence of Lusitanean species, such as the big-scale sand smelt (*Atherine boyeri*), the undulate ray (*Raja undulata*) and boreal species such as the Norway haddock (*Sebastes viviparous*) in assemblages along the English coast during periods, coincided with changes in temperture; they also found that diversity was higher along the southeast and east English coasts (Shannon's Diversity H'= 1.78 and 1.93) compared to the northeast (H'=1.58), and that these communities showed lower inter-annual variability. In recent years, there has been some concern over the loss of certain species due to human activities. Elasmobranchs appear to be particularly sensitive: two species of dogfish (*Scyliorhinus canicula* and *Mustelus mustelus*) have decreased as have rays, skates, conger (*Conger conger*), and the thornback ray (*Raja clavata*); sturgeon (*Acipenser sturio*) have virtually disappeared (Walker and Heessen 1996) and the greater weaver (*Trachinus draco*) is now extinct (Bergman *et al.* 1996). The anchovy, however, has returned to the Wadden Sea, after it disappeared following the construction of a dam in 1932 and the build-up of phosphates between 1955 and the early 1980s (Boddeke and Vingerhoed 1996).

Bird communities

About 110 species of birds utilise the North Sea; they can be divided into those which feed primarily intertidally (50 species), those using nearshore shallow waters (30 species) and those feeding offshore (30 species). The nearshore and intertidal groups feed mainly on benthic invertebrates, while those offshore feed on fish and zooplankton. Patterns of usage vary, with some species visiting only in summer or winter; in many cases the immature birds differ from adults in their occurrence. Many of the inshore and intertidal species move to the North Sea coasts from colder areas to the north and east; they not only breed along the coasts but also move inland. The coasts of north Britain in the northwestern North Sea hold the largest numbers and greatest diversity; the Channel and southwest Norway hold the fewest birds of any of the regions. Reasons for these differences are likely to be complex, and involve direct predation by humans and other birds, food supplies and pollution.

During the 20th century most species of seabirds in the North Sea have greatly increased in numbers, established new colonies and/or expanded their breeding range (Lloyd *et al.* 1991; Camphuysen and Garthe 2000). Before the beginning of the century, seabirds were heavily exploited, both as animals and eggs, and persecuted (Croxall *et al.* 1984; Mearns and Mearns 1998), and in some areas the practice persists (Beatty 1992). Some have argued that the increase observed in species such as the black-legged kittiwake (*Rissa tridactyla*) is due to the relaxation of exploitation (Camphuysen 1995); however, the increase in species such as the northern fulmar (*Fulmarus glacialis*) throughout the area is thought to have occurred because they have benefited from the offal produced in various fisheries (Fisher 1952; Camphuysen and Garthe 1997). Some authors have also suggested that seabirds have benefited from changes in the abundance of small fish arising from the activities of commercial fisheries (Sherman *et al.* 1981; Dunnet *et al.* 1990; Hunt and Furness 1996). There have also been increases in the northern fulmar, northern gannets (*Morus bassanus*), great skuas (*Catharacta skua*), and most *Larus* gulls. Similar trends have been reported for non-scavenging seabirds such as common guillemots (*Uria aalge*), razorbills (*Alca torda*) and puffins (*Fratercula arctica*).

Today, most populations of seabirds are currently at their highest historical levels. In the eastern and southern half of the North Sea, several *Larus* gulls have established populations that have shown exponential growth; for example, in the German Bight the number of pairs has increased from 18,000 in 1950 to 130,000 in 1995. Herring and lesser-backed gulls nesting in the Netherlands have gone from 2500 pairs in 1900 to nearly 120,000 pairs in 1996. Northern fulmars and gannets have established breeding colonies on the island of Helgoland, alongside increases in kittiwakes and guillemots. In the northeastern North Sea, similar patterns have been reported along the Norwegian west coast for kittiwakes, northern fulmars and gannets. It is now estimated that the total number of offshore seabirds in the North Sea ranges from c. 4 million (June) to over 8 million (February and late summer and early autumn). In summer, more than 50% are found in the northern and northwestern North Sea. The estimated annual energy requirements of the more common species are 3900 x 10^9 kJ (Anon. 1994; Camphuysen and Garthe 2000). Most of this (89%) is used by the populations of northern fulmar, gannet, herring gull, great black-backed gull, kittiwake, and guillemot. Results from a number of detailed surveys and studies of scavenging conducted between 1985 and 1998 to examine the effect of fisheries on population abundance (Camphuysen *et al.* 1995; Stone *et al.* 1995) showed that a substantial part of this energy demand comes from the discards of fishing vessels. By contrast, the energy requirements of the three most common terns (with over 300,000 breeding pairs), and the cormorant and shag (*Phalacrocorax carbo* and *P. aristotelis*) are almost negligible.

Nearly ten milllion seaducks feed on shellfish in northwest Europe during the winter, as do over one million waders and several thousands of gulls. The most detailed information exists for the southeastern North Sea, where key species include the eider (*Somateria spp.*), common scoter (*Melanitta spp.*), oystercatcher (*Haematopus ostralegus*), and herring gull (*Larus argentatus*). Natural fluctuations in some of these species are directly related to the spatfall of invertebrates in their diet, and birds such as oystercatchers, that are restricted to intertidal flats, are vulnerable to shellfish fisheries as it is difficult for them to switch prey items.

Fisheries also have an effect on the abundance and breeding success of scavenging species via competition for food items (Camphuysen and Garthe 2000). For example, the breeding success of several species of seabirds in the Shetland Islands in the northern part of the North Sea declined markedly during the 1980s, at a time when there was a noticeable drop in landings of sandeels (*Ammodytes marinus*) from an industrial fishery operating in the waters nearby (Huebeck 1988). The seabirds most affected were surface feeders, i.e. Arctic terns, kittiwakes and great skuas. After a few years of poor breeding success, significant declines occurred in breeding populations, compounded by the fact that some birds, such as the great skua and great black-backed gull which also depend on sandeels, switched to direct predation of adults and chicks of smaller seabirds. Numbers rose dramatically in 1991 coinciding with the emergence of a large sandeel year-class, giving rise to the suggestion that these birds are directly competing with industrial fisheries for food supplies. The controversy surrounding the size and hence the effect of the industrial fishery in this area on seabirds still persists (Anon. 1994).

Marine mammals

The marine mammal populations in the North Sea have been assessed via the SCANS survey (Hammond *et al.* 1995a,b), the BYCARE programme (Harwood 2000), and various land-based sightings and strandings (Tregenza 1992, 2000); the sixteen species include porpoises, dolphins, whales, and seals. In the eastern North Atlantic including the Baltic, five populations of harbour porpoise (*Phocoena phocoena*) have been identified through sightings, strandings, genetic analyses, and by-catch data (Gaskin 1984, 1985; Berggren 1994; Andersen *et al.* 1995, 1997; Hammond *et al.* 1995a, Tiedemann *et al.* 1996; Harwood 2000; McGlade and Metuzals 2000): these occur in the southern North Sea, and in inshore waters off the Netherlands and Danish coasts. The abundance of harbour porpoise, *Phocoena phocoena*, in the North Sea is estimated to be approximately 340,000 (Hammond *et al.* 1995a,b), with by-catches in the Danish gillnet fishery potentially as high as 10,000 (Vinther, 1994; McGlade and Metuzals 2000). As harbour porpoises are known to migrate between these areas, such population integrity is thought to arise mainly from female philopatry, the males being more highly dispersed (Tiedemann *et al.* 1996).

Bottlenose dolphins (*Tursiops truncatus*) are resident in the Moray Firth and in the English Channel. The resident groups are small, with the total number estimated at 250 animals. There is evidence that these dolphins were more common historically. Whitebeaked dolphins (*Lagenorhynchus albirostrirs*) are the most commonly sighted in the northern and western North Sea, with several sightings of whitesided dolphins (*L. acutus*) in mixed herds in the central area; the total number of both was estimated at 10,900 in 1994. Rissoís dolphins (*Grampus griseus*) occur in small numbers in the northern North Sea and common dolphins (*Delphinus delphis*), although rarely seen in the North Sea, do occur in the Channel.

Minke whales (*Balaenoptera acutorostrata*) migrate through the North Sea and are often seen in the northern and central areas during the summer. They belong to the northeast Atlantic stock that was estimated in 1996 to be 112,125 animals (International Whaling Commission/48/4 1996). Pilot whales (*Globicephala melaena*) frequently occur in the northern and northwestern North Sea; the stock is not restricted to the North Sea but rather occurs in the whole central and northeast Atlantic where it has been estimated to be comprised of 778,000 animals. Large whales are also seen from time to time, including sperm whales (*Physeter catodon*), fin whales (*Balaenoptera physalus*) and humpback whales (*Megaptera novaeangliae*).

Trends in small cetacean numbers in Europe show a historical decline in many areas, with notable drops in porpoises in the Baltic, southern North Sea and English Channel. A retrospective survey of sightings in the southern part of the Channel showed a decline of over 95% in sighting rates from the coast over the last 50 years (Treganza 1992, 2000). This appeared to involve both porpoise and bottlenose dolphins and precedes the use of monofilament gill-nets.

There are six species of seal in the North Sea: the grey (*Halichoerus grypus*), harbour (*Phoca vitulina*), ringed (*P. hispida*), harp (*P. groenlandica*), hooded (*Cystophora cristata*), and bearded (*Erignathus barbatus*). Only the grey and harbour seals breed in the North Sea. The North Sea has 40% of the European population of grey seals and 15% of the world population; the total number has been estimated at 52,000 animals. The majority breed around the coasts of Britain, although some pups are observed in the Wadden Sea, the Skagerrak/Kattegat and off the French Channel coast. The largest colonies are on Orkney where numbers of pups being born are c. 11,500. Harbour seals are found throughout the North Sea, although the largest concentrations occur in the Wadden Sea (8900 in 1994), Orkney (7900 animals in 1993), Shetland (6200 in 1993), and the Skagerrak/Kattegat (5200 in 1994). The total number (c. 33,000) represents 45% of the European population and 5% of the world's population. All the marine mammals contribute to the dynamics of the marine ecosystem in so far as they consume large quantities of fish to survive.

Coastal habitats

The periphery of the North Sea is made up of open and subtidal sandbanks and mudflats, reefs, partially submerged sea caves and rocky areas, intertidal sedimentary areas, and estuaries (Elliott *et al.* 1998); there is also a variety of cliffs, shingle beaches, pebble areas, saline lagoons, and large shallow inlets. Offshore, there are coralline banks and open seabanks with hydrocarbon-rich substrata having methane-seep pockmarked areas, where high levels of chemosynthesis have been observed (Dando *et al.* 1991). All these habitats have been affected over the past century by changes in climate, pollution and anthropogenic activities; as such habitat size and condition are widely threatened (North Sea Task Force 1993). Salt-marshes in particular are highly sensitive to coastal development. Fortunately, many of these habitats have specific floral and faunal assemblages associated with them, and as such are designated as Natura 2000 sites under Annexes I and II of the European Union's Habitat Directive (92/43/EEC) on the conservation of natural habitats and of wild fauna and flora, and the Birds Directive (79/409/EEC) on the conservation of wild birds. The intention of these Directives is to protect and maintain them for the foreseeable future.

Ecosystem models and food-web dynamics

Longhurst (1998) has proposed a series of biomes to reflect the pelagic ecology of the oceans. The North Sea falls into the Northeast Atlantic Shelves Province (NECS), which comprises the continental shelf of western Europe from northern Spain to Denmark and then into the Baltic Sea. The edge of the deep Faeroe-Shetland channel and the Norwegian Trench forms the separation between the NECS province and the Atlantic Subarctic province. The NECS province can be rationally subdivided into seven parts: the North Sea from the straits of Dover to the Shetlands; the English Channel from Dover west to Ushant; the southern outer shelf from Spain to Ushant; the northern outer shelf, including the Celtic Sea and the Irish, Malin and Hebrides shelves off Britain; the Irish Sea; the central Baltic; and the Gulfs of Bothnia and Finland.

There are four ecological seasons in the province: i) mixed conditions and light limitation in winter, ii) a nutrient-limited spring bloom, iii) stratified conditions during summer with localized zones of high chlorophyll along fronts, and iv) a second general bloom when autumn gales break down the summer stratified condition. As described above, local conditions make this schematic more complex, but generally the phytoplankton cycle with a large spring diatom and weaker autumn dinoflagellate bloom seems to occur most typically in the central North Sea. In the southern North Sea the spring and autumn peaks are about the same magnitude, except in the mixed region off the Dutch coast where a single bloom is more common. In the northern North Sea and western English Channel, the spring bloom is stronger compared to the autumn bloom than elsewhere.   Seasonal succession, as observed in the English Channel, follows a pattern of: a nearsurface spring bloom (<4mg chl m^{-3}, 0-15m, April), a summer sub-surface bloom in the thermocline (2-4mg chl m^{-3}, 20-25 m, May-September) fuelled by regenerated NH4, and an autumn near-surface bloom (<2mg chl m^{-3}, 0-15m, late September-October). The spring bloom is dominated by diatoms, which are progressively replaced by dinoflagellates and flagellates, and in the autumn diatoms become important again. Autotrophic production by pico and nano-plankton are a vital component in the North Sea: Joint and Williams (1985) calculated that 36% of primary production is by 0.2 to 1.0μm cell fraction and that 77% is by the 0.2-5.0μm fraction. Before the spring bloom the autotrophic picoplankton account for 50% of production. In the summer in the western English Channel, two species of chaetognaths, *Sagitta elegans* and *S. setosa*, partition the water column, with the former in the lower and the latter in the upper layer; similarly, in the North Sea proper *Calanus finmarchicus* dominates, but *C. helgolandicus* is restricted to the upper layer and the euphausiids make migrations between layers. All these elements typify the province in Longhurst's classification and have been used in a number of ecosystem models to simulate the seasonal changes in the microplankton (e.g. Tett and Droop 1988; Fasham *et al.* 1990; Fasham 1993; Yool 1998; Tett and Wilson, in press).

Despite the wealth of data from the North Sea, few large-scale trophic models have been published. Steele (1974) estimated the energy flows among 10 compartments for the whole North Sea: these comprised primary producer, pelagic herbivores, bacteria, macrobenthos, meiobenthos, invertebrate carnivores, pelagic fish, demersal fish, large fish, and other carnivores. He then calculated the straight chain trophic transfers inherent in this web by combining the components into trophic levels. Some smaller-scale trophic models have also been developed (e.g. for the Fladen Ground, Radach 1982; and the Ems-Dollard estuary, Baretta and Ruardij 1988; Baird *et al.* 1990), in which the most prominent feature has been shown to be the benthic-pelagic coupling. In general, the benthic system shows the importance of allocthonous organic matter as a food supply: a large proportion of this is consumed by the bacteria, which itself is an important food source for the meiofauna; but as the total demand of benthic consumers is lower than the supply of energy, the benthos exports energy to the adjacent sea areas.

The pelagic food web in coastal areas indicates the role of microbial grazers as the major link between planktonic primary production and the mesoplankton. Similar energetics exist in the shallow and intertidal areas, although the microbial part of the food-web appears to be more important (Billen *et al.* 1991). In these areas, shrimps are a key food item for fishes, especially gadoids and flatfishes; conversely, adult shrimps and crabs predate on juvenile fish. Suspension feeding bivalves are effective converters of phytoplankton in the water column, and together with shrimps are keystone species in inshore and estuarine areas. The significance of euphausiids in the Norwegian Trench as a food source for fish means that the food web has more energy in the pelagic phase compared to those in estuarine and shallower areas (North Sea Task Force 1993).

FISH and FISHERIES

North Sea fishing fleets

Fishing is one of the most extensive and historically well-established activities in the North Sea. Roman merchants in Utrecht conducted an active fishing business, and over the intervening nineteen centuries, the North Sea has become home to many small-scale local fishing operations. In the fourteenth century conflicts over access to herring arose between the British and Europeans (Coull 1988). The Dutch were the first to establish an open-sea fishery, and by the fifteenth century had pioneered the basic technique of catching herring in drift-nets and then curing them with salt to sell at home and for export. Conflicts continued to increase eventually leading to war between the British and the Dutch over herring in the North Sea. Ultimately the issue of access was resolved by the publication and acceptance of Hugo Grotius' 1609 essay *Mare Liberum* that set out the principles of the freedom of the seas and with it rights to harvest fish resources (McGlade 2001a).

In 1549, Fridays, Saturdays and Lent were given over to being "fish" days, and by 1563 Wednesdays had been added. Although this added a boost to fishing, the British industry gradually declined until the introduction of the fishing smack in 1830, followed by steam drifters and trawlers and the widespread development of the railways linking the ports to markets at the beginning of the 1900s (Dyson 1977). The new steam vessels transformed the North Sea fisheries, which had previously relied on hook-and-gill nets, so that between 1900 and the First World War herring fishing reached its peak of prosperity: during November 3000 miles of nets were shot every night off the coast of east England, and Lowestoft became the busiest fishing port in the world. In 1913, 854 million herring worth nearly £1million were landed at Great Yarmouth on the east coast in just 14 weeks, and 436 million at Lowestoft (Dyson 1977). British boats were thought to have caught approximately two-thirds of all herring and nearly all of the other pelagic fish (Morley 1968).

During World War I, fishing declined and although British industry picked up afterwards it was never as prominent. Indeed, it was the German fleets that began to compete most heavily, and by 1925 the British catches and exports were almost the same. The German fleets also introduced drift-nets with trawlers, instead of drifters. After World War II, during which fishing was again brought to a standstill, the need for fish oil increased, as demand for agricultural feedstuffs increased; the type of species caught switched to Norway pout and other pelagic species, in what is now known as the industrial fishery. The change in demand led to a reduction in the overall British catch because the United Kingdom did not find it economical to engage in industrial fishing. But in parallel, the growing pressure to find resources led to dramatic improvements in non-industrial fishing equipment, and in turn increased total catches of North Sea herring by the distant-water fleets from the former USSR and eastern Europe. The advent of the purse-seine net in the mid-1960s, plus the increased strength of synthetic fibres and the use of the power-block, meant that surface herring fisheries soon collapsed. The Norwegians also changed fishing grounds from the North Atlantic to the North Sea. By 1972, the enlargement of the European Economic Council to include the two major fishing nations - the United Kingdom and Denmark - led to the development of the Common Fisheries Policy; Norway, however, did not join and when Exclusive Economic Zones were extended to 200 nm in 1977, a median line was drawn between the EEC and the Norwegian EEZ.

Today, fishing and aquaculture are important economic activities in the European Union. Although the fishing sector's contribution to the gross national product of many European Union Member States is generally less than 1%, its impact is highly significant in terms of employment in all countries surrounding the North Sea, especially in areas where there are few alternatives. This is well-documented in the anthropological literature on north European fishing communities in studies from the United Kingdom (Tunstall 1972), Portugal (Johnson 1979; Mendonsa 1982; Brogger 1990; Cole 1991 including accounts by a woman skipper), the Netherlands (van Ginkel 1989), Spain (Sanmartin 1982), and Brittany in France (Jorion 1982).

Currently, the European Union fishing industry comprises 97,000 vessels, which vary greatly in size and capacity, and approximately 260,000 fishermen directly involved in catching fish. The industry also supports significant numbers of jobs in processing, packing, transportation, and marketing, plus shipbuilding, fishing gear manufacture, chandlers, and servicing. These jobs form the backbone of many remote coastal areas. Restrictions on fishing due to overcapacity and modernisation of the fleet, arising from the multi-annual guidance programmes (see below), are thus likely to have a significant effect on many communities surrounding the North Sea. The United Kingdom has the greatest number of fishermen; however, as a proportion of total national employment, fishing is of greatest importance in Denmark. The United Kingdom's fleet is the largest, both in terms of number and capacity, followed by the Danish, Dutch and German; however, the majority of vessels in the fleet are less than 10m, generating mainly part-time employment, whereas the average capacity per vessel in the Netherlands is much larger.

The North Sea fishery, with a production of over 2.5 million mt of fish and shellfish, supplies products to the European Union market, one of the largest in the world. Yet in Europe as a whole, 1.6 million mt of fish products were exported in 1995 and 4.3 million mt were imported, generating a significant economic imbalance. Denmark is the largest exporter of fish and fish products, with Dutch exports 60% and the United Kingdom and Germany less than half those of Denmark; the latter two are, in fact, net importers of fish. Fisheries in the North Sea are either aimed at human consumption markets or provide fish for reduction to meal and oil (industrial fisheries). Fish landed for human consumption include pelagic species such as herring, mackerel and horse mackerel; demersal roundfish species including cod, haddock, whiting, and saithe; and benthic flatfish species such as plaice and sole. Landings from the industrial fishery mainly consist of sandeels, Norway pout and sprat. There are by-catches of younger age groups of commercially exploited gadids in each of these fisheries. The Danish fleet is the largest in terms of catch, with approximately 70% by volume and 25% by value for the production of fish meal and oil. The most important species landed for human consumption in the United Kingdom are mackerel, cod, haddock, and herring; in Germany cod; and in Denmark cod and herring.

Aquaculture, although relatively minor compared to fishing, is an important economic sector in Scotland and Norway where salmon and trout (*Salmo salar* and *Onchorynchus mykiss*) are produced. France and the Netherlands also produce quantities of oysters, scallops and blue mussels, and the United Kingdom, France and Norway produce seaweeds, such as *Laminaria*, for pharmaceuticals and fertilizers. However, the greater public awareness of the need to protect natural resources and increase food safety has resulted in a more heavily regulated approach, with stringent controls on new and existing facilities and operations. This, combined with technical problems and risk from disease, has added to the vulnerability of many aquaculture enterprises around the North Sea.

General assessment of fish stocks

The North Sea is a highly productive area supporting landings of about 2.5 million mt of fish and shellfish every year, plus an equivalent amount as food for predatory fish species and 0.75 million mt as food for birds and mammals, from a total biomass of approximately 10 million mt. This does not include the amount potentially eaten by birds and mammals in the Skagerrak/Kattegat and Channel. Landings for human consumption and industrial purposes are currently on average 1.6 million mt and 1.75 million mt, and are valued at 1,282 and 134 million Euros respectively.

Whilst the North Sea ecosystem does not fluctuate wildly in terms of main species, it is not entirely stable with regard to individual species. Changes in the abundance of commercially important fish stocks in the North Sea have been monitored since the 1950s; all are heavily exploited and the majority of those landed for human consumption are considered to be in a seriously depleted condition, either outside Safe

Biological Limits or below their Minimum Biologically Acceptable Level (i.e. a level of spawning stock size below which the stock may be in danger of severe depletion if it is not allowed to rebuild as quickly as possible). Results from surveys also suggest that there has been a change in the size composition of North Sea fish, with the quantity of larger fish declining and the numbers of small fish increasing (Rice and Gislason 1996).

Analytical assessments of all commercially important species are carried out by the International Council for the Exploration of the Seas (http://www.ices.dk). The results are used to establish a system of total allowable catches (TACs), and from there national catch quotas, which are the main instruments for attempting to control fishing mortality rates. About 40 fish and crustacean stocks are managed by quotas: in the European Union analytical TACs are set for stock where scientific knowledge is of sufficient quality as to be able to support the prediction of the next year's stock size. Precautionary TACs are set where knowledge is insufficient; in these cases the levels are set according to past catches so as to curtail unregulated expansion.

The TAC system has generally suffered from problems of enforcement, and as such there is a general recognition that it has failed to control fishing mortality in most North Sea fisheries. This is because without sufficient direct controls on the amount of fishing effort, fish can be caught in excess of the TAC and either discarded or landed illegally. Thus in reality, even though there are no discrepancies between the advice from ICES for TACs, the agreed TAC and the reported amount of fish landed for many North Sea stocks, large differences do exist. The result is that not only are the high levels of fishing mortality lowering the long-term yield, but also the quality and reliability of data have deteriorated, causing further problems in formulating advice (McGlade and Shepherd 1992; McGlade 1999a,b).

Other more direct means to control fishing mortality require restrictions to be placed on effort, via technical measures such as closures, restrictions on the number of vessels, capacity of the vessels, fishing gear, and time, but these have not been used or enforced systematically throughout the North Sea. At present, European Union regulations on size selectivity are mainly concerned with the definition of minimum mesh sizes; however, in some cases reference is made to voluntary use of square-mesh netting, separator trawls and aspects of the geometry of towed gear (Community Regulation 3094/86). Separation of fish and crustaceans and the avoidance of non-target species have been achieved to some extent in mobile gear using separator trawls, by escape panels and by inclusion of veil nets (Community Regulation 284/93). Closed areas are also important elements in the regulation 3094/86: within the European Community legislation a number of areas are defined within which fishing activities are limited, generally for part of a year. Examples include the plaice box in the southeastern North Sea and the Norway pout box. In the United Kingdom, a system of 34 nursery areas has been designated for juvenile bass, and both Sweden and the United Kingdom have closed areas for salmon and sea trout. Other fisheries such as shellfish have a variety of measures designed to protect them under national regulations.

The status of key pelagic, demersal and flatfish fisheries

Herring in the North Sea consist of a mixture of populations, most of which spawn in the summer or winter. In the eastern North Sea, Skagerrak and Kattegat, and along the coast of Norway there are spring-spawning herring that migrate into the North Sea to feed. Herring for human consumption are caught in the North Sea, eastern Channel and in the Skagerrak/Kattegat; juveniles are also taken in mixed fisheries with sprat. Landings peaked in 1965 at 1.2 million mt when purse seining was introduced; but landings and stock size declined rapidly until 1977 when a total ban was imposed. In 1981 the fishery was reopened and landings increased to a peak of 875,000 t in 1988, since then they have decreased. Spawning stock biomass declined from an estimated 4-5 million mt after World War II to approximately 50,000 mt in 1977; the minimum biologically acceptable level in 1995 was estimated to be 500,000 mt, outside Safe Biological Limits (Intermediate Ministerial Meeting 1997).

Mackerel has an extensive migration and changing distribution in the North Sea; individuals spawn in a number of areas, but it is not yet clear whether the North Sea stock is separate from the stock to the west of the United Kingdom and Ireland. In winter mackerel are almost absent in the North Sea; in spring they enter via the Channel and from the northwest; and in summer adults are found throughout the area, and juveniles in the southeast and Channel. They are mainly caught in directed fisheries with pelagic trawls and purse seines or as by-catch in demersal fisheries. The North Sea fishery was intensively exploited in the 1960s: landings peaked in 1967 at 930,000 mt and the spawning stock biomass was estimated to be over 3 million mt. Since then the North Sea component has collapsed, with the last good year-class occurring in 1969; currently the population spawning in the North Sea is estimated to be between 50,000 and 100,000 mt. It is therefore outside Safe Biological Limits.

Horse mackerel is also a highly migratory species. Few are found in the North Sea in winter, although some juveniles overwinter in the Kattegat. The majority are found in the Channel from where they migrate in spring into the North Sea; in summer juveniles are mostly found in the shallow southeastern area and adults are mainly in the northernmost and southern parts. Although the North Sea and western stocks are distinguishable, the data for the North Sea are not reliable enough for stock assessments. Landings of the North Sea stock have fluctuated since 1987 between 6,000 and 33,000 mt; they are taken as by-catch in the small-mesh industrial trawl fishery and the directed fishery for human consumption. Spawning stock biomass for 1989-1991 was estimated to be over 200,000 mt.

The stock of North Sea cod is found widely over the North Sea, Skagerrak and Kattegat and in the Channel; cod in the Kattegat are thought to belong to a different stock, as are those in the western Channel. Cod are mainly caught in mixed demersal fisheries, either together with haddock and whiting or with plaice and sole. There are also important fisheries targeted at cod using gillnets and lines (McGlade and Metuzals 2000). From the beginning of the last century until the late 1960s, landings fluctuated between

50,000 and 100,000 mt. In the 1960s landings increased and reached a maximum of 350,000 mt in 1972; they slowly declined from 1981-1991 and then rose again in 1995 to 140,000 mt. Over the past four decades the average yield has been 200,000 mt. Fishing mortalities have been very high and the spawning stock biomass has declined from a peak in 1962 of 280,000 mt to 75,000 mt in 1997, well below the minimum biologically acceptable level of 150,000 mt (Daan *et al.* 1994). Nowadays, fewer than 1% of the 1-year old fish survive to maturity and the catch is mainly constituted of immature fish younger than 3 years old. Given the low level of recruitment, it is now thought that the collapse of this stock may be imminent. There is also evidence that the low recruitment has been exacerbated by environmental effects related to changes in sea temperature in the North Sea, the latitudinal limit of this species distribution (Pope and Macer 1996; O'Brien *et al.* 2000).

Haddock have a northern distribution in the North Sea; they are likely to be from one stock, with juveniles and adults sharing the same distribution. Haddock are caught for human consumption in mixed demersal fisheries, together with cod and whiting. Catches are uneven because haddock tend to produce exceptionally strong year-classes at irregular intervals: landings increased in the early 1960s and peaked to a record high of over 900,000 mt in 1969. Since then landings gradually decreased and reached a record low in 1990, but have shown a slight increase due to recruitment. Fishing mortality gradually increased after World War II and has been high since the 1960s. The minimum biologically acceptable level is 100,000 mt; the spawning stock size has followed more or less the fluctuations in catches, but in the years 1990-1992 it was close to or below the limit. Exploitation has been sub-optimal with only 2% of the age 0 recruits surviving to maturity. Maintenance of the stock above safe limits depends on strong year-classes.

Whiting is one of the most abundant and widely distributed gadoid species in the North Sea. Separate assessments are carried out for the North Sea and eastern Channel, Skagerrak/Kattegat and the western Channel/Celtic Sea. Most whiting are sexually mature at two years old; they are caught for human consumption in mixed demersal fisheries and to a lesser extent in directed fisheries in the southern North Sea; however, large quantities are discarded at sea. Whiting are also taken as part of the industrial fisheries, partly in the Norway pout fishery and partly in the southern North Sea. Landings were stable until 1960 at a level of 75,000 mt; they then rose to a high in the late 1970s of around 175,000 mt; since then they have decreased to around 100,000 mt. The spawning stock biomass peaked in 1969 and 1976 at about 600,000 mt but declined between 1980 and 1985. Since then it has been stable varying between 250,000 and 300,000 mt.

Saithe are mainly found in the northern part of the North Sea and the edges of the Norwegian deep. They all belong to the same stock, maturing at 5 years and spawning in spring and winter; the nursery areas occur along the Scottish and Norwegian coasts. Landings of saithe fluctuated between virtually nil to 50,000 mt up until the 1960s; they then increased to over 300,000 mt in the mid-1970s, declined and then increased again to 200,000 mt in 1985. Since 1988 average landings have been 100,000 mt. Saithe are taken in both directed and mixed fisheries. The spawning stock biomass reached a maximum of about 450,000 mt in the early 1970s but gradually declined to an historical low in 1990; the stock is considered to be close to its safe biological limits.

Five species of sandeel occur in the North Sea; landings, however, consist almost entirely of *Ammodytes marinus*. There are three different stocks that are assessed: the North Sea, the Skagerrak/Kattegat and the Shetland. The fishery for sandeels, using small-mesh trawls, occurs in the spring and summer; sandeels are also caught in the industrial fisheries directed at sprat and Norway pout (Robertson *et al.* 1996). Landings peaked at 1,000,000 mt in 1989 and averaged 800,000 mt in the 1990s. Sandeels are thus the largest component of total fish landings from the North Sea. The spawning stock biomass has fluctuated between 400,000 and 1,700,000 mt with a peak in 1987-88. Although no downward trend has been reported for the past 20 years (Bannister 2001), recent evidence from the fishing industry, on the absence of sandeels in the stomachs of cod, suggests that there may in fact be a genuine decline in sandeel populations in the North Sea. However, given the official analysis, they are considered to be within safe biological limits. Sandeels are a major prey species in the North Sea ecosystem, supporting birds (38% of diets from 1974-1995), other fish (8-25%) and mammals (Bannister 2001).

Norway pout is a small, short-lived species, seldom living beyond three years in the North Sea. It is mainly found in deeper waters in the northern and central areas and in the Skagerrak/Kattegat: it is considered to be one stock. Small-meshed trawls are used in the Norway pout fishery; the catch is mostly used for reduction purposes. The annual landings rose from nil in the 1950s to over 750,000 mt in 1974; since then they have fluctuated between 100,000 and 300,000 mt. In the fishery there is a by-catch of juveniles of haddock and whiting: thus the Norway pout box in the northwestern North Sea was introduced in the late 1970s as an area where fishing for this species was prohibited, in order to prevent the by-catch of protected species. In deeper waters it is caught with juvenile blue whiting. The Norway pout stock is considered to be within safe biological limits: spawning stock size has increased in recent years and is currently at a high level. Recruitment, however, is variable and depends heavily on strong year-classes.

Sprat is widely distributed in the North Sea and although it is caught for human consumption in the Skagerrak/Kattegat and coastal areas, it is primarily used for the production of fish meal and oil (Roberston *et al.* 1996). Three stocks of sprat are assessed: the North Sea, Skagerrak/Kattegat and the Channel. Landings rose to 700,000 mt between 1970-1975, decreased until 1986 and have now increased. Although the size of the stock is not known it is thought to be within safe biological limits.

Four plaice stocks are assessed separately: the North Sea, the western and eastern Channel, and Skagerrak/Kattegat; these stocks are composed of several sub-populations which separate during spawning and partly overlap during feeding. Nurseries are found in shallow coastal areas. Adult plaice have been tagged and shown to migrate many hundreds of kilometres over short periods of time (Metcalfe *et al.* 1994; see: http://www.cefas.co.uk); females mature between three and four years of age, males somewhat younger. Plaice are fished in directed fisheries, often with sole. Since the 1960s the fishery has been largely prosecuted as a beam trawl fishery. Landings in the first half of the last century were around 50,000 mt; having recovered during World War II, landings remained between 75,000 and 170,000 mt from the mid-1950s to 1989. Since then they have declined to under 100,000 mt; fishing mortality levels are high and well above the level estimated for maximum sustainable yield. Recruitment has been very stable, perturbed only by occasional strong year-classes, particularly two in the 1980s. Spawning stock biomass peaked at 500,000 mt in the 1960s, decreased to 300,000 mt in the 1970s, rose again to over 400,000 mt, and declined to 215,000 mt at the beginning of 1995, well below the minimum biologically acceptable level. The plaice box was introduced in 1989 to protect juvenile plaice along the continental coast. Originally the box was only closed to heavy beam trawlers (>300horse power) for half the year, but since 1994 a year-round closure has been in place. Surveys have indicated that adult as well as undersize plaice have increased in abundance since the box was established. However, fleets of small beam trawlers doubled their effort in the box between 1989 and 1993, somewhat negating the effect and reducing the increase in stock from the 25% expected to 14%.

Sole are at their northern limit in the North Sea. Several stocks are distinguished including two in the Channel, one in the North Sea and one in the Skagerrak/Kattegat. Most females mature at three years and spawning takes place in coastal areas in April-May; nursery areas occur along the shallow waters of the southern North Sea. Sole are exploited with plaice in mixed flatfish fishery by beam trawlers; they are also taken by otter trawlers and fixed nets in inshore areas. Landings increased during the second half of the last century, and as with plaice were strongly influenced by strong year-classes; they fluctuated between 10,000 and 35,000 mt, with fishing mortality increasing by a factor of four from the mid-1950s to the mid-1980s. Recruitment is stable, but since it is a southern species it is affected by cold winters. In the severe winters of 1962-63 about 60% of the adult sole did not survive the low temperatures, and similarly in 1995-96. The spawning stock size peaked in the 1960s at 150,000 mt; the current level of 35,000 mt is above the minimum biologically acceptable level and the stock is considered to be within safe biological limits.

Status of shellfish fisheries

There are several commercially important shellfish species of molluscs and crustaceans. The most important crustaceans are the deep-sea shrimp (*Pandalus borealis*), brown shrimp (*Crangon crangon*), edible crab (*Cancer pagurus*), spider crab (*Maja squinado*), Norway lobster (*Nephrops norvegicus*), and lobster (*Homarus gammarus*). Important molluscan species are squid (*Loligo sp.*), cuttlefish (*Sepia officianalis*), oyster, mussels (*Mytilus spp.* and *Modiolus spp.*), Great scallop (*Pecten maximus*), Queen scallop

(*Chlamys opercularis*), trough shell (*Spisula spp.*), cockle (*Cerastoderma*), razor clam (*Ensis spp.*), winkle (*Littorina littorea*), and whelk (*Buccinum undatum*). Information on these fisheries is limited and, apart from Norway lobster and deep-sea shrimp, is largely restricted to landings data.

Mussel, whelk, winkle, cockle, crab, lobster, and shrimp fishing activities are concentrated in the coastal zones and estuaries, whereas Norway lobster and deep-sea shrimp are international fisheries. The cultivation of mussels is a major activity in the Wadden Sea, the eastern Scheldt and along the northern coast of France; in the Wadden Sea it occurs on both subtidal and intertidal beds. The fisheries, especially for seed mussels, have contributed to the disappearance of mature beds. As a result, no mature beds remain in the Dutch part of the Wadden Sea and numbers have strongly declined in the Niedersachsen part of the Wadden Sea. The recovery of mature beds and the opportunities for the development of new ones are seriously disrupted by the seed mussel fisheries, and several areas within the Wadden Sea have been closed. Mussel production, especially in this area, fluctuates strongly, and the availability of seed restricts production; more recently, production in the Dutch sector of the Wadden Sea has been augmented by half-grown mussels from the German sector. In the 1990s, annual landings from the Dutch, German and Danish Wadden Sea have totalled 110,000 mt. In Danish waters, fishing is restricted to natural mussel beds in the subtidal areas inside the islands, of which c. 50% are closed to fishing, giving rise to annual landings of c. 5-10,000 mt. The French production in the Channel is 26,500 mt, and in the Wash on the east coast of the United Kingdom annual landings are 5-10,000 mt despite the persistent spatfall failure since 1990.

The French production of Pacific oyster (*Crassostrea gigas*) in the Channel is c. 40,000 mt; there is also a small-scale production of oysters in several estuaries in southeast England, and mariculture in the northern isles off Scotland in the northern North Sea, where oysters, scallops and mussels are produced. In the Wadden Sea, the European oyster (*Ostrea edulis*) is extinct; however, in Limfjord the species has re-established natural beds. Cockle fishing is significant in the Moray Firth (Scotland), the Wash (England), the Dutch Wadden Sea and Delta area, and in some Danish waters. Annual landings in the Netherlands range from zero to c. 7000 mt of flesh (i.e. 50,000 mt wet weight). The proportion of total stocks harvested by the Dutch fisheries is below 10% in most years. In Germany, the cockle fishery has been closed for bird protection. Fisheries for trough shells (*Spisula spp.*) began in Denmark and the Netherlands in 1990 and in Germany in 1992 and have increased since. Biological information and statistics on the two species that are harvested are limited; off Germany alone the stock size was estimated to be 200,000 mt. However, the extreme low temperatures in the winter of 1995-96 led to mass mortalities of *Spisula solida*. Razor clams are found in large concentrations off the Wadden Sea coast, but as yet are not exploited. Scallops have been fished off Scotland for many years; maximum annual landings are c. 2500 mt. Queen scallops are also fished around the northern isles of Scotland and northeast England, with annual landings of c. 330 mt and 600 mt respectively.

The deep-sea shrimp fishery differs across the North Sea. In some areas, such as the Norwegian Deep and Fladen Ground in the northern North Sea, it tends to be an opportunistic fishery, strongly influenced by stock abundance and market prices. In some other places, such as off the Swedish coast, the fishery is year-round and the catch landed for the domestic market. Deep-sea shrimp stocks are short-lived and dependent on recruiting year-classes. The spawning stock biomass in the Skagerrak and the Norwegian Deep increased in 1994 to a level of c. 15,000 mt due to the strong 1992 year-class. The stock is currently considered to be within Safe Biological Limits; however, the state of the stocks on the Fladen Ground and in the Farn Deep is not known.

The main commercial fishery for crustacea in the North Sea is for Norway lobster; important fisheries exist in the Skagerrak/Kattegat, Moray Firth, Farn Deep, Firth of Forth, and the Fladen Ground. Landings from the Skagerrak/Kattegat have been stable over the past 10 years at c. 3,000 mt. Total landings for the rest of the North Sea increased from c. 8,000 mt in 1985 to 14,000 mt in 1994, due to the increased fishery on the Fladen Ground. The landings are mainly comprised of males, because the egg-bearing females remain in their burrows; this behaviour means that the lobster is restricted to areas with particular sediment types. Spawning stock biomass and recruitment appears to be stable; however, there are signs of growth overfishing in a number of stocks.

Governance of European fisheries

The primary aim of the North Sea management is to ensure sustainable, sound and healthy ecosystems, maintain biodiversity and ensure sustainable exploitation of the living resources in order to achieve economically viable fisheries. The Esbjerg Declaration 1997

Management of the North Sea fisheries falls within the European Union's Common Fisheries Policy. Overfishing and competition due to the globalisation of the market in fish products have become major threats to the future of fish stocks and the industry. As a result, European Union Member States have adopted a policy of restructuring the fleet, in order to respond to the constraints imposed both by the resource and market demand. In particular, financial support has been made available to the sector, in line with its objectives of promoting economic and social cohesion in less well-off regions. Specifically, support has been aimed at reducing fleet sizes: progress is monitored through the multi-annual guidance programmes (MAGPs), which have been used to set conditions since 1983. Modernisation has been necessary to increase safety, improve hygiene conditions on board vessels, adapt vessels to new fisheries, and facilitate selective fishing methods.

The idea behind the MAGPs was quite simply one of devising programmes to show how fleets needed to evolve in relation to resource status and market demand. MAGP I

(1983-1986) aimed to prevent further increases in fleet capacity; although it did not entirely succeed, the idea was sown that fleets could not continue to increase. MAGPII (1987-91) introduced a modest cut in fleet tonnage and engine power; however, good returns from fishing and the fear of increased competition from two new Community Members (Spain and Portugal) encouraged Member States to disregard their targets. Moreover, there was no check on whether funds given to scrap vessels were being used to build new, more efficient ones. In contrast, MAGP III (1992-96) did not express cuts in terms of capacity reductions but rather as decreases in fishing effort. To ensure that stocks in most danger were protected, the totality of stock was divided into three groups and matched against segments in the community fleet. The reductions included 30% for demersal stocks, 20% for benthic species (flatfish) and no change for pelagic stocks. Overall the fleet was cut by 7%. Before preparing the MAGP IV (1997-2001), the European Commission asked a group of independent scientists to evaluate the state of fish stocks (the Lassen report). This showed that several commercial stocks were still under too much pressure. Thus the European Council decided to cut fishing effort in 1997 by 30% on stocks in danger of collapse and 20% on overfished stocks. There was also a focus on the impact of technological progress. Unfortunately, much more still needs to be done, not only to conserve stocks but also to make the Common Fisheries Policy structural policy more effective.

In 1993, all the fishing sector budgets were combined to create the Financial Instrument for Fisheries guidance: together with funds from socio-economic measures for fishermen losing their jobs and by an initiative called PESCA, which provided access to structural funds for regions dependent on fishing, alternative activities and jobs were created and developed. While PESCA has not been renewed, opportunities provided by the initiative will be retained.

The common organisation of the market in fisheries and aquaculture products was set up in the European Community 30 years ago; since then the free circulation of goods has been achieved for all products in line with the framework regulated by the World Trade Organisation. The common organisation has several components, including: common marketing standards for fresh products on quality grades, packaging and labelling; producer organisations, which are voluntary associations of fishermen set up to stabilise markets, whose role is to protect fishermen from sudden changes in market demand; a price support system which sets minimum prices below which fish products cannot be sold; and rules for trade with non-European Union countries. The aim is to balance the needs of the European Union and the interests of the European Union fishermen and to ensure that the rules on fair competition are respected.

As with other fisheries policies, the Common Fisheries Policy has had to adapt to major changes since its inception in 1970. Supply often does not match the needs of the market in terms of quantity, regularity or quality, mainly as a result of poor conservation and overexploitation. In addition, supermarket chains, now the main buyers of fish products, expect steady supplies and turn to imports to meet their demands. In late 1997, the Commission started a process of consultation on the future of the Community market in fish products. Three major issues have emerged beyond the need for better conservation: these are the role of producer organisations, responsible fishing and consumer protection.

Producer organisations are formed by fishermen or fish farmers freely associated in order to take measures to ensure the best marketing conditions for their products. While membership is voluntary, producer organisation members have to respect rules in their production and marketing operations. Producer organisations have to meet a number of conditions before they are recognised: they must represent a minimum level of economic activity in the area they cover, not operate any discrimination in terms of nationality or geographical location of members, and meet the necessary legal requirements of member states. The European Union has created mechanisms to correct the worst effects of fluctuations in supply and demand: the Council of Ministers sets guidance prices for a number of species and these are then used as a reference to set withdrawal prices. To ensure a minimum revenue for fishermen, producer organisations can enforce these prices by taking products off the market when prices fall. Members then receive an indemnity from their producer organisations, which in turn receives compensations from the European Union. Other measures include carry-over operations, and aid schemes in case of market collapse; producer organisations can also apply for extensions to their discipline. Greater involvement by fishermen in producer organisations and closer co-operation with the market is envisaged in the next Common Fisheries Policy.

In a number of Member States co-management has also been introduced (McGlade 2001a). Central to the idea of co-governance is the concept that society should participate more thoroughly in governing resources, including an active involvement in the formulation and implementation of public policies. In 1993, the Dutch government introduced a co-management scheme, which has subsequently been very successful (Kooiman *et al.* 1999).

The Common Fisheries Policy does not exist in isolation, and its international dimension has increased in recent years to include both bilateral and multilateral agreements. Through these the concept of responsible fishing has become more important; the idea of responsible fishing refers to an assurance that conservation measures have been adopted for a particular fishery. Consumers can contribute to this process by buying products that do not come from fisheries in breach of the regulations; to this end there is a growing effort to set up certification schemes. An example of such a scheme is that of the Marine Stewardship Council (see http://www.msc.org).

ECOSYSTEM HEALTH and POLLUTION

No one single definition of aquatic ecosystem health exists, although a range of indicators and criteria have been proposed including biodiversity, stability, yields, productivity, and resilience. Aquatic ecosystems do not exist in a steady state, but exhibit continuous changes in production and species composition. To be healthy and sustainable, an ecosystem needs to be able to maintain its internal functioning in the face of external stresses. What happens when an ecosystem is subjected to short, sharp shocks, and how quickly it recovers from transient shocks determine its resilience. In aquatic food webs, resilience has been shown to be linked to the number of trophic levels present (Pimm 1999) and the degree to which the biomass is altered: the greater the alteration, the faster the change. Resistance in ecosystems is the degree to which it remains unchanged when its component parts are altered: this is related to the presence and maintenance of keystone organisms (Paine 1966, 1974). Keystone organisms are often overlooked until such time as they are lost or displaced by the unintentional introduction of non-indigenous organisms, when dramatic changes in community structure or collapse of the trophic cascade can occur (Carpenter *et al.* 1985).

The health of marine ecosystems can be altered indirectly by a large number of human-derived processes, including excessive nutrient loading from agriculture, industrial pollution, changes in freshwater fluxes, and sedimentation. Because biogeochemical cycles and the availability of organisms to make the best use of them are tightly coupled, variability in the biogeochemistry of a water body can also lead to sudden trophic changes. Detecting the occurrence of such changes, and understanding their origin for multiple-state comparisons, is thus an important part of any assessment.

Measures to reduce pollution in the North Sea have been in effect since the 1970s, with the advent of the Convention for the Prevention of Marine Pollution from Land-Based Sources (the Paris Convention 1974), which was merged with the Convention for the Prevention of Marine Pollution by dumping from ships and aircraft (the Oslo Convention 1972) to form the Convention for the Protection of the Marine Environment of the North-east Atlantic (the OSPAR Convention). The OSPAR Convention has been ratified by all signatories and entered into force in March 1998. At Sintra, Portugal, in July 1998, a new agreement was signed under the auspices of the OSPAR Convention. Signatories agreed to continue to reduce emissions of hazardous substances with the ultimate aim of achieving background concentrations for naturally occurring substances and close to these concentrations for synthetic substances within one generation. By 2003, programmes for the prioritisation of hazardous substances and the control of emissions are to be established.

Inputs of dangerous substances are controlled through European Community and national legislation via limitations on discharges where they may enter the North Sea, together with restrictions on use. In addition, inputs from urban wastewater treatment plants, from agriculture in the form of nitrate, from industry, from indirect inputs of nitrates from motor vehicle exhausts, and from pesticides are regulated *inter alia* via specific controls. At each Ministerial Conference various measures have been adopted, including undertakings to reduce aqueous discharges and atmospheric emissions of substances that are toxic, persistent and liable to bio-accumulate; phase out certain harmful practices; adopt a precautionary principle and best available technology and environmental practice; and enhance habitats and species protection (International

Ministerial Conference on the Protection of the North Sea 1984, 1987, 1990, 1995; Intermediate Ministerial Conferences 1993, 1997).

Ecological impacts of pollution

Both offshore and land-based activities have a significant effect on the North Sea ecosystem. Often the cause and effect of specific pollutants on the ecosystem are difficult to assess, as changes in the environment are often obfuscated by poor sampling and natural variability. Nevertheless, some key adverse effects have been identified in the North Sea: these include a) the impact of tributyltin (TBT) on the development of male sexual characteristics in the female dogwhelk (*Buccinum undatum*) - the imposex effect (Ten Hallers-Tjabbes *et al.* 1994), and on the shell structure, growth and condition of Pacific oysters; b) a prevalence of tumours in flatfish in areas exposed to polycyclic aromatic hydrocarbons (PAHs); c) the effects of organochlorine compounds on hatching success of fish eggs; and d) the impact of polychlorinated biphenyls (PCBs) and mercury on seafood (Intermediate Ministerial Conferences 1997). Key areas of interest in relation to pollution are nutrients, eutrophication and bacterial contamination, hazardous substances, oil and oil wastes, radioactive substances, and litter.

Nutrients, eutrophication and bacterial contamination
The general increase in nutrient discharges from rivers, run-off and the atmosphere, largely resulting from sewage effluents, leaching from agricultural land, contributions from rural populations and atmospheric nitrogen deposition, is such that eutrophication is now a major environmental issue at both national and international levels (Gerlach 1990; North Sea Task Force 1993; Asman and Berkowicz 1994). Although both phosphorus and nitrogen contribute to eutrophication, nitrogen is most important in estuarine and coastal waters. In 2000, five estuaries were designated as Sensitive Areas (eutrophic) under the EC Urban Wastewater Treatment Directive (91/271/EEC): for each designated area, there is now a requirement for nutrient reduction in discharges from sewage works serving a population of 10,000 or more, unless it can be shown that this will not result in improvements in water quality.

Eutrophication in the North Sea manifests itself in a variety of ways, both inshore and in offshore deposition areas such as the Dogger Bank (Kroncke and Knust 1995). Inshore, there has been an increased growth of *Enteromorpha* and *Ulva* on intertidal mudflats, particularly in estuaries, generally indicative of excessive nutrient inputs (Piriou *et al.* 1991; Den Hartog 1994; Reise *et al.* 1994; Scott *et al.* 1999). There has also been a shift from long-lived macrophytes to short-lived algae; 50,000 t of these nuisance algae are removed from North Sea beaches each year (North Sea Task Force 1993). These algal mats can also create anaerobic conditions preventing birds from feeding on the intertidal mudflats, and leading to fish kills (Desprez *et al.* 1992).

In the North Sea, later stages of eutrophication are associated with algal blooms. Communities of phytoplankton build up and reduce water clarity and block sunlight, resulting in a reduction in biodiversity, as sensitive plants and animals are replaced by more nutrient tolerant species (Dippner 1998). Algal blooms in the North Sea can cause large fluctuations in dissolved oxygen concentrations as they photosynthesize during daylight hours, adding oxygen to the water, but consume oxygen at night, leading to invertebrate and fish mortalities (Richardson 1989). Blooms are often spectacular events evident as large discolourations of the water by pigments in the algae involved (often referred to as "red tides"), and excessive accumulations of foam on the shore, producing unpleasant odours. In the North Sea, marine algal blooms most commonly result in an aesthetic nuisance when large accumulations, foams or scums are washed inshore. The most common taxa include *Phaeocystis* and *Noctiluca* which can discolour the water brownish red.

High concentrations of algae can affect water treatment for public supply by blocking filters and affecting taste and odour. The high nutritional quality of algal carbon also leads to bioaccumulation of hydrophobic organic contaminants, such as polychlorinated biphenyls and polycyclic aromatic hydrocarbons, in benthic communities and their predators (Gunnarsson *et al.* 2000). Hydrophobic organic compounds have a high affinity for organic matter and will thus sorb to phytoplankton cells, humic acids and colloids in the pelagic system, making them readily available as food for organisms such as blue mussels (*Mytilus edulis*), brittle stars and polychaetes, hence providing a benthic-pelagic coupling for these compounds. Bioturbation then enhances the release of these contaminants from the sediment to overlying water.

Although many marine algal blooms that discolour the water are harmless, some, such as *Alexandrium* and *Dinophysis*, can cause poisoning through the food chain when shellfish become contaminated with toxins from ingested algae and are subsequently consumed by fish, birds and humans. This can occur even in non-bloom conditions, as filtering by shellfish will concentrate the toxin. Such events give rise to paralytic, diarrhetic, neurotoxic, and amnesic shellfish poisoning (PSP, DSP and ASP), depending on the type of marine algae involved. The main toxic species in the North Sea are *Alexandrium spp.*, the dinoflagellate species thought to be the cause of PSP, *Dinophysis spp.* and *Prorocentrum lima,* thought to produce DSP, and *Pseudo-nitzschia spp.* associated with ASP. The first documented human fatalities from algal toxins occurred in Germany in the 1880s. From 1828-1968, only 10 outbreaks of paralytic shellfish poisoning were documented in the United Kingdom; however, in 1968 an outbreak occurred in the northeast of England, and 85 people were affected after eating locally collected mussels. In the last 30 years all three types of poisoning have been detected, with PSP and DSP occurring in mussels in Danish, Norwegian and French waters in the late 1980s and 1990s, DSP in Dutch mussels from 1976 onwards, and PSP regularly in the United Kingdom. Some types of algae, although harmless to humans, may still be toxic or harmful to marine organisms. *Gyrodinium aureolum* has been associated with shellfish and fish mortalities in addition to causing a red discolouration of the water (a red tide). Another alga, *Chaetoceros*, has spines that may clog the gills of fish, particularly in fish farms where migration by fish away from a bloom is not possible.

Nowadays, the presence of marine algal blooms can be inferred from chlorophyll-*a* concentrations, when a value in excess of 10µg/l is indicative of the presence of an algal bloom, using both spaceborne and aerial surveillance techniques. Spaceborne sensors such as the Sea-viewing Wide Field of View Sensor (see SeaWiFS http:// seawifs.gfsc.nasa.gov) now provide a good spatial overview of a large part of the North Sea, allowing continuous assessments to be made of the extent and characteristics of algal blooms, such as those of the haptophyte (*Emiliania huxleyi*), and red tides (*Alexandrium tamarense*) (see http://www.environment-agency.gov.uk).

The quality of bathing waters in the North Sea is monitored by each Member State against standards laid down in the EC Bathing Water Directive (76/160/EEC). As the regulations have been implemented, the number of beaches to be monitored has increased: in England and Wales for example, only 27 bathing waters were included in 1979, then a further 333 were identified in 1987, and the number has since grown to 473 coastal bathing waters in 2000. The consistency of monitoring practices changed during this period but is now strictly standardised.

The assessment of compliance against the standards of the Directive are rather complex. Each sample is analysed for total coliform bacteria and for faecal coliform bacteria. Faecal coliform bacteria are an indicator of the presence of traces of human sewage. The *mandatory* (or imperative) standards, which should not be exceeded, are 10,000 total coliforms per 100 millilitres (ml) of water and 2,000 faecal coliforms per 100ml of water. In order for a bathing water to comply with the Directive, 95% of the samples (i.e. at least 19 out of the 20 taken) must meet these standards, plus other criteria. The *guideline* standards, which should be achieved where possible, are no more than 500 total coliforms per 100ml of water and no more than 100 faecal coliforms per 100ml of water in at least 80% of the samples (i.e. 16 or more out of 20) and no more than 100 *faecal streptococci* per 100ml of water in at least 90% of the samples (i.e. 18 or more out of 20).

Other standards that must be met include the presence of enteroviruses and *Salmonellae*. Consistent compliance with the imperative and guideline standards of the Directive (for coastal and freshwater sites combined) has been steadily increasing throughout the North Sea in recent years as a result of a substantial and continuing investment by the water service companies, and the cessation of disposal of sewage sludge in 1998. The results of the various bathing water analyses are used to assist in giving "awards" to bathing beaches, such as the European Blue Flag Scheme, organised by the Federation of Environmental Education in Europe.

Hazardous substances
Member States surrounding the North Sea have an obligation to monitor and control the discharge of hazardous substances to estuaries and the marine environment. Hazardous

substances include persistent organic contaminants, such as polychlorinated biphenyls, dioxins and pesticides, heavy metals (e.g. lead, copper, mercury, cadmium), antifouling agents (e.g. tributyltin), and polycylic aromatic hydrocarbons. Legislation exists that is concerned with inputs to the North Sea and Atlantic including both riverine and direct discharges. Estuarine and coastal waters and sediments are monitored downstream of major industrial and sewage discharges containing List I substances, as classified in the EC Directive on Dangerous Substances (76/464/EEC).

Standard methods for estimating the inputs of chemicals have been provided by OSPAR: there are two principal means, both of which are based on the product of two numbers: the flow rate of the river, sewage or industrial discharge and the concentration of the substance. Where the substance cannot be detected, a calculation can be made assuming that it is not present at all (low-load estimate), or a calculation can be made assuming that the substance is present exactly at the limit of detection (high-load estimate). The limit of detection is the lowest concentration of a substance that can be reliably measured - any real concentration lower than this level, including zero, is reported as present at less than the limit of detection concentration. Monitoring takes place at sites remote from these discharges, known as national network sites. Sampling is carried out monthly at sites close to the discharges and quarterly at the national network sites. Quarterly sampling is also carried out in the three nautical mile zone of coastal waters, in order to establish a baseline concentration for a range of substances. Total riverine inputs to the sea can also be estimated. The results are reported annually and are compared against List I and List II statutory environmental quality standards (EQSs). Sources of organic inputs into estuarine sediments can be determined by pollutants and biogenic markers (Readman *et al.* 1986).

Long-term trends in heavy metal concentrations in the North Sea are difficult to determine, as data collected in the North Sea before 1975 are generally not considered to be very reliable because values were close to the analytical limits (Hempel 1978; Pedersen 1996). Since then, however, improvements in analytical methods have stimulated the organisation of a number of international programmes to monitor pollution in the marine environment; of particular importance is the Joint Monitoring Programme of the Oslo and Paris Commissions (1990). However, despite reductions in emissions of many metals during the last 20 years as a result of European and international regulations, it has only been possible to detect significant changes in the environment in a very few cases: for example, in heavily contaminated coastal waters (Pedersen 1996). Nevertheless, overall compliance has improved over the last six years.

Compliance for List II substances has been much lower; List II failures include tributyltin and endosulphan from industries and sewage-treatment works, copper derived from antifouling paints, and titanium dioxide for use as a white pigment which can result in a highly acidic, metal rich effluent. The three plants producing titanium dioxide in England are monitored under the EC Directive on titanium dioxide (78/176/EEC, 82/883/EEC, 89/428/EEC, 92/112/EEC). These plants discharge into the

Tees and Humber estuaries. In the 1980s there was evidence of localised pollution near the two discharges into the Humber, resulting in acidic water far from the discharge and damage to life inhabiting the bed of the estuary. The discharge point was moved further offshore, allowing for greater dilution and thereby minimising the impact on the environment. The Teesside plant has had no significant environmental impact due to the cleaner process technology employed at the site. This can be seen by the diverse fish and invertebrate communities found in the vicinity of the plant and a small seal population on the nearby Seal Sands. However, results of sediment monitoring show elevated metal concentrations at outfall sites compared to reference sites, and biological monitoring results show localised effects at all three discharges, suggesting that the Humber estuary is a trap for fine, metal contaminants (Millward and Glegg 1997). In the German Bight, Radach and Heyer (1997) showed that less cadmium left compared to the amount that entered: they found that it was concentrated in deposition areas where it was absorbed by phytoplankton and ingested by the benthos (Hall *et al.* 1996), leading to high levels of cadmium and mercury in the kidneys and liver of predators.

Data on the biological effects of hazardous wastes are still rather limited. Specific effects that have been observed include feeding and endocrine disruption, reduced "scope for growth," decreased reproductive success, embryo malformation, increased susceptibility to diseases, and lower survival rates in fish, mammals and birds (Readman *et al.* 1993; Dethlefsen *et al.* 1996; Donkin *et al.* 1997; Lowe 1998). The North Sea Task Force (1993) concluded that the greatest risks to the North Sea ecosystem came from copper, mainly in lower trophic levels; from cadmium and mercury, in top predators and birds (Furness *et al.* 1994); and from lead, mainly in shellfish. In a survey of fish diseases in the North Sea, Vethaak and Rheinallt (1992), suggested that a range of ætiologies were associated with direct effects of contaminants (e.g. skeletal deformities in many species associated with heavy metals and toxicants; liver nodules and tumours in flatfish with polycylcic aromatic hydrocarbons; papilloma with heavy metals such as chromium; and skin ulcers in flatfish and eels and ulcus syndrome in cod associated with nutrient enrichment), together with a variety of ætiologies associated with stress.

Polycyclic aromatic hydrocarbons occur naturally, but are enhanced in the North Sea by inputs from human activities, such as discharges from oil and oil cuttings, incomplete combustion processes such as flaring and industrial effluents. Background hydrocarbon concentration in biota is strongly dependent on species and age (Little 1997). In mussels, the "scope for growth" associated with PAHs, tributyltin and other hydrocarbons declined from north to south, possibly reflecting the inflow of clean water from the North Atlantic via the north of Scotland (Widdows *et al.* 1995). Many other synthetic organic compounds, such as hexachlorobenzene, octachlorostyrene, hexachlorocyclohexanes (lindane), PCBs, DDT, polychlorinated dibenzodioxins, polychlorinated dibenzofurans, polybrominated diphenylethers, and polybrominated biphenyls, are also found in estuaries and in the open North Sea in sediments and concentrated in the tissues of organisms, such as the blubber of marine mammals

(Jensen 1996). The fact that relatively high levels of DDT concentration still occur in biota and in many of the major river estuaries running into the North Sea (e.g. Thames, Weser, Elbe, Western Scheldt) suggests that more work is still required to understand its sources. Other organics, such as pesticides (e.g. dichlorvos) and antibiotics used in aquaculture, have also come under closer monitoring in recent years, but the long-term effects of these on the ecosystem are as yet not known.

Exposure to tributyltin, derived from anti-fouling paints, produces imposex in dogwhelks (*Nucella lapillus*), leading to sterility and destructive effects on the population at very low levels (2ng/l). Contamination now occurs primarily from leaching from small boats; where statutory controls have been implemented the rates have been falling, especially where copolymers have been used. Results from the Dutch surveys in 1991-92 concluded that the whelk (*Buccinum undatum*) was no longer present in the shipping lanes in coastal areas, and that imposex was present in the deepwater shipping lanes and near the Dogger Bank. In a survey of the North Sea, Harding *et al.* (1992) showed that the majority of dogwhelk populations around the North Sea had marked reductions in the production of egg capsules. TBT also effects fish reproduction and shellfish growth; thus its persistence in the North Sea is still seen as a major problem.

Oil and oily wastes
The quantity of oil entering the North Sea annually is primarily from land-based sources including ports and harbours (24-76 kt), general shipping, illegal discharges and oil spills (15-60kt), and sewage sludge and waste (4-22 kt), rather than from oil and gas operations (12kt) (North Sea Task Force 1993; Mathiesen 1994; Olson 1994). Oil and oily wastes are an important source of polycyclic aromatic hydrocarbons in the North Sea; they accumulate in the sediment mainly from drill-cuttings and activities around platforms. The associated effects on the benthic communities include smothering and chronic pollution, which in turn can cause a reduction in the number of sensitive species, an increase in opportunistic species, increased mortality, overall reduction in macrobenthos abundance, and reduced diversity of the whole macrobenthos community (Gray *et al.* 1990). Other effects include reduced "scope for growth" in shellfish such as mussels, reduced algal concentrations, increased mortality in early life history stages in fish, and a number of distorted physiological processes. For example, data on mono-oxygenase activity, indicated by ethoxyresorufin-O-deethylase (EROD), shows elevated levels in fish near oil and gas drilling sites: EROD is a measure of biotransformation of environmental contaminants. After drilling has ended, recovery of the macrobenthos takes place over two to three years, although in heavily contaminated areas recovery may take longer. Options for the long-term treatment of cuttings piles are likely to be costly, and need to set against some 400km^3 of seabed showing biological effects (Environment Agency 1998).

In response to the impact of drilling mud oil, OSPARCOM introduced tighter standards on the quantity of oil that can be discharged and banned oil-based mud systems. More recently water-based muds have been used for drilling; in a study of their effects in the

Dutch sector, no adverse effects were observed in benthic communities, even within 25m of the discharge site (Daan and Mulder 1996). By comparison, one year after drilling, biological effects of oil-based mud discharges were detectable up to ≥1000m by reduced abundances of a few very sensitive species (particularly *Echinocardium cordatum* and *Montacuta ferruginosa*). Closer to the well sites, increasing numbers of species showed effects, even after 8 years; but, in the longer term, the macrofauna seemed to recover at distances ≥500m (Daan and Mulder 1996). The chemicals used in drilling muds are controlled by the OSPARCOM Harmonised Notification Scheme, which requires data on ecotoxicity of all substances used to be monitored. This is run in conjunction with the Chemical Hazard Assessment and Risk Management scheme, which predicts the risk of a particular chemical in a particular location. Given the remoteness of most platforms, oil companies are largely self-regulating; thus oil spills must be reported and permission sought for gas flaring.

Oil in production water is currently the largest direct input of oil to the sea; the industry is working to reduce this and concentrations of less than 20mg/l have been reached at some sites. Oil concentrations in the water itself are generally low; however, large oil spills from tankers can have major impacts on coastal ecosystems as seen from the *Braer*, which lost 85,000 tonnes of light crude oil, and the *Sea Empress*, which lost 72,000 tonnes of crude oil. In the case of the Braer, 35% of the oil was deposited in the subtidal sediments, 50% dispersed to the sea and 14% evaporated into the atmosphere (Edgell 1994; Davies *et al.* 1997).

The presence of oil as slicks is mainly a problem for birds and marine mammals; not only can these arise from drilling operations and dumping of bilgewater from ships, but also as concentrations of lithophilic substances such as alkylphenols, as in the case of the Wadden Sea where 10,000 birds died at one time (Skov 1991). Despite the number of controls on shipping, there does not appear to be a reduction in the trend of oiling in birds throughout the North Sea. The illegal discharge of oily waste by ships is less obvious, but over an annual period amounts to a quantity similar to that of a large oil spill such as from the *Sea Empress*. The designation of North West European Waters as a Special Area for the discharge of oily waste and the requirement for port authorities to prepare port waste management plans is therefore an important step in containing oil pollution. Large oil spills can cause aesthetic and economic impacts. There is no established scheme for assessing the aesthetic quality of beaches which adequately takes oil into account; however, Coastwatch UK found oil and petrol in the intertidal and upper shore areas in four to six percent of 0.5km units surveyed between 1991 and 1996 (Pond and Rees 1996), and Beachwatch 1995 found an average of more than one oil drum per kilometre on the 187km of beaches surveyed (Marine Conservation Society 1995).

Decommissioning of oil and gas platforms has presented a number of technical problems, as well as social debate, evidenced in the disposal of *Brent Spar* (Natural Environment Research Council 1996). For example, there is currently no tested technology to bring drill cuttings safely to the surface without dispersing any contamination into the environment. However, rules for decommissioning were agreed upon by OSPARCOM in 1998 (OSPAR Decision 98/3 on the Disposal of Disused Offshore Installations), with the result that the topsides of all installations plus all the steel jacket structures must now be brought ashore for onshore dismantling, with the possible exception of the footings which go deep into the seabed. The decision allows for exceptions in cases where the regulating government assesses that an alternative disposal method is preferable, following consultation with other OSPAR contracting parties: all installations put in place after 9th February 1999 must be completely removed when decommissioned. It still remains for the European Union to clarify the legal status of the waste arising in international waters.

Radioactive substances
Natural and artificial radionuclides occur in the North Sea. The most important natural nuclide is polonium-210, which emits alpha radiation to humans and biota such as shrimps and mussels. It occurs at enhanced levels in areas affected by waste discharges from phosphate-ore processing. It has been estimated that the dosage from consuming natural nuclides in marine food products is approximately 2 mSv per year (North Sea Task Force 1993). Artificial nuclides, the most important of which is caesium-137, are discharged from reprocessing plants at Cap de la Hague and Sellafield; the dose is two orders of magnitude less than that of the natural nuclides. The effects of Chernobyl can still be observed in the Kattegat and Skagerrak as a result of outflows from the Baltic.

Litter and dumping
Litter can be found throughout the North Sea and all along its coasts. The main sources are shipping and tourists on beaches, rivers, dumping of sewage sludge, and aggregate waste. Disposal of dredged material amounted to approximately 70 million mt per year in the 1990s; however, the release of inert material has stopped and vessel and other vehicle scrap dumping will be prohibited by 2005. Discharge of sewage sludge was supposed to have ceased in 1998, but organic pollution from fish processing still remains high. Fisheries also generate significant amounts of litter in the form of lost and damaged gear that, in the case of ghost nets, can cause mortalities by entanglement for birds, fish and mammals. Plastics are also a major hazard. In 1988, during a survey in the southern North Sea and German Bight, 350-700kg/km^2 were caught, and on a beach survey there was a mean density of 9000kg/km^2 of litter on the Dutch shore (North Sea Task Force 1993). Ammunition and various types of canisters used in warfare still continue to wash up on beaches around the North Sea.

Ecological effects of other marine industries

One of the most obvious industries to impact the marine ecosystem is fishing (Gislason 1994; Kaiser and de Groot 2000). At present, between 30 and 40% of the biomass of commercially exploited fish species is taken annually, by industrial fishing (58%), purse seine (16%), otter trawl (8%), beam trawl (5%), seines (5%), pelagic trawl and gillnet (3%), pair trawl (3%), and other (3%). As well as its direct effect on fish populations, each fishery has a different impact on non-target species taken as by-catch.

In 2000, ICES reviewed the effects of different types of fisheries on North Sea benthic ecosystems (International Council for the Exploration of the Sea 2000). The report concluded that the effects of bottom trawling on both habitats and species could be identified over a range of temporal and spatial scales.   More especially, it was concluded that low-energy environments (i.e. those with consolidated sediments such as muds, gravel and boulders) were more vulnerable than high energy sites (i.e. those frequently disturbed by natural processes - storms, tidal scour etc. - such as the southern North Sea) where sensitive species had already been depleted or locally extirpated. Secondly, the impact of bottom trawling on benthic communities could have permanently compromised their resilience, i.e. their ability to return to their original condition; thus cessation of trawling would not always result in a return to a pre-impacted state.

The report cited evidence from the North Sea and surrounding coasts that trawling had removed some physical features (e.g. cobbles and boulders) and altered the physical structure of the seafloor by making sediments more homogeneous; reduced structural biota such as colonial bryozoans, *Sabellaria*, hydroids, seapens, sponges, mussel and oyster beds; caused reduction in habitat complexity; caused loss of species from parts of their normal range; decreased populations with low turnover rates; caused some fragmentation of populations; altered relative abundance of species; impacted surface-living species more than deep-burrowing species; increased sub-lethal effects on individuals (e.g. via slower growth, disease, predation, longevity); caused an increase in populations with a high turn-over rate; and favoured scavenging species such as seabirds. In ecosystem terms, fishing had added an extra super-predator, thereby altering the trophic status of the North Sea; and by applying fishing mortality differentially amongst species, fishing had altered the dynamics of the food web. However, given the significant changes that have occurred in the natural environment as well as in fishing effort distribution over the past two to three decades, the magnitude and seriousness of the consequences of direct effects on the overall food web are still uncertain (Greenstreet *et al.* 1999; Jennings *et al.* 1999a).

The marine aggregate extraction industry in the North Sea contributes approximately 15% of some North Sea coastal nations' demands for sand and gravel (de Groot 1996). The methods most commonly used include anchor and trailer suction hopper dredging. Estimates of marine sand requirements are 32×10^6 t yr^{-1} for the Netherlands and 21-35 $\times 10^6$ t yr^{-1} for the United Kingdom. Approximately four times the amount of sand extracted is put into suspension; thus many of the effects on habitats are caused by sedimentation. In the Wadden Sea, extraction of aggregates has had a much greater impact on the intertidal areas, especially on long-lived bivalves.

Disposal of dredged materials amounts to nearly 70 million mt per year, although this is decreasing as stricter environmental regulations come into effect. Where areas have been extensively dredged, concomitant alterations to infauna and epifaunal biota have been observed (de Groot 1996). In Belgian and Dutch coastal areas, more than 1000km^2 of seabed has been changed, leading to high nematode/copepod ratios; in the Wadden Sea, large-scale dredging has interfered with hydrological and geomorphological processes; and in the German Bight, disposal of mining and industrial wastes has had significant effects on the benthic fauna.

Impoundment of rivers, construction of extensive coastal structures such as dykes, underwater barriers, conventional and tidal power stations, harbours and ports, plus recreational facilities have led to high levels of disturbance in coastal and specialised habitats. Shoreline protection schemes and beach replenishment are now common features, especially in the low-lying coastal areas of the United Kingdom and the Netherlands. However, recent debates on climate-induced sea-level rise in relation to the fall in extreme-event horizons of coastal structures are being met by soft-engineering solutions, which are often less detrimental to sensitive habitats and species.

Nearshore developments, including wind-turbines off the east coast of England, Rotterdam, Zeebruge, and Tunø Knob, and structures to support aquaculture and mariculture along the coasts of Norway, Scotland, France, and the Netherlands are also increasing. In the absence of monitoring data, planning guidelines for these are generally restricted to shipping and individual species rather than any long-term effects on sub-surface habitats.

Ecological impacts of non-indigenous species

Introductions of alien organisms into the North Sea have arisen from escapes and from the transferral of water from a variety of sources, such as ships and vessels in rivers and coastal waters; as their impact on indigenous species can be significant, their introduction should be prevented where at all possible (Hiscock 1997). The most important vectors are shipping (ballast water and fouling) and aquaculture. Secondary dispersal into new regions from introductions in other parts of Europe has also played a part, as well as through aquatic products as human food. It is through this route that the swim bladder nematode *Anguillicola crassus* was introduced along with Japanese eels: this parasite is now spreading and causing severe problems in wild and farmed stocks of European eels. Another parasite of note is the oyster parasite (*Bonamia ostrea*). In the North Sea various alien species of seaweed have been introduced, including *Sargassum muticum* and *Unadria pinnatifida* (1973), *Grateloupia doryphora and G. filcina* (1990), and *Heterosigma akashiwo* (1996). Introduced invertebrates include the worm *Marenzelleria viridis,* the slipper limpet *Crepidula fornicata* which is having a negative impact on scallop and other shellfish culture, and the Chinese mitten crab which now infests a number of ports and rivers. In coastal areas, there has been hybridisation of salt-marsh plants between *Spartina alterniflora* from North America and *S. maritime* to produce the polyploid *S. anglica.*

The International Maritime Organisation has issued guidelines for the control and management of ballast water (http://www.imo.org) but the problem is already extensive. It has been estimated that the major 40,000 cargo vessels of the world transfer 10 billion tonnes of ballast water globally each year, and it has been demonstrated that on average 3,000-4,000 species are transported daily, with severe consequences for ecosystem as well as public health through the spread of human disease agents (http://members.aol.com.sgollasch/sgollasch/index.html; http://invasions.si.edu).

Aquaculture is another obvious source of introductions as it is generally accompanied by the escape of individuals. Most of the stocks used in the North Sea have been developed for economically important characteristics: thus the triploid Pacific oyster, rainbow trout and salmon (*Onchorhynchus mykiss*) were introduced on the basis that they had good growth and were putatively sterile. However, the ICES Working Group on Introduction and Transfer of Marine Organisms has reported that the triploidy in oysters may be incomplete, resulting in changes to biosafety. Escaped farmed salmon are also known to cause harm to natural populations by destroying spawned eggs and breeding with wild individuals. In Norway, the overall relative abundance of escapees from aquaculture has varied between 20 and 30% in 1992-1997, with the highest proportion in regions with a high density of aquaculture installations, although in some rivers proportions of 80-90% have been recorded (Intermediate Ministerial Conferences 1997). There is also increasing focus on genetically modified fish for mariculture, including Atlantic salmon with anti-freeze genes and greatly enhanced growth potential. But before these types of organisms are introduced into aquaculture installations, there will need to be a greater understanding of the risks and impacts on wild populations.

SOCIO-ECONOMIC TRENDS AND GOVERNANCE

Development and sectoral trends

The North Sea plays a key role in the development of the regions that border it. It acts as a source of economic resources, a transport highway as well as a sink for waste and pollution. Increasing pressures and conflicts over these various elements have historically epitomized the region (e.g. House 1986; Sanger 1987; Schama 1991). In addition, major changes in the international environment, such as German unification, political and economic liberalisation of Central and Eastern Europe, and completion of the single market, are having far-reaching effects on the whole northern sea-board. The enlargement of the European Community is also a powerful force for change, as it affects the peripherality of the North Sea states. The coastal regions of southern and eastern England, Germany, Denmark, and Belgium are likely to become increasingly focused on the Rotterdam-Berlin corridor, with coastal areas in Scotland becoming increasingly peripheral and dependent on transport links via the haven ports serving the north of the United Kingdom. Moreover, the existing development pattern in the

Table 12-3. Population in countries in the North Sea catchment area (millions).

Country	Population in North Sea catchment area (m)
Belgium & Luxembourg	10
Czech & Slovak Republics	5
Denmark	2
France	20
Germany	70
Netherlands	16
Sweden	3
Switzerland	5
United Kingdom	30

Highlands and Islands of Scotland is costly to support, so that despite strong social and cultural drivers, a move towards private and public investment along the more favourable east coast areas around Inverness and Aberdeen is anticipated.

Approximately 164 million people use the coastline and marine environment of the North Sea catchment (Table 12-3). Extra-European Community migration and migration between member states are expected to add over 9 million to the total population of the northern seaboard by 2020. But given the decline in population growth rates, there will be a considerable shift towards the elderly age groups, leading to an increase in dependency rates, a fall in the labour force, increasing labour costs, with retirement migration becoming more important. To this end, exploitation patterns in the North Sea are likely to change significantly over the next 20 years.

The fisheries sector is expected to decrease as it comes under increasing pressure to allow fish stocks to recover. It is also expected that there will be increasing restrictions on fishing effort, extension of the use of structural elements into fishing vessel decommissioning and land-based job-creation schemes, stabilization of fish prices, a decline in the fortunes in small fishing ports and a concentration of fishing effort on fewer but larger ports (e.g. the Scottish fleet is now centred on major population centres such as Fraserburgh), and a growth in the number of distressed communities where alternative jobs do not exist or are insufficient to replace those once provided by fishing and allied activities. Aquaculture, as an alternative to fishing, is still developing, with financial support from the European Union. Stabilization rather than growth is now expected, as environmental concerns are likely to lead to restrictions on the number of sites and size of farms.

The tourism sector will see its dominant flows become intra-European and predominantly domestic; however, world tourism is also projected to become an increasingly significant element in total demand, putting high pressure on certain coastal areas. To support this sector, it is expected that there will be particularly high levels of development in the hinterland of the main ports around the North Sea, underlining the strong east-west transport corridors such as Rotterdam-Berlin. The pattern and hierarchy of urban centres are not likely to exhibit significant change, but there will be an increase in new nodal points for transport and trans-shipping. The present pattern of port development is set to continue with increases in traffic growth in short-sea and containerised cargo markets and a concentration of shipping services in fewer, larger ports. The single market makes port areas attractive for distribution facilities, light industry and assembly activities.

The northern seaboard will continue to supply at least 50% of the total energy requirements of the European Union, with increases in natural oil and gas production from the North Sea and off Scotland, and significant generation of renewable energy from offshore wind farms and wave and tidal power plants.

Political and legal regimes

The exploitation of the natural marine resources of the North Sea follows a number of conventions, declarations and regulations including the Geneva Convention on the Continental Shelf (1958), the joint declaration of the Commission of the European Union on the co-ordinated extension of jurisdiction in the North Sea through the establishment of Exclusive Economic Zones (1992), and European Commission Directives and Regulations within the Common Fisheries Policies. A large number of instruments from international bodies, such as the United Nations and the International Maritime Organization, and the European Community, also exist to conserve natural resources, protect the environment and ensure health and safety standards.   They include:

- 1971 Ramsar Convention on Wetlands of International Importance, in which parties are required to conserve wetlands, *inter alia*, as habitats of distinctive ecosystems, that constitute a resource of great economic, cultural, scientific, and recreational value, the loss of which would be irreparable;
- 1972 London Convention to ban sea dumping of radioactive waste;
- 1973/78 International Convention for the Prevention of Pollution from Ships (MARPOL);
- 1973 Convention on International Trade in Endangered Species of Wild Flora and Fauna (CITES) which recognises the important impact that introductions could have on marine ecosystems;
- 1979 Bonn Convention on the Conservation of Migratory Species of Wild Animals which recognises the importance of managing habitats for their support;
- 1982 UN Convention on the Law of the Sea, in which states agreed to a set of conditions for the extraction of living marine resources within EEZs, which

included their conservation and management and the promotion of optimum utilisation;

- 1992 Convention for the Prevention of marine pollution from land-based sources (Paris Convention), aimed at preventing pollution of the North-East Atlantic and Arctic Oceans, the North Sea and the Baltic Sea from pollution arising in rivers, estuaries, pipelines, or man-made structures and emissions to air from land or man-made structures. This has now been replaced by OSPAR, the merged Oslo and Paris Conventions, which extends pollution to include dumping and incineration;
- 1992 Helsinki Convention on the Protection and Use of Transboundary Watercourses, the first to codify on a regional basis rules governing the protection and use of international watercourses down to where they flow directly into the sea, thereby linking human activities to these orphan ecosystems;
- Convention on the Conservation of European Wildlife and Natural Habitats which was negotiated under the auspices of the Council of Europe to overcome the inadequacies of the piecemeal and outdated European Conventions and to provide for co-operation among States. The aims of this Convention are to conserve natural habitat, especially those which require the co-operation of several States, and to promote co-operation;
- UN Framework Convention on Climate Change (UNFCC) is also concerned with ecosystem-based management. The objective of the Convention is that Parties should reduce greenhouse gas emissions to allow ecosystems, including marine ecosystems, to adapt naturally to climate change. Further, Article 4.1(d) commits Parties to promote sustainable management and cooperate in the conservation of sinks and reservoirs of greenhouse gases, including oceans as well as other coastal and marine ecosystems; and
- 1992 Declaration of the UN Conference on Environment and Development, involving the action plans of Agenda 21, the Convention on Biological Diversity (CBD) and the non-legally binding Statement of Principles on Forests;
- Bonn Agreement for co-operation in dealing with pollution of the North Sea by oil and other harmful substances;
- Convention on protection of the marine environment of the Baltic;
- Convention for the protection of the Rhine against chemical pollution. Limits on discharges of substances in annex I of the convention are laid down by the International Commission for the Protection of the Rhine against Pollution, which was set up to implement the convention. Annex II substances are to be controlled by governments under the commission's supervision;
- UN Conference on Straddling Fish Stocks and Highly Migratory Fish Stock. This gives the general provisions relating to the precautionary approach;
- 1995 FAO Code of Conduct for Responsible Fisheries.

The EC laws for the protection of the environment can be grouped into six main categories: general, air and noise, chemicals and industrial risks, nature conservation, waste, and water. From these it can be seen that EC environmental law is now a pervasive part of the legislation affecting business, government agencies, the voluntary sector, and individual citizens across the whole of the European Union: matters affecting the North Sea are now frequently dealt with under an environmental umbrella. Enforcement has also become stronger, with many individuals as well as organisations taking cases to the European Courts.

The main institutions involved in adopting European Community laws dealing with the North Sea are the European Commission, the European Parliament, the Council of the European Union, and the European Court and the Court of First Instance. Amongst the 20 Commissioners, at least 7 have a direct involvement with the North Sea and its environment (i.e. environment, industry, agriculture, transport, fisheries, energy, and consumer policy). Various types of environmental cases will come to the European Court, including actions by the Commission against member states that are breaking EC environmental laws. The European Environment Agency (EEA), which was established in 1994, provides the Community and member states with information relating to the environment and aims to ensure that the public is informed about the state of the environment. It does not have a role in enforcing compliance. Participation in the EEA is open to non-EC members, such as countries of the European Free Trade Association and the Central and East European States.

When the treaty to set up the European Commission was signed in 1957, environmental degradation was not generally recognised as an important problem. However, since 1972 five action programmes for the environment have been adopted. These are not binding measures in themselves but are useful indicators of the framework within which the Commission is working to bring about new environmental laws. Recent priorities have been to improve the enforcement of legislation, the integration of environmental considerations into other policies, use of a wide range of policy instruments (such as environmental charges and liability), the raising of public awareness, and reinforcement of the Community's international actions to protect the environment. In the environment field, there are now three forms of legally binding measures - *directives, regulations and decisions.*

Directives are the most common form of legislation and to comply, states have to pass national laws within the timetable laid down in the directive (normally within two years). *Regulations* are directly binding and applicable in all Member States; i.e. no further national legislation is necessary or permissible. *Decisions* are directly binding on the persons to whom they are addressed, including member states, individuals and legal persons. International agreements are entered into by the EC and they then become part of EC law. This has three consequences: first, it means that the international agreement can give rise to rights and duties that may be relied upon by individuals in national courts. Secondly, decisions of organisations created by the agreement will also become part of community law. And finally, the European Court is able to interpret and apply the agreement and decisions of the organisation created by the agreement. Judgements of the European courts are important as they affect the way in which EC environmental laws are applied.

An integrated ecosystem approach to North Sea management should seek to consider all significant factors that affect species and populations within the ecosystem. Indeed, in legal terms the use of the term *ecosystem* is embedded in many of the international and regional conventions, and declarations supporting its use have been recognised and ratified by European Member States, and hence are operational within the North Sea and its catchment. However, achieving a more integrated environmental approach from a legislative perspective will require a shift from sector-based instruments to one that integrates issues across a range of socio-economic and physical scales. In the management of the North Sea, ecosystem considerations have only played a minor role until now. For example, in fisheries an ecosystem approach has been restricted to looking at multi-species interactions between commercially important fish stocks. The result has been that fisheries management measures have been unable to obtain an appropriate balance between fishing effort and available fish resources (Olsen 1994).

By comparison, over the years since the first International Conference on the Protection of the North Sea in 1984, the North Sea states have achieved good results in their efforts to reduce pollution and enhance environmental health, helped in large part by the North Sea Task Force which has provided comparable data on the level and distribution of contaminants plus surveys of biological impacts (Tromp and Wieriks 1994). For example, there are commitments, fully or partly implemented, for a ban on the dumping and incineration of waste at sea, reduction of inputs of nutrients by 50%, cessation of all inputs of hazardous substances by 2020, a ban on dumping of offshore installations, and a ban on the application of tributyltin. The ambitious target set for phosphorus has almost been met. In 1990 it was agreed to set ambitious targets to reduce inputs of hazardous substances by 50% and substances that cause threats by 70%. Several pesticides have already been phased out: the agreement in 1990 to phase out and destroy in an environmentally safe manner all identifiable PCBs by 1995 and at the latest by the end of 1999 (International Ministerial Conference on the Protection of the North Sea 1995). After an initiative at the Esbjerg Conference, the North Sea states have also taken a concerted action within the IMO to introduce a global ban on the use of harmful anti-fouling agents by 2003. In addition, biosafety protocols on ballast water transfers are being introduced under the aegis of the IMO.

The OSPAR Convention is the obvious key to co-operation in the North Sea, acting as the interface between the Ministerial and regional meetings (Tromp and Wieriks 1994). However, in the growing debate about the nature and role of biodiversity in the environment, the Convention on Biological Diversity also occupies central stage (Holdgate 1994). As with the 1972 Stockholm Declaration, the Rio Declaration is not formally binding, but it introduces important new principles that are highly relevant to the North Sea Large Marine Ecosystem. These include the adoption of a precautionary approach, the polluter pays principle, environmental impact assessment, and public participation. Agenda 21 sets out a basis of actions to provide for an integrated policy and decision-making process, including all involved sectors, to promote compatibility and a balance of uses, identify existing and projected uses, apply preventive and precautionary approaches to project planning and implementation, promote development and application of techniques to value loss of environmental services, and provide access to information for all concerned individuals.

Both the twelve CBD principles (UNEP/CBD/SBSTTA/5/11) and Agenda 21 play an important role in determining much of current European Union legislation regarding the North Sea. Since its inception, the Conference of the Parties to the CBD has worked to establish the principles and operational guidance for the application of an ecosystem approach (see UNEP/CBD/CPO/5/3, page 78). There are now moves to encourage the take up of the Jakarta Mandate on Marine and Coastal Biological Diversity for the North Sea (see Statement of Conclusions from the Intermediate Ministerial Meeting 1997). The Esbjerg Declaration specifically states that the precautionary principle is the guiding principle to achieve the objective of a sustainable, sound and healthy ecosystem. Thus, in the North Sea the scope of the precautionary principle has been successively broadened from toxic substances to natural substances (e.g. nutrients) to all emissions responsible for global warming, and since the 1990s it has been widened to encompass the management of fisheries. By implication this means that for the North Sea: management boundaries must reflect a wide range of environmental issues including ecological integrity; monitoring programmes must be supported in order to ensure a robust basis for management decisions and to track the results of actions; interagency and/or transboundary co-operation is essential; structural changes in many national and regional resource management agencies will be necessary; humans must be recognised as a component of the ecosystem; and human values must play a dominant role in the establishment of management goals, but must become or remain the primary forcing agent. In the case of the North Sea, the precautionary approach concept, rather than the precautionary principle, still needs development, i.e. where there are threats of serious or irreversible damage, lack of full scientific certainty shall not be used as a reason for postponing cost-effective measures to prevent environmental degradation.

GENERAL ASSESSMENT OF THE NORTH SEA ECOSYSTEM

In the absence of long-term data series and detailed statistics, comparative analyses are often the only basis upon which the status of an ecosystem can be evaluated. Various approaches are available, but key to them all is the need to identify the interlinkages existing between features (natural and social), activities and processes and the effects of these through time on the dynamics of the ecosystem, as well as the outputs, or goods and services, that it provides (Anon 1999; McGlade 1999a).

For this assessment of the health of the North Sea ecosystem, two principle criteria have been used: socio-economic performance and ecosystem health (McGlade 1989). To properly undertake the assessment, management objectives, system complexity and forms of governance have been examined. A set of attributes have been selected: for ecosystem health, biodiversity, level of pollution and trophic stability are used; for socio-economic performance, sectoral outputs of good and services, cohesion and institutional strength/governance are used (McGlade 1989). Each attribute may have more than one measure associated with it. The time periods chosen were pre-1957 and 1958-present, to coincide with the Treaty of Rome, the establishment of the European Commission and the extension of industrial activities in the North Sea.

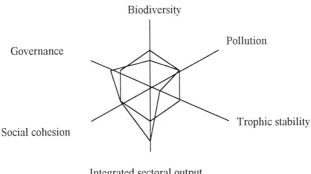

Figure 12-2. Kite diagram to indicate changes in the North Sea Large Marine Ecosystem, with pre-1957 values set to zero.

To enable numeric and non-numeric attributes with different numeraires to be compared, observations are placed in fuzzy classes (i.e. small, medium, large; positive or negative) depending on the extent of the information available (Anon. 1999; McGlade 2001b). The value for each measure lies within the range -5 to +5, with pre-1957 values set at zero. A value for each attribute for the post-1957 period is obtained by combining different measures values of each measure into one distribution and obtaining the mean with a central weight function (McGlade 1999a). The resulting values of this defuzzification are then placed into a kite diagram to indicate the changes that have occurred. Wherever possible the evaluation has been undertaken for the entire North Sea (Figure 12-2; Table 12-4).

As the combined results show, the ecological attributes indicate a general decline, and the socio-economic attributes a notable increase. In other words, the outputs derived from the ecosystem have been arrived at via some cost to the environment, albeit not commensurate on the scale used. This is unsurprising. However, the measures also suggest that: i) the changes observed in trophic structure are indicative of a trend towards decreasing resilience, ii) the trend is not only a response to fishing pressure and resource exploitation, but also to inter-annual changes in the physical oceanography of

Table 12-4. Attributes for general ecosystem assessment.

Attribute	Measure
Biodiversity (all cited communities)	- 2
Pollution levels (Class I and II)	- 1
Trophic stability (abundance, size-classes and life-span)	- 3
Integrated sectoral outputs (fisheries, energy, tourism, transport)	+ 4
Social cohesion (conflicts; migration)	+ 1
Institutional strength (legal & political regimes)	+ 2

the North Atlantic, and iii) traditional economic measures of sectoral outputs (e.g. GNP) are not a true reflection of the true value of the North Sea ecosystem to the states involved. Rather, the measure reflecting social cohesion and institutional strengths are also of significance. Overall, despite several decades of increasing exploitation, the North Sea large marine ecosystem has provided and continues to provide a high level of goods and services to the human and biological communities that rely on it.

ACKNOWLEDGEMENTS

The author would like to thank Dr. K. Zwanenberg for his efforts as a reviewer and in providing many useful comments and suggestions.

REFERENCES

AAAS (American Association for the Advancement of Science). 1986. Variability and Management of Large Marine Ecosystems. AAAS Selected Symposium 99. Westview Press, Inc. Boulder, CO. 319 p.

Adams, J.A. 1987. The primary ecological sub-divisions of the North Sea: some aspects of their plankton communities. In: R.S. Bailey and B.B. Parrish (eds.). Developments in Fisheries Research in Scotland. Fishing News Books, Farnham. 165-181.

Aebischer, N.J., J.C. Coulson, and J.M. Colebrook. 1990. Parallel long-term trends across four marine trophic levels and weather. Nature 347:753-755.

Alexander, L.M. 1986. Large marine ecosystems as regional phenomena. In: K. Sherman and L.M. Alexander (eds.). Variability and Management of Large Marine Ecosystems. AAAS Selected Symposium 99. Westview Press, Inc. Boulder, CO. 239-240.

Andersen, L.W., L.E. Holm, B. Clausen, and C.C. Kinze. 1995. Preliminary results of a DNA microsatellite study of the population and social structure of the harbour porpoise. In: A. Blix, L. Walloe, and Ø. Ulltang (eds.). Whales, Seals, Fish, and Man. Elsevier, Amsterdam. 117-127.

Andersen, L.W., L.E. Holm, H. Siegismund, B. Clausen, C.C. Kinze, and V. Loeschcke. 1997. A combined DNA-microsatellite and isozyme study of the population structure of the harbour porpoise in Danish waters and West Greenland. Heredity 78:270-276.

Andersson, L. and L. Rydberg. 1988. Trends in nutrient and oxygen conditions within the Kattegat: effects of local nutrient supply. Estuar. Coast. Shelf. Sci. 26:559-579.

Anon. 1994. Report of the study group on seabird/fish interactions. ICES CM 1994/L:3.

Anon. 1999. Fuzzy logic to manage coastal regions. SimCoast™ solves planning and development problems. International Ocean Systems Design 3:4-7.

Asman, W.A.H. and R. Berkowicz. 1994. Atmospheric nitrogen deposition to the North Sea. Mar. Poll. Bull. 29:426-434.

Austen, M.C., J.B. Buchanan, H.G. Hunt, A.B. Josefson, and M.A. Kendall. 1991. Comparison of long-term trends in benthic and pelagic communities of the North Sea. J. Mar. Biol. Assoc. UK 71:179-190.

Bacon, S. and D.J.T. Carter. 1991. Wave climate changes in the North Atlantic and North Sea. Int. J. Clim. 11:545-558.

Backhaus, J.O., I.H. Harms, M. Krause, and M. Heath. 1994. An hypothesis concerning the space-time succession of *Calanus finmarchicus* in the northern North Sea. ICES J. Mar. Sci. 51:169-180.

Baird, D., J.M. McGlade, and R.E. Ulanowicz. 1990. The comparative ecology of six marine ecosystems. Philosophical Transactions of the Royal Society London B. 333:15-29.

Bannister, C. 2001. Questions and answers on North Sea sandeels. Fishing News 6th April, p 7.

Baretta, J. and P. Ruardij (eds.). 1988. Tidal Flat Estuaries, Simulation and Analysis of the Ems Estuary. Springer, Berlin.

Basford, D.J., D.C. Moore, and A.S. Eleftheriou. 1996. Variations in benthos in the north-western North Sea in relation to the inflow of Atlantic Water, 1980-1984. ICES J. Mar. Sci. 53:957-963.

Beatty, J. 1992. Sula - the Seabird-Hunters of Lewis. Michael Joseph, London.

Becker, G.A. and D. Pauly. 1996. Sea surface temperature changes in the North Sea and their causes. ICES J. Mar. Sci. 53:887-898.

Berggren, P. 1994. Bycatches of the harbour porpoise (*Phocoena phocoena*) in the Swedish Skaggerak, Kattegat and Baltic Seas: 1973-1993. Rep. Int. Whal. Comm. 16:211-215.

Bergman, M.J.N., M. Fonds, S.J. de Groot, and J.W. Van Santbrink. 1996. Direct effects of beam trawl fishery on bottom fauna in the Southern North Sea. In: J. Andersen, H. Kraup, and U.B. Nielsen (eds.). Scientific Symposium on the North Sea Quality Status Report, 1994. Danish Protection Agency, Copenhagen. 204-209.

Beukema, J.J. 1986. Long-term and recent changes in the benthic macrofauna living on tidal flats in the western part of the Wadden Sea. Neth. Inst. Sea Res. 20:135-141.

Beukema, J.J. 1992a. Expected changes in the Wadden Sea benthos in a warmer world: lessons from periods with milder winters. Neth. J. Sea Res. 30:73-79.

Beukema, J.J. 1992b. Expected changes in winter temperatures on benthic animals living in soft sediments in coastal North Sea areas. In: J.J. Beukema, W.J. Wolff, and J.W.M. Brouns (eds.). Expected Effects of Climatic Change on Marine Coastal Ecosystems, Kluwer, Dordrecht. 83-92.

Billen, G., C. Lancelot, and M. Meybeck. 1991. N, P and Si retention along the aquatic continuum from land to ocean. In: R.F.C. Mantoura, J.M. Martin, and J. Wollast (eds.). Ocean margin processes in global change. John Wiley & Sons, London. 19-44.

Blackburn, T.H. and K. Henriksen. 1983. Nitrogen cycling in different types of sediments from Danish waters. Limnol. Oceanogr. 28:477-493.

Boddeke, R. and P. Hagel. 1995. Eutrophication, fisheries and productivity of the north Sea continental zone. In: N.B. Armantrout (ed.). Procs. World Fisheries Congress. Oxford and IBH Publishing Co. Pvt. Ltd, New Delhi. 290-325.

Boddeke, R. and B. Vingerhoed. 1996. The anchovy returns to the Wadden Sea. ICES J. Mar. Sci. 53:1003-1007.

Bot, P.V.M. and F. Colijn. 1996. A method for estimating primary production from chlorophyll concentrations with results showing trends in the Irish Sea and the Dutch coastal zone. ICES J. Mar. Sci. 53:945-950.

Bouws, E., D. Jannick, and G.J. Komen. 1996. On increasing wave height in the North Atlantic. Bull. Amer. Met. Soc. 77:2275-2277.

Brockmann, U., G. Billen, and W.W.C. Gieskes. 1988. North Sea nutrients and eutrophication. In: W. Salomons, B.L. Bayne, E.K. Duursma, and U. Förstner (eds.). Pollution of the North Sea: An Assessment. Springer Verlag, Berlin. 348-389.

Broekhuizen, N. and E. McKenzie. 1995. Patterns of abundance for *Calanus* and smaller copepods in the North Sea: time series decomposition of two CPR data sets. Mar. Ecol. Prog. Ser. 118:103-120.

Brogger, J. 1990. Pre-Bureaucratic Europeans. Norwegian University Press, Oslo.

Buchanan, J.B. 1993. Evidence of benthic pelagic coupling at a station off the Northumberland coast. J. Exp. Mar. Biol. & Ecol. 172:1-10.

Buchanan, J.B. and J.J. Moore. 1986a. A broad view of variability and persistence in the Northumberland benthic fauna - 1971-1985. J. Mar. Biol. Assoc. UK 66:641-657.

Buchanan, J.B. and J.J. Moore. 1986b. Long-term studies at a benthic station off the coast of Northumberland. Hydrobiologica 142:121-127.

Burton, J.D., G.E. Althaus, G.E. Millward, A.W. Morris, P.J. Statham, A.D. Tappin, and A. Turner. 1994. Processes influencing the fate of trace metals in the North Sea. In: H. Charnock, K.R. Dyer, J.M. Huthnance, P.S. Liss, J.H. Simpson, and P.B. Tett (eds.). Understanding the North Sea system. The Royal Society. Chapman and Hall, London. 179-190.

Cadée, G.C. 1992. Trends in Marsdiep phytoplankton. In: N. Danker, C.J. Smit, and M. Scholl (eds.). Present and future conservation of the Wadden Sea. Neth. Inst. Sea Res. Publ. Ser 20. 143-149.

Camphuysen, C.J. 1995. Kittiwakes *Rissa tridactyla* in the North Sea: pelagic ecology, fisheries relationships and feeding strategies. Limosa 68:123.

Camphuysen, C.J. and S. Garthe. 1997. Distribution and scavenging habits of northern fulmars in the North Sea. ICES J. Mar. Sci. 54:654-683.

Camphuysen, C.J. and S. Garthe. 2000. Seabirds and commercial fisheries population trends of piscivorous seabirds explained? In: M.J. Kaiser and S.J. de Groot (eds.). Effects of Fishing on Non-target Species and Habitats. Blackwell Science, Oxford. 163-184.

Camphuysen, C.J., B. Calvo, J. Durinck, K. Ensor, A. Follestad, R.W. Furness, S. Garthe, G. Leaper, H. Skov, M.L. Tasker, and C.J.N. Winter. 1995. Consumption of discards by seabirds in the North Sea. EC Contract BIOECO/93/10, Brussels.

Carpenter, S.R., J.F. Kitchell, and J.R. Hodgson. 1985. Cascading trophic interactions and lake productivity. BioScience 35:634-639.

Chambers, C., A. Matthews, S. Lavender, and T. Sawyer. 2000. Temporal variability in chlorophyll-a concentrations in the coastal environment. Final Report, National Centre for Environmental Data and Surveillance, Environment Agency, Twerton, UK.

Chester, R., G.F. Bradshaw, C.J. Ottley, R.M. Harrison, J.L. Merrett, M.R. Preston, A.R. Rendell, M.M. Kane, and T.D. Jickells. 1994. The atmospheric distributions of trace metals, trace organics and nitrogen sources over the North Sea. In: H. Charnock, K.R. Dyer, J.M. Huthnance, P.S. Liss, J.H. Simpson, and P.B. Tett (eds.). Understanding the North Sea system. The Royal Society. Chapman and Hall, London. 165-178.

Coeling, J.P., A.J.M. Vanwijk, and A.A.M. Holtslag. 1996. Analysis of wind speed observations over the North Sea. J. Wind Engineering 61:51-69.

Cole, S. 1991. Women of the Praia. Princeton University Press, Princeton, NJ.

Colebrook, J.M. 1960. Continuous plankton records: methods of analysis, 1950-59. Bull. Mar. Ecol. 5:51-64.

Corten, A. 1990. Long-term changes in pelagic fish stocks of the North Sea and adjacent waters and their possible connection to hydrographic changes. Neth. J. Sea Res. 25:227 - 235.

Corten, A. and G. van de Kamp. 1996. Variation in the abundance of southern fish species in the southern North Sea in relation to hydrography and wind. ICES. J. Mar. Sci. 53:1113-1119.

Coull, J.R. 1988. The North Sea herring fishery in the twentieth century. Ocean Yearbook 7:115-131.

Croxall, J.P., P.G.H. Evans, and R.W. Schrieber. 1984. Status and conservation of the world's seabirds. Technical Publication 2. International Council for Bird Preservation, Cambridge.

Cushing, D.H. 1984. The gadoid outburst in the North Sea. J. Cons. Int. pour l'Explor. Mer 41:159-166.

Daan, N., P.J. Bromley, J.R.G. Hislop, and N.A. Nielsen. 1990. Ecology of North Sea fish. Neth. J. Sea. Res. 26:343-368.

Daan, N., H.J.L. Heessen, and J.G. Pope. 1994. Changes in the North Sea cod during the twentieth century. ICES Mar. Sci. Symp. 197:228-243.

Daan, R. and M. Mulder. 1996. On the short-term and long-term impact of drilling activities in the Dutch sector of the North Sea. ICES J. Mar. Sci. 53:1036-1044.

Dando, P.R., M.C. Austen, M.A. Burke Jr., M.A. Kendall, M.C. Kennicut II, A.G. Judd, D.C. Moore, S.C.M. O'Hara, R. Schmalljohan, and A.J. Southward. 1991. Ecology of a North Sea pockmark with an active methane seep. Mar. Ecol. Prog. Ser. 70:49-63.

Danielssen, D.S., L. Davidsson, L. Edler, E. Folquist, S.H. Fonselius, L. Føyn, L. Hernroth, B. Håkansson, I. Olsson, and E. Svendsen. 1991. SKAGEX: some preliminary results. ICES CM 1991/C:2.

Danielssen, D.S., E. Svendsen, and M. Ostrowski. 1996. Long-term hydrographic variation in the Skaggerak based on the section through Torungen-Hirtshals. ICES J. Mar. Sci. 53:917-925.

Davies, J.M., A.D. McIntosh, R. Stagg, G. Topping, and J. Rees. 1997. The fate of the *Braer* oil in the marine and terrestrial environments. In: J.M. Davies and G. Topping (eds.). The impact of an oil spill in turbulent waters: the *Braer*. The Stationery Office, London. 26-41.

Davies, A.M. and J. Xing. 1997. Towards dynamic coupling of open ocean and shelf sea models. In: J.H. Stiel, H.W.A. Behrens, J.C. Borst, L.J. Droppert and J. van der Meulen (eds.). Operational Oceanography. The Challenge for European Co-operation. Elsevier Science B.V., Amsterdam. 455-462.

de Groot, S.J. 1996. The physical impact of marine aggregate extraction in the North Sea. ICES J. Mar. Sci. 53:1051-1053.

Dehairs, F., W. Baeyens, and D. van Gansbeke. 1989. Tight coupling between enrichment of iron and manganese in North Sea suspended matter and sedimentary redox processes: evidence for seasonal variability. Est. Coast. Shelf Sci. 29:457-471.

Den Hartog, C. 1987. "Wasting disease" and other dynamic phenomena in *Zostera* beds. Aquat. Bot. 47:21-28.

Den Hartog, C. 1994. Suffocation of a littoral *Zostera* bed by *Enteromorpha radiata*. Aquat. Bot 27:3-14.

Desprez, M., H. Rybarczyk, J.G. Wilson, J.-P. Ducrotoy, and R. Olivesi. 1992. Biological impact of eutrophication in the Bay of Somme and the induction and impact of anoxia. Netherlands J. Sea Res. 30:149-159.

Dethlefsen, V., H. von Westernhagen, and P. Cameron. 1996. Malformations in North Sea pelagic fish embryos during the period 1984-1995. ICES J. Mar. Sci. 53:1024-1035.

De Wolf, P. and J.J. Zijlstra. 1988. The ecosystem. In: W. Salomons, B.L. Bayne, E.K. Duursma, and U. Förstner (eds.). Pollution of the North Sea: An Assessment. Springer Verlag, Berlin. 118-151.

Dickson, R.R., D.S. Kirkwood, G. Topping, A.J. van Bennekom, and W. Schreurs. 1988a. A preliminary trend analysis for nitrate in the North Sea west of 3°E. ICES CM 1988/C:4.

Dickson, R.R., J. Meincke, S.A. Malmberg, and A.J. Lee. 1988b. The "Great Salinity Anomaly" in the northern North Atlantic 1968-1982. Progress in Oceanography 20:103-151.

Dippner, J.W. 1998. Competition between different groups of phytoplankton for nutrients in the southern North Sea. J. Mar. Systems 14:181-198.

Donkin, P., J. Widdows, S.V. Evans, F.J. Staff, and T. Yan. 1997. Effects of neurotoxic pesticides on feeding rate of marine mussels (*Mytilus edulis*). Pesticide Sci. 49:196-209.

Doody, J.P., C. Johnston, and B. Smith (eds.). 1993. Directory of the North Sea Coastal Margin. UK Joint Nature Conservation Committee, Peterborough.

Dooley, H.D., J.H.A. Martin, and D.J. Ellett. 1984. Abnormal hydrographic conditions in the Northeast Atlantic during the 1970s. Rapports et Procès-Verbaux des Réunions du Conseil International pour l'Exploration de la Mer 185:179.

Ducrotoy, J.-P. 1999. Indication of changes in the marine flora of the North Sea in the 1990s. Mar. Poll. Bull. 38:646-654.

Dunnet, G.M., R.W. Furness, M.L. Tasker, and P.H. Becker. 1990. Seabird ecology in the North Sea. Neth. J. Sea Res. 26:387-425.

Dyson, J. 1977. Business in Great Waters. The Story of British Fishermen. Angus & Robertson, London.

Edgell, N. 1994. The *Braer* tanker incident: some lessons from the Shetland Islands. Mar. Poll. Bull. 29:361-367.

Eisma, D. 1987. The North Sea: an overview. Phil. Trans. R. Soc. London B. 316:416-485.

Eisma, D. and G. Irion. 1988. Suspended matter and sediment transport. In: W. Salomons, B.L. Bayne, E.K. Duursma, and U. Förstner (eds.). Pollution of the North Sea: An Assessment. Springer Verlag, Berlin. 20-33.

Eisma, D. and J. Kalf. 1987a. Dispersal, concentration and deposition of suspended matter in the North Sea. J. Geol. Soc. Lond. 144:161-178.

Eisma, D. and J. Kalf. 1987b. Distribution, organic content and particle size of suspended matter in the North Sea. Neth. J. Sea. Res. 21:265-285.

Ekman, M. 1988. The world's longest continued series of sea level observations. Pure Appl. Geophys. 127:73-77.

Eleftheriou, A. and D.J. Basford. 1989. The macrobenthic infauna of the offshore northern North Sea. J. Mar. Biol. Assoc. UK 69:123-143.

Ellett, D.J. and W.R Turrell. 1992. Increased salinity levels in the NE Atlantic. ICES CM 1992/C:20.

Elliott, M., S. Nedwell, N.V. Jones, S. Read, N.D. Cutts, and K.L. Hemingway. 1998. Intertidal Sand and Mudflats and Subtidal Mobile Sandbanks. Scottish Association for Marine Science, UK Special Area of Conservation Project.

Environment Agency. 1998. Oil and gas in the environment. The Stationary Office, London.

Fasham, M.J.R. 1993. Modelling the marine biota. In: M. Heinemann (ed.). The Global Carbon Cycle, Springer-Verlag, Berlin.

Fasham, M.J.R., H. Ducklow, and S.M. McKelvie. 1990. A nitrogen-based model of plankton dynamics in the oceanic mixed layer. J. Mar Res. 48:591-639.

Fisher, J. 1952. The Fulmar. Collins New Naturalist Series, London.

Flather, R. 1987. Estimates of extreme conditions of tide and surge using a numerical model of the north-west continental shelf. Estuar. Coastal Shelf. Sci. 24:69-93.

Flather, R., R. Proctor, and J. Wolf. 1991. Oceanographic forecast models. In: Computer modelling in the environmental sciences. Clarendon, Oxford. 15-30.

Fonds, M. and S. Groenewold. 2000. Food subsidies generated by the beam-trawl fishery in the southern North Sea. In: M.J. Kaiser and S.J. de Groot (eds.). Effects of Fishing on Non-target Species and Habitats. Blackwell Science, Oxford. 130-150.

Fransz, H.G., J.M. Colebrook, J.C. Gamble, and M. Krause. 1991. The zooplankton of the North Sea. Neth. J. Sea Res. 28:1-52.

Frid, C.L.J. and R.A. Clark. 2000. Long-term changes in North Sea benthos: discerning the role of fisheries. In: M.J. Kaiser and S.J. de Groot (eds.). Effects of Fishing on Non-target Species and Habitats. Blackwell Science, Oxford. 198-216.

Frid, C.L.J. and N.V. Huliselan. 1996. Far-field control of long-term changes in Northumberland (NW North Sea) coastal zooplankton. ICES J. Mar. Sci. 53:972-977.

Frid, C.L.J., J.B. Buchanan, and P.R. Garwood. 1996. Variability and stability in benthos: twenty-two years of monitoring off Northumberland. ICES J. Mar. Sci. 53:978-980.

Frid, C.L.J., L.C. Newton, and J.A. Williams. 1994. The feeding rates of *Pleurobrachia* (Ctenophora) and *Sagitta* (Chaetognatha) with notes on the potential seasonal role of planktonic predators in the dynamics of North Sea Zooplankton communities. Neth. J. Aquat. Ecol. 28:181-191.

Furness, R.W., D.R. Thompson, and P.H. Becker. 1994. Spatial and temporal variation in mercury contamination in seabirds in the north Sea. Helgoländer Meeresuntersuchungen 49:605-615.

Gaskin, D.E. 1984. The harbour porpoise *Phocoena phocoena* (L.): regional population status and information on direct and indirect catches. Rep. Int. Whaling Comm. 34:569-586.

Gaskin, D.E. 1985. The ecology of whales and dolphins. Heinemann, London.

Gerlach, S.A. 1990. Nitrogen, phosphorus, plankton and oxygen deficiency in the German Bight and in Kiel Bay. Kieler Meeresforsch., Sonderh 7:1-341.

Gislason, H. 1994. Ecosystem effects of fishing activities in the North Sea. Mar. Poll. Bull. 29:520-527.

Glover, R.S. 1967. The continuous plankton recorder survey of the north Atlantic. Symp. Zool. Soc. London 19:189-210.

Gray, J.S., K.R. Clarke, R.M. Warwick, and G. Hobbs. 1990. Detection of initial effects of pollution in marine benthos: an example from the Ekofisk and Eldfisk oil-fields, North Sea. Mar. Ecol. Prog. Ser. 66:285-299.

Greenstreet, S.P.R., A.D. Bryant, N. Broekhuizen, S.J. Hall, and M.R. Heath. 1997. Seasonal variation in the consumption of food by fish in the North Sea and implications for food web dynamics. ICES J. Mar. Sci. 54:243-266.

Greenstreet, S.P.R., F.E. Spence, and J.A. MacMillan. 1999. Fishing effects in northeast Atlantic shelf seas: patterns in fishing effort, diversity and community structure. II Trends in fishing effort in the North Sea by UK registered vessels landing in Scotland. Fisher. Res. 40:107-124.

Greve, W., F. Reiners, and J. Nast. 1996. Biocoenotic changes of the zooplankton in the German Bight: the possible effects of eutrophication and climate. ICES J. Mar. Sci. 53:951-956.

Guillen, J. and P. Hoekstra. 1997. Sediment distribution in the nearshore zone. Grain size evolution in response to shore-surface nourishment. Estuar. Coast. Shelf Sci. 45:639-652.

Gunnarsson, J., M. Bjork, M. Gilek, M. Granberg, and R. Rosenberg. 2000. Effects of eutrophication on contaminant cycling in marine benthic systems. Ambio 29:252-259.

Gunther, H., W. Rosenthal, M. Stawarz, J.C. Carretero, M. Gomez, I. Lozano, O. Serrano, and M. Reistad. 1998. The wave climate of the Northeast North Atlantic over the period 1955-1994: the WASA wave hindcast. Norwegian Meteorological Institute (DNMI), Bergen.

Hall, S.J. 1994. Physical disturbance and marine benthic communities: life in unconsolidated sediments. Oceanogr. & Mar. Biol.: Ann. Rev. 32:179-239.

Hall, J.A., C.L.J. Frid, and R.K. Proudfoot. 1996. Effects of metal contamination on macrobenthos of two North Sea estuaries. ICES J. Mar. Sci. 53:1014-1023.

Hammond, P.S., H. Benke, P. Berggren, D.L. Borchers, S.T. Buckland, A. Collet, M.P. Heide-Jorgensen, S. Heimlich-Boran, A.R. Hilby, M. Leopold, and N. Øien. 1995a. Distribution and abundance of the harbour porpoise and other small cetaceans in the North Sea and adjacent waters. SCANS Final Report, LIFE 92-2/UK/027, European Commission, Brussels.

Hammond, P.S., H. Benke, P. Berggren, A. Collet, S. Heimlich-Boran, M. Leopold, and N. Øien. 1995b. The distribution and abundance of harbour porpoises and other small cetaceans in the North Sea and adjacent waters. ICES C.M. 1995/N:10.

Harding, M.J.C., S.K. Bailey, and I.M. Davies. 1992. UK Department of the Environment TBT Imposex Survey of the North Sea. PECD 7/8/214. Scottish Fisheries Working Paper 9/92. 26 p.

Hardisty, J. 1990. The British Seas. Routledge, London.

Harwood, J. (ed.). 2000. Assessment and reduction of the by-catch of small cetaceans. BYCARE Final Report, FAIR CT95 0523, European Commission, Brussels.

Hays, G.C., M.C. Carr, and A.H. Taylor. 1993. The relationship between Gulf Stream position and copepod abundance derived from the Continuous Plankton Recorder Survey: separating biological signal from sampling noise. J. Plankton Res. 15:1359-1373.

Heessen, H.J.L. 1996. Time-series data for a selection of forty fish species caught during the International Bottom Trawl Survey. ICES J. Mar Sci. 53:1079-1084.

Heessen, H.J.L. and N. Daan. 1996. Long-term trends in ten non-target North Sea fish species. ICES J. Mar Sci. 53:1062-1078.

Heip, C. and J.A. Craeymeersch. 1995. Benthic community structures in the North Sea. Helgoländer Meeresuntersuchungen 49:313-328.

Heip, C., D. Basford, J.A. Craeymeersch, J.M. Dewarumez, J. Dorjes, P. De Wilde, G. Duineveld, A. Eleftheriou, P.M. Hermann, U. Nierman, P. Kingston, P. Künitzer, A. Rachor, H. Ruhmor, K. Soetaert, and T. Soltwedel. 1992. Trends in biomass, density and diversity of North Sea macrofauna. ICES J. Mar. Sci. 49:13-22.

Hempel, G. (ed.). 1978. North Sea fish stocks - recent changes and their causes. Rapports et Procès-Verbaux des Réunions du Conseil International pour l'Exploration de la Mer 172.

Hickel, W., P. Mangelsdorf, and J. Berg. 1993. The human impact in the German Bight: eutrophication during three decades (1962-1991). Helgoländer Meeresuntersuchungen 47:243-263.

Hickel, W., E. Bauerfeind, U. Niermann, and H. von Westernhagen. 1989. Oxygen deficiency in the South-eastern North Sea: sources and biological effects. Ber. Boil. Anst. Helgoland 4:1-148.

Hill, A.E., I.D. James, P.F. Linden, J.P. Matthews, D. Prandle, J.H. Simpson, E.M. Gmitrowicz, D.A. Smeed, K.M.M. Lwiza, R. Durazo, A.D. Fox, and D.G. Bowers. 1994. Dynamics of tidal mixing fronts in the North Sea. In: H. Charnock, K.R. Dyer, J.M. Huthnance, P.S. Liss, J.H. Simpson, and P.B. Tett (eds.). Understanding the North Sea System. The Royal Society. Chapman and Hall, London. 53-68.

Hiscock, K. 1997. Conserving biodiversity in north-east Atlantic marine ecosystems. In: R.F.G. Ormond, J.D. Gage, and M.V. Angel (eds.). Marine Biodiversity, Patterns and Processes. Cambridge University Press, Cambridge. 415-427.

Hislop, J.R.G. 1996. Changes in North Sea gadoid stocks. ICES J. Mar Sci. 53:1146-1156.

Hogben, N. 1994. Increases in wave heights over the North Atlantic: a review of evidence and some implications for the naval architect. Trans. R. Inst. Naval Arch. 93-101.

Holdgate, M. 1994. Environmental challenges and priorities for action. Mar. Poll. Bull. 29:258-261.

Holme, N.A. 1983. Fluctuations in the benthos of the western English Channel. In: Proceedings of the 17th European Marine Biology Symposium, Brest, France. 121-124.

House, J.D. (ed.). 1986. Fish vs. Oil. Resources and Rural Development in North Atlantic Societies. Social and Economic Papers 16. Institute of Social and Economic Research, Memorial University of Newfoundland.

Howarth, M.J., K.R. Dyer, I.R. Joint, D.J. Hydes, D.A. Purdie, H. Edmunds, J.E. Jones, R.K. Lowry, T.J. Moffat, A.J. Pomeroy, and R. Proctor. 1994. Seasonal cycles and their spatial variability. In: H. Charnock, K.R. Dyer, J.M. Huthnance, P.S. Liss, J.H. Simpson, and P.B. Tett (eds.). Understanding the North Sea system. The Royal Society. Chapman and Hall, London. 5-25.

Huebeck, M. 1988. Shetland's seabirds in dire straits. British Trust for Ornithology News 158:1-2.

Hunt, G.L. and R.W. Furness (eds.). 1996. Seabird/fish interactions, with particular reference to seabirds in the North Sea. ICES Coop. Res. Rept. 216, Copenhagen.

Huntley, D.A., J.M. Huthnance, M.B. Collins, C.-L. Liu, R.J. Nicholls, and C. Hewitson. 1994. Hydrodynamics and sediment dynamics of North Sea sand waves and sand banks. In: H. Charnock, K.R. Dyer, J.M. Huthnance, P.S. Liss, J.H. Simpson, and P.B. Tett (eds.). Understanding the North Sea system. The Royal Society. Chapman and Hall, London. 83-96.

Huys, R., P.M.J. Herman, C.H.R. Heip, and K. Soetaert. 1992. The meiobenthos of the North Sea, density, biomass trends and distribution of copepod communities. ICES J. Mar. Sci. 49:23-44.

International Council for the Exploration of the Sea. 2000. Report of the Working Group on ecosystem effects of fishing activities. ICES CM 2000/ACME:2. 95 p.

International Ministerial Conference on the Protection of the North Sea 1984, 1987, 1990, 1995. The North Sea Secretariat, Ministry of the Environment, PO Box 8013 Dep., N-0030 Oslo, Norway. http://odin.dep.no/md/html/conf

Intermediate Ministerial Conferences 1993, 1997. The North Sea Secretariat, Ministry of the Environment, PO Box 8013 Dep., N-0030 Oslo, Norway. http://odin.dep.no/md/html/conf

Jago, C.F., A.J. Bale, M.O. Green, M.J. Howarth, S.E. Jones, I.N. McCave, G.E. Milward, A.W. Morris, A.A. Rowden, and J.J. Williams. 1994. Resuspension processes and seston dynamics, southern North Sea. In: H. Charnock, K.R. Dyer, J.M. Huthnance, P.S. Liss, J.H. Simpson and P.B. Tett (eds.). Understanding the North Sea system. The Royal Society. Chapman and Hall, London. 96-113.

Jennings, S., J. Alvsvåg, A.J. Cotter, S. Ehrich, S.P.R. Greenstreet, A. Jarre-Teichmann, N. Mergardt, A.D. Rijnsdorp, and O. Smedstad. 1999a. Fishing effects in northeast Atlantic shelf seas: patterns in fishing effort, diversity and community structure. III International fishing effort in the North Sea: an analysis of spatial and temporal trands. Fisher. Res. 40:125-134.

Jennings, S., J. Lancaster, A. Woolmer, and J. Cotter. 1999b. Distribution, diversity and abundance of epibenthic fauna in the North Sea. J. Mar. Biol. Assoc. UK 79:385-399.

Jensen, B.M. 1996. An overview of exposure to, and effects of, petroleum oil and organochlorine pollution in grey seals *Halochoerus grypus*. Sci. Total Envir. 186:109-118.

Joint, I. 1997. Phytoplankton production in the North Sea. LOICZ Report 29.

Joint, I. and A. Pomeroy. 1993. Phytoplankton biomass and production in the North Sea. Mar. Ecol. Prog. Ser. 99:169-182.

Joint, I.R. and R. Williams. 1985. Demands of the herbivore community on phytoplankton production in the Celtic Sea in August. Mar. Biol. 87:297-306.

Johnson, T. 1979. Work together, eat together. In: R. Anderson (ed.). North Atlantic Maritime Cultures. Mouton de Gruyter, The Hague.

Jorion, P. 1982. The priest and the fisherman. Man 17:275-286.

Josefson, A.B., J.N. Jensen, and G. Aertebjerg. 1993. The benthos community structure in the late 1970s and early 1980s - a result of a major food pulse. J. Exp. Mar. Biol. Ecol. 172:31-45.

Kaiser, M.J. and S.J. de Groot. 2000. The Effects of Fishing on Non-Target Species and Habitats. Blackwell Science, Oxford.

Knijn, R.J., T.W. Boon, H.J.L. Heessen, and J.R.G. Hislop. 1993. Atlas of North Sea fishes. ICES Cooperative Research Report 194. 268 p.

Kooiman, J., M. van Vliet, and S. Jentoft (eds.). 1999. Creative Governance: Opportunities for Fisheries in Europe. Ashgate, Aldershot.

Krauss, W. 1986. The North Atlantic Current. J. Geophys. Res. C4 91:5061-5074.

Kröncke, I. 1990. Macrofauna standing stock of the Dogger Bank - a comparison. 2. 1951-1952 versus 1985-1987 are changes in the community of the northeastern part of the Dogger Bank due to environmental changes. Neth. J. Sea Res. 25:189-198.

Kröncke, I. 1992. Macrofauna standing stock of the Dogger Bank - a comparison. 3. 1950-54 versus 1985-87 - a final summary. Helgoländer Meeresuntersuchungen 46:137-169.

Kröncke, I. and R. Knust. 1995. The Dogger-Bank - a special ecological region in the central North Sea. Helgoländer Meeresuntersuchungen 49:335-353.

Kröncke, I. and E. Rachor. 1992. Macrofauna investigations along a transect from the inner German Bight towards the Dogger Bank. Mar. Ecol. Prog. Ser. 91:269-276.

Künitzer, A., D. Basford, J.A. Craeymersch, J.-M. Dewarumez, J. Dorjes, G.C.A. Duineveld, A. Eleftheriou, C. Heip, P. Herman, P. Kingston, U. Niermann, E. Rachor, H. Rumohr, and P.A. De Wilde. 1992. The benthic fauna of the North Sea, species distributions and assemblages. ICES J. Mar. Sci. 49:127-144.

Kushnir, Y. 1994. Interdecadal variation in North Atlantic sea surface temperature and associated atmospheric conditions. J. Climate 7:141-157.

Kushnir, Y., V.J. Cardone, J.G. Greenwood, and M.A. Cane. 1997. The recent increase in North Atlantic wave heights. J. Climate 10:2107-2113.

Laane, R.W.P.M., A.J. Southwood, D.J. Slinn, J. Allen, G. Groeneveld, and A. de Vries. 1996. Changes and causes of availability in salinity and dissolved inorganic phosphate in the Irish Sea, English Channel, and Dutch coastal zone. ICES J. Mar. Sci. 53:933-944.

Law, C.S. and N.P.J. Owens. 1990. Dentrification and nitrous oxide in the North Sea. Neth. J. Sea Res. 25:65-74.

Le Fèvre, J. 1986. Aspects of the biology of frontal systems. Adv. Mar. Biol. 23:164-299.

Lindeboom, H., W. van Raaphorst, J. Beukema, G. Cadée, and G. Swennen. 1995. (Sudden) changes in the North Sea and Wadden Sea: oceanic influences underestimated. Deutsche Hydrographische Zeitschrift, Supplement 2. 87-100.

Lindley, J.A., J.C. Gamble, and H.G. Hunt. 1995. A change in the zooplankton of the central North Sea (55° to 58° N): a possible consequence of changes in the benthos. Mar. Ecol. Prog. Ser. 119:299-303.

Little, A.D. 1997. Benchmarking of hydrocarbons in the area affected by the Sea Empress oil spill. Final Report to the Sea Empress Evaluation Committee, A.D. Little Ltd., Cambridge.

Liss, P.S., A.J. Watson, M.I. Liddicoat, G. Malin, P.D. Nightingale, S.M. Turner, and R.C. Upstill-Goddard. 1994. Trace gases and air-sea exchanges. In: H. Charnock, K.R. Dyer, J.M. Huthnance, P.S. Liss, J.H. Simpson and P.B. Tett (eds.). Understanding the North Sea system. The Royal Society. Chapman and Hall, London. 153-163.

Lloyd, C., M.L. Tasker, and K. Partridge. 1991. The Status of Seabirds in Britain and Ireland. T & A.D. Poyser, London.

Lomstein, B. and T.H. Blackburn. 1992. Sediment nitrogen cycling in Aarhus Bay, Denmark. Havforskning fra Miljøstyrelsen 16. Miljøstyrelsen, Copenhagen.

Longhurst, A. 1998. Ecological geography of the sea. Academic Press, London.

Lowe, D.M. 1998. Alterations in cellular structure in *Mytilus edulis* resulting from exposure to environmental contaminants under filed and experimental conditions. Mar. Ecol. Prog. Ser. 46:91-100.

Luff, R. and T. Pohlmann. 1995. Calculation of the water exchange times in the ICES-Boxes with an Eulerian dispersion model using a half-life time approach. Deutsche Hydrographische Zeitschrift 47:287-299.

Luff, R., A. Moll, and T. Pohlmann. 1996. Program animates North Sea model results. American Geophysical Union. http://www.agu.org/eos_elec/96197e.html

Marine Conservation Society. 1995. Nationwide beach-clean and survey report. Marine Conservation Society, Ross On Wye.

Martin, J.H.A., H.D. Dooley, and W. Shearer. 1984. Ideas on the origin and biological consequences of the 1970s salinity anomaly. ICES CM 1994/Gen:18.

Mathiesen, T.-C. 1994. The human element in environmental protection. Mar. Poll. Bull. 29:375-377.

McCartney, M.S. and L.D. Talley. 1982. The sub-polar mode water of the North Atlantic Ocean. J. Phys. Oceanogr. 12:1169-1188.

McGlade, J.M. 1989. Integrated marine resource management: understanding the limits to exploitation. Amer. Fish. Soc. Symp. 6:139-165.

McGlade, J.M. 1999a. Ecosystem analysis and the governance of natural resources. In: J.M. McGlade (ed.). Advanced Ecological Theory. Blackwell Science, Oxford. 309-341.

McGlade, J.M. 1999b. Bridging disciplines: the role of scientific advice, especially biological modelling. In: J. Kooiman, M. van Vliet, and S. Jentoft (eds.). Creative Governance: Opportunities for Fisheries in Europe. Ashgate, Aldershot. 175-185.

McGlade, J.M. 2001a. Governance and sustainable fisheries. In: B. von Bodungen and R.K. Turner (eds.). Science and Integrated Coastal Management. Dahlem University Press, Dahlem. 339-361.

McGlade, J.M. 2001b. SimCoast™: a fuzzy logic expert system for planning and management (CD-ROM and User Manuals). Version 2.0. University College London, UK.

McGlade, J.M. and K.I. Metuzals. 2000. Options for the reduction of by-catches of harbour porpoises (*Phocoena phocoena*) in the North Sea. In: M.J. Kaiser and S.J. de Groot (eds.). Effects of Fishing on Non-target Species and Habitats. Blackwell Science, Oxford. 332-353.

McGlade, J.M. and J. Shepherd (eds.). 1992. Techniques for biological assessment in fisheries management. Berichte aus der Ökologischen Forschung Vol. 9. 63 p. ISBN 3-89336-091-3

Mearns, B. and R. Mearns. 1998. The Bird Collectors. Academic Press, San Diego.

Mendonsa, E.L. 1982. Benefits of migration as a personal strategy in Nazare, Portugal. Int. Migr. Rev. 16:635-645.

Metcalfe, J.D., G.P. Arnold, and B.H. Holford. 1994. The migratory behaviour of plaice in the North Sea as revealed by data storage tags. ICES CM 1994: Mini 11. 8 p.

Millner, R.S. and C.L. Whiting. 1996. Long-term changes in growth and population abundance of sole in the North Sea from 1940 to the present. ICES J. Mar. Sci. 53:1185-1195.

Millward, G.E. and G.A. Glegg. 1997. Fluxes and retention of trace metals in the Humber Estuary. Est. Coast. Shelf Sci. 46:175-184.

Moll, A. 1997. Modeling primary production in the North Sea. Oceanography 10.

Morley, G. 1968. The North Sea. Frederick Muller, London.

Natural Environment Research Council. 1996. Report of the Scientific Group on Decommissioning Offshore Structures. NERC, Swindon.

North Sea Task Force. 1993. North Sea Quality Status Report. Oslo and Paris Commissions, London. Olsen & Olsen, Fredensborg.

O'Brien, C.M., C.J. Fox, B. Planque, and J. Casey. 2000. Climate variability and North Sea cod. Nature 404:142.

Olsen, J.H.T. 1994. New challenges for the international management of ocean resources. Mar. Poll. Bull. 29:262-265.

Olson, P.H. 1994. Handling of waste in ports. Mar. Poll. Bull. 29:284-295.

Oslo and Paris Commissions. 1990. Temporal trend studies in biota, 1983-1991. Oslo and Paris Commissions, London.

Ottley, C.J. and R.M. Harrison. 1992. The spatial distribution and particle size of some inorganic nitrogen, sulphur and chlorine species over the North Sea. Atmos. Environ. A 26:1689-1699.

Otto, L., J.T.F. Zimmerman, G.K. Furness, M. Mork, R. Seatre, and G. Becker. 1990. Review of the physical oceanography of the North Sea. Neth. J. Sea Res. 26:161-238.

Paine, R.T. 1966. Food web complexity and species diversity. American Naturalist 100:65-75.

Paine, R.T. 1974. Intertidal community structure: experimental studies on the relationship between a dominant competitor and its principal predator. Oecologia 15:93-120.

Pedersen, B. 1996. Metal concentration in biota in the North Sea: changes and causes. ICES J. Mar. Sci. 53:1008-1013.

Philippart, C.J.M., H.J. Lindeboom, J. van der Meer, H.W. van der Veer, and J. IJ. Witte. 1996. Long-term fluctuations in fish recruit abundance in the western Wadden Sea in relation to variation in the marine environment. ICES. J. Mar. Sci. 53:1120-1129.

Pimm, S. 1999. The dynamics of the flows of matter and energy. In: J.M. McGlade (ed.). Advanced Ecological Theory. Principles and Applications. Blackwell Science, Oxford.

Pingree, R.D., P. Holligan, and G.T. Mardell. 1978. The effects of vertical stability on phytoplankton distribution in summer on the northwestern European shelf. Deep-Sea Res. 25:1011-1028.

Pingree, R.D. and D.K. Griffiths. 1978. Tidal fronts on the shelf seas around the British Isles. J. Geophys. Res. 83:4615-4622.

Pingree, R.D. and D.K. Griffiths. 1980. Currents driven by a steady uniform wind stress on the shelf seas around the British Isles. Oceanol. Acta 3:227-236.

Pingree, R.D. and L. Maddock. 1977. Tidal residuals in the English Channel. J. Mar. Biol. Assoc. 58:965-973.

Piriou, J.-Y., A. Menesquen, and J.-C. Salomon. 1991. Les marées vertes à ulves: conditions nécessaries, évolution et comparison de sites. In: M. Elliott and J.-P. Ducrotoy (eds.). Estuaries and coasts: spatial and temporal intercomparisons. Olsen & Olsen, Fredensborg, Denmark. 117-122.

Pohlmann, T. 1996. Predicting the thermocline in a circulation model of the North Sea. Part I Model description, calibration and verification. Cont. Shelf. Res. 16:131-146.

Pohlmann, T. and W. Puls. 1993. Currents and transport in water. In: J. Sündermann (ed.). Circulation and contaminant fluxes in the North Sea. Springer Verlag, Berlin.

Pollard, R.T. and S. Pu. 1985. Structure and circulation of the upper Atlantic Ocean north-east of the Azores. Prog. Oceanograph. 14:443-462.

Pond, K. and G. Rees. 1996. Coastwatch UK 1996 Survey Report. International Environmental and Public Health Centre, University of Surrey. ISBN 1 899090 70 8

Pope, J.G. and C.T. Macer. 1996. An evaluation of the stock structure of North Sea cod, haddock, and whiting since 1920, together with a consideration of the impacts of fisheries and predation effects on their biomass and recruitment. ICES J. Mar. Sci. 53:1157-1169.

Prandle, D., C.F. Jago, S.E. Jones, D.A. Purdie, and A. Tappin. 1994. The influence of horizontal circulation on the supply and distribution of tracers. In: H. Charnock, K.R. Dyer, J.M. Huthnance, P.S. Liss, J.H. Simpson, and P.B. Tett (eds.). Understanding the North Sea system. The Royal Society. Chapman and Hall, London. 27-44.

Prandle, D., D.J. Hydes, J. Jarvis, and J. MacManus. 1997. The seasonal cycles of temperature salinity, nutrients and suspended sediments in the southern North Sea in 1988 and 1989. J. Coastal & Shelf Sci. 45:669-680.

Proctor, R. and I.D. James. 1996. A fine resolution 3D model of the Southern North Sea. J. Mar. Syst. 8:285-295.

Puls, W. and J. Sündermann. 1990. Simulation of suspended sediment dispersion in the North Sea. In: R. Cheng (ed.). Physics of estuaries and bays. Springer Verlag, Berlin.

Quality Status Report. 1987. Second International Conference on the Protection of the North Sea: a Report by the Scientific and Technical Working Group. HMSO, London.

Radach, G. 1982. Dynamics interactions between lower trophic levels of the marine foodweb in relation to the physical environment during the Fladen Ground experiment. Neth. J. Sea Res. 16:23-246.

Radach, G. and K. Heyer. 1997. Cadmium budget for the German Bight in the North Sea. Mar. Poll. Bull. 34:375-381.

Radach, G. and H.J. Lenhart. 1995. Nutrient dynamics in the North Sea - Fluxes and budgets in the water column derived from ERSEM. Neth. J. Sea Res. 33:301-335.

Readman, J.W., L. Liong Wee Kwong, D. Grondin, J. Bartocci, J-P. Villeneuve, and L.D. Mee. 1993. Coastal water contamination from a triazine herbicide used in antifouling paints. Env. Sci. & Tech. 27:1940-1942.

Readman, J.W., R.F.C. Mantoura, M.R. Preston, A.D. Reeves, and C.A. Llewellyn. 1986. The use of pollutant and biogenic markers as source discriminants of organic inputs to estuarine sediments. Int. J. Enviro. Analy. Chem. 27:29-54.

Reid, P.C., C. Lancelot, W.W.C. Gieskes, E. Hagmeier, and G. Weichart. 1990. Phytoplankton of the North Sea and its dynamics: a review. Neth. J. Sea Res. 26:295-331.

Reise, K., E. Herre, and M. Sturm. 1989. Historical changes in the benthos of the Wadden Sea around the island of Sylt in the North Sea. Helgoländer Meeresunters 43:417-433.

Reise, K., K. Kolbe, and V.N. De Jonge. 1994. Makroalgen und Seegrasbestände im Wattenmeer. In: J.L. Lozán, E. Rachor, K. Reise, P. von Westernhagen, and W. Lenz (eds.). Warnsignale aus dem Wattenmeer - Wissenschaftliche Fakten, Blackwell, Oxford. 90-100.

Rice, J. and H. Gislason. 1996. Patterns of change in the size spectra of numbers and diversity of the North Sea fish assemblage, as reflected in surveys and models. ICES J. Mar. Sci. 53:1214-1225.

Richardson, K. 1989. Algal blooms in the North Sea, the good, the bad and the ugly. Dana 8:83-94.

Richardson, K. and J.P. Heilmann. 1995. Primary production in the Kattegat: past and present. Ophelia 41:317-328.

Rijnsdorp, A.D. and P.I. van Leeuwen. 1996. Changes in growth of North Sea plaice since 1950 in relation to density, eutrophication, beam-trawl effort, and temperature. ICES J. Mar Sci. 53:1199-1213.

Rijnsdorp, A.D., P.I. van Leeuwen, N. Daan, and H.J.L. Heessen. 1996. Changes in abundance of demersal fish species in the North Sea between 1906-1909 and 1990-1995. ICES J. Mar. Sci. 53:1054-1062.

Robertson, J., J.M. McGlade, and I. Leaver. 1996. Ecological effects of the North Sea industrial fishing industry on the availability of human consumption species. Unilever Commissioned Report, Univation, Aberdeen. 47 p.

Rogers, S.I. and R.S. Millner. 1996. Factors affecting the annual abundance and regional distributions of English inshore demersal fish populations: 1973-1995. ICES J. Mar. Sci. 53:1094-1112.

Sanger, C. 1987. Ordering the Oceans. The Making of the Law of the Sea. University of Toronto Press, Toronto.

Sanmartin, R. 1982. Marriage and inheritance in a Mediterranean fishing community. Man 17:664-672.

Schama, S. 1991. The Embarrassment of Riches. An Interpretation of Dutch culture in the Golden Age. Fontana Press, London.

Schroeder, A. and R. Kunst. 1999. Long-term changes in the benthos of the German Bight (North Sea) - possible influences of fisheries? ICES J. Mar. Sci.

Scott, C.R., K.L. Hemingway, M. Elliott, V.N. De Jonge, J.S. Pethick, S. Malcolm, and M. Wilkinson. 1999. Impact of nutrients in estuaries. Environment Agency & English Nature Report, Cambridge Coastal Research Unit, University of Cambridge & Centre for Environment, Fisheries and Aquaculture Science. 250 p.

Serchuk, F.M., E. Kirkegaard, and N. Daan. 1996. Status and trends of the major roundfish, flatfish and pelagic fish stocks in the North Sea: thirty-year overview. ICES J. Mar. Sci. 53:1130-1145.

Shennan, I. and P.L. Woodworth. 1992. A comparison of late Holocene and twentieth century sea-level trends from the UK and North Sea region. Geophys. J. Int. 109:96-105.

Sherman, K. 1994. Sustainability, biomass yields, and health of coastal ecosystems: an ecological perspective. Mar. Ecol. Prog. Ser. 112:277-301.

Sherman, K., C. Jones, L. Sullivan, W. Smith, P. Berrien, and L. Ejsymont. 1981. Congruent shifts in sand-eel abundance in western and eastern North Atlantic ecosystems. Nature 291:487-489.

Simpson, J.H. 1981. The shelf-sea fronts: implications of their existence and behaviour. Phil. Trans. R. Soc. Lond. A 302:531-546.

Skov, H. 1991. Trends in the oil contamination of seabirds in the North Sea. SULA 5: 22-23.

Steele, J.H. 1974. The Structure of Marine Ecosystems. Harvard University Press, Cambridge, MA.

Stein, O. and A. Hense. 1994. A reconstructed time series of the number of low pressure events since 1980. Z. Meteor. 3:43-46.

Stone, C.J., A. Webb, C. Barton, N. Ratcliffe, T.C. Reed, M.L. Tasker, C.J. Camphuysen, and M.W. Pienkowski. 1995. An Atlas of Seabird Distribution in North-west European Waters. Joint Nature Conservation Committee, Peterborough.

Sündermann, J. 1994. Suspended particulate matter in the North Sea: field observations and model simulations. In: H. Charnock, K.R. Dyer, J.M.

Huthnance, P.S. Liss, J.H. Simpson, and P.B. Tett (eds.). Understanding the North Sea system. The Royal Society. Chapman and Hall, London. 45-52.

Svansson, A. 1975. Physical and chemical oceanography of the Skaggerak and the Kattegat. I. Open sea conditions. Fishery Board of Sweden, Institute of Marine Research, Report 1. 88 p.

Taylor, A.H. 1995. North-south shifts of the Gulf Stream and their climatic connection with the abundance of zooplankton in the UK and its surrounding seas. ICES J. Mar. Sci. 52:711-721.

Taylor, A.H. 1996. North-south shifts of the Gulf Stream: ocean-atmosphere interactions in the North Atlantic. Int. J. Climatology 16:559-583.

Taylor, A.H., J.M. Colebrook, J.A. Stephens, and N.G. Baker. 1992. Latitude displacements of the Gulf Stream and the abundance of plankton in the north-east Atlantic. J. Mar. Biol. Assoc. UK 72:919-921.

Taylor, A.H. and J.A. Stephens. 1980. Latitudinal displacements of the Gulf Stream (1966 to 1977) and their relation to changes in temperature and zooplankton abundance in the NE Atlantic. Oceanologica Acta 3:145-149.

Ten Hallers-Tjabbes, S.C.C., J.F. Kemp, and J.P. Boon. 1994. Imposex in whelks (*Buccinum undatum*) from the open North Sea - relation to shipping traffic intensities. Mar. Poll. Bull. 28:311-313.

Tett, P. and M.R. Droop. 1988. Cell quota models and plankton primary production. In: J.W.T. Wimpenny (ed.). Handbook of laboratory model ecosystems for microbial ecosystems. CRC Press, Boca Raton. 177-233.

Tett, P.B., I.R. Joint, D.A. Purdie, M. Baars, S. Oosterhius, G. Daneri, F. Hannah, D.K. Mills, D. Plummer, A.J. Pomeroy, A.W. Walne, and H.J. Witte. 1994. Biological consequences of tidal stirring gradients in the North Sea. In: H. Charnock, K.R. Dyer, J.M. Huthnance, P.S. Liss, J.H. Simpson, and P.B. Tett (eds.). Understanding the North Sea system. The Royal Society. Chapman and Hall, London. 115-130.

Tett, P. and H. Wilson. In press. From biogeochemical to ecological models of marine microplankton. J. Mar. Syst.

Tiedemann, R., J. Harder, C. Gmeiner, and E. Haase. 1996. Mitochondrial DNA sequence patterns of harbour porpoises (*Phocoena phocoena*) from the North and Baltic Sea. Zeitschrift für Säugetierkunde 61:104-111.

Tregenza, N.J.C. 1992. Fifty years of cetacean sightings from the Cornish coast, SW England. Biol. Cons. 57:65-71.

Tregenza, N.J.C. 2000. Fishing and cetacean by-catches. In: M.J. Kaiser and S.J. de Groot (eds.). Effects of Fishing on Non-target Species and Habitats. Blackwell Science, Oxford. 269-280.

Tromp, D. and K. Wieriks. 1994. The OSPAR Convention: 25 years of North Sea protection. Mar. Poll. Bull. 29:622-626.

Tunstall, J. 1972. The Fishermen (3rd ed.). MacGibbon & Kee, London.

Turrell, W.R., E.W. Henderson, G. Slesser, R. Payne, and R.D. Adams. 1992. Seasonal changes in the circulation of the northern North Sea. Cont. Shelf Res. 12:257-286.

Turrell, W.R., G. Slesser, R. Payne, R.D. Adams, and P.A. Gillibrand. 1996. Hydrography of the East Shetland Basin in relation to decadal North Sea variability. ICES J. Mar. Sci. 53:899-916.

van den Hoek, C., W. Admiral, F. Colijn, and V.N. de Jonge. 1979. The role of algae and seagrasses in the ecosystems of the Wadden Sea: A review. In: W.J. Wolff (ed.). Flora and vegetation of the Wadden Sea. Wadden Sea Working Group Report 3, Balkema, Rotterdam. 9-118.

van Dijk, H.W.J. 1994. Integrated management and conservation of Dutch coastal areas: premises, practice and solutions. Mar. Poll. Bull. 29:609-616.

van Ginkel, R. 1989. Plundered into planters. In: J. Boissevain and J. Verrips (eds.). Dutch Dilemmas. Van Gorcum, Maastricht.

van Leussen, W., G. Radach, W. van Raaphorst, F. Colijn, and R.W.P.M. Laane. 1996. The North-West European shelf project (NOWESP): integrated analyses of shelf processes based on existing data and models. ICES J. Mar. Sci. 53:926-932.

Varela, R.A., A. Gonzado, and J.E. Gabaldan. 1995. Modelling primary production in the North Sea. Neth. J. Sea Res. 33:337-361.

Verboom, G.K., J.G. de Ronde, and R. P. van Dijk. 1992. A fine grid tidal flow and storm surge model of the North Sea. Cont. Shelf. Res. 12:213-233.

Vethaak, A.D. and T. Rheinallt. 1992. Fish disease as a monitor for marine pollution: the case of the North Sea. Rev. Fish Biol. & Fisheries 2:1-32.

Vinther, M. 1994. Investigations on the North Sea gillnet fisheries. DFU Report 485-95. Landbrugs-og Fiskeriministeriet. Danmarks Fiskeriundersogelser.

von Storch, H. 1996. The WASA project. Changing storm and wave climate in the North East Atlantic and adjacent seas. GKSS 96/E/61.

von Westernhagen, H., W. Hickel, E. Bauerfeind, U. Niermann, and I. Kröncke. 1986. Sources and effects of oxygen deficiencies in the South-eastern North Sea. Ophelia 26:457-473.

Voorrips, A.C., H. Hersbach, F.B. Koek, G.J. Komen, V.K. Makin, and J.R.N. Onvlee. 1997. Wave prediction and data assimilation at the North Sea. In: J.H. Stel, H.W.A. Behrens, J.C. Borst, L.J. Dropper, and J. v.d. Meulen (eds.). Operational Oceanography. The Challenge for European Co-operation. Elsevier Science B.V., Amsterdam. 463-471.

Walker, P.A. and H.J.L. Heessen. 1996. Long-term changes in ray populations in the North Sea. ICES J. Mar. Sci. 53:1085-1093.

Warner, A.J. and G.C. Hays. 1994. Sampling by the continuous plankton recorder survey. Prog. Oceanog. 34:237-256.

Watson, R.T., Z.C. Zinyowera, and R.H. Moss (eds.). 1997. The Regional Impact of Climate Change: an Assessment of Vulnerability. Cambridge University Press, Cambridge.

Wheeler, A.C. 1978. Key to the fishes of Northern Europe. Frederick Warner, London.

Widdows, J., P. Donkin, M.D. Brinsley, S.V. Evans, P.N. Salkeld, A. Franklin, R.J. Law, and M.J. Waldock. 1995. Scope for growth and contaminant levels in North Sea mussels *Mytilus edulis*. Mar. Ecol Prog. Ser. 127:131-148.

Wieking, G. and I. Kröncke. 1999. Long term comparison of macrofaunal communities on the Dogger Bank between 1996 and 1985-1987. ICES J. Mar Sci.

Wilson, W.H. 1990. Competition and predation in marine soft-sediment communities. Ann. Rev. Ecol. & Syst. 21:221-241.

Witbaard, R. 1996. Growth variations in *Arctica islandica* L. (Mollusca): a reflection of hydrography related food supply. ICES J. Mar. Sci. 53:981-987.

Woodworth, P.L. 1999. High waters at Liverpool since 1768: the UK's longest sea level record. Geophys. Res. Letts. 26:1589-1592.

Yool, A. 1998. The dynamics of open-ocean plankton ecosystem models. PhD thesis University of Warwick.
http://www.oikos.warwick.ac.uk/ecosystems/ThesisArchive/yool_thesis.html

Zuhlke, R. and K. Reise. 1994. Response of macrofauna to drifting tidal sediments. Helgoländer Meeresuntersuchungen 48:277-289.

Zweip, K. v.d. 1990. The Wadden Sea: a yardstick for a clean North Sea. In: D. Freestone and T. IJ. Ijlstra (eds.). The North Sea: Perspective on Regional Environmental Co-operation. International Journal of Estuarine & Coastal Law, Graham & Trottman / Martinus Nijhoff, London. 201-212.

V
Summary and Comments

Large Marine Ecosystems of the North Atlantic
K. Sherman and H.R. Skjoldal (Editors)
© 2002 Published by Elsevier Science B.V.

13

Changing States of the Large Marine Ecosystems in the North Atlantic: Summary and Comments

Gotthilf Hempel

This summary is based on notes taken during the "Symposium on Changing States of the Large Marine Ecosystems of the North Atlantic and Global Environmental Trends," held in Bergen, Norway, 17-19 June, 1999. The summary is selective and might not always reflect what the various authors consider the salient points of their papers. *My personal comments are shown in italics.*

INTRODUCTION AND GENERAL PAPERS

In his welcome address, the Research Director of the Institute of Marine Research in Bergen, Åsmund Bjordal, stressed the need for ecosystem research as the basis for any medium and long-term prediction of fish catches, as five years is about the upper limit for short-term predictions based on stock and catch data. *In fact, this time-span applies to long-lived fish and a relatively stable temperate ocean climate, while in many other species and systems the time-span for sensible predictions is even shorter. Unfortunately the questions of predictability were not much addressed by the speakers of the Symposium.*

Introducing the theme of the Symposium, Hein-Rune Skjoldal pointed to the global scales of climatic driving forces and to the regional scales of biological reactions on all trophic levels of an ecosystem. Correspondingly, the Global Ocean Observing System (GOOS) takes care of monitoring the ocean environment while Large Marine Ecosystem (LME) research projects address the regional phenomena, particularly the interaction of nature and man when exploiting the resources or using the sea as a sink for all kinds of waste. GOOS and LME assessment activities have to be linked to process oriented research like GLOBEC. ICES would be the right body for this, as it has been active in assessment and research for many years.

Kenneth Sherman reminded the audience of the paradigm shift in LME activity from being primarily fish stock oriented programmes to a broad ecological and socio-economic concept, as reflected in the five modules of modern LME assessments: Productivity, Fish and Fisheries, Pollution and Ecosystem Health, Socio-economics, and Governance. The LME approach links science-based assessments of the changing states of the systems to the potential socio-economic benefits gained from long-term sustainability of their resources and from environmental integrity. Marine scientists and administrators should be aware of the enormous economic and societal value of the coastal ecosystems. K. Sherman referred to a recent study by economists, estimating the overall annual output value of US$ 330 billion for the waters off the NE coast of the United States, while the most conservative figure amounts to US$ one billion. Up to 44 million people in the region are in one way or another dependent on the adjacent marine ecosystem.

The call for good and continued data series was introduced by K. Sherman as a Leitmotiv of the Symposium. Later Kees Zwanenburg stated that "the lack of long-term data for many of the trophic levels within the system (primary/secondary production, benthic invertebrate composition and abundance) and the rudimentary understanding of the connections both between these levels and with higher trophic levels indicates the need for significant expenditures on monitoring and research."

Regarding time series of environmental data, the biologists are catching up with the physical oceanographers by making more use of towed systems. K. Sherman advocated the advanced versions of the Continous Plankton Recorder (CPR) carrying packages of sensors for nutrients, particle counts and pigments. Advanced towed sensor and sampling devices are on the horizon. Those tools will provide environmental data of more direct relevance to estimating productivity and trophodynamics of the system. Remote sensing, including SeaWIFS of sea surface temperature and pigments, has become widely used but requires further ground truthing for the translation of near-surface pigment concentrations into primary production.

To my mind, any good director of a fisheries institute has two prime obligations: 1. He/she should support new research for the understanding of the functioning of fish populations and of man's impact on them. 2. He/she should continuously ensure the generation of reliable data files for present and future analysis of fish stocks and their environment. The quality of fisheries data has deteriorated in recent years for various reasons related to the global trade as well as to the multitude of catch and tax regulations which lead to false reporting of fishing sites and of figures of catch and discards. Trends in certain catch statistics might reflect shifts in the quality of the reporting rather than in the stocks themselves. Another source of error is the improvement in searching power and positioning of the vessels, giving hidden rise to fishing efficiency without augmenting the fleet. Those difficulties regarding data derived from commercial fisheries call for fishery-independent data files and make the continuation of the conventional trawling surveys and other sampling programmes imperative.

The large sets of environmental and fisheries data in LME assessment call for statistical treatment of various kinds and invite a multitude of spectra analyses and all kinds of possible combinations, correlations and ratios. Modern computer programmes make data treatment rather easy. Some answers are surprising and some not meaningful.

The critical dialogue about the reliability of environmental and fisheries data and about the appropriateness of the statistical analysis and its interpretation in ecological terms is crucial for any LME project and should be fostered in symposia and workshops where different LMEs are compared.

Fish, benthic invertebrates, and other biological indicator species are used in the pollution and ecosystem health module to measure pollution effects on the ecosystem at the levels of individual species, populations, and communities following the protocols of NOAA's "National Status and Trends Program." Multiple indices used to describe different ecosystem states of health are under investigation by NOAA, including: biodiversity, stability, resilience, productivity, and yield levels.

K. Sherman told the audience the success story of making LME research a central theme of the activities of the Global Environmental Facility (GEF) with its US$ 2 billion trust fund. Four GEF funded LME projects are underway (Black Sea, Gulf of Guinea, Yellow Sea, South China Sea), while several others are in various stages of planning and preparation (e.g. Baltic Sea, Benguela System). Funding and implementation of each of these projects are based on partnership with UNIDO, UNDP, UNEP, and World Bank as well as with regional and national organisations. Further integrated LME programmes for research, monitoring and management are in place or preparation for LMEs under the jurisdiction of a single coastal state, like USA or Canada, or on a regional basis like in the North Sea and in the Gulf of Mexico.

Jacqueline McGlade provided a generic look at the LME approach. She asked: Is the LME concept based on sound theory, or is it a compromise or a practicality? J. McGlade analysed the state of knowledge on several criteria of an LME:

- Regarding biogeochemistry and the scales of natural forcings, various classes of LMEs can be distinguished (oceanic, strong upwelling, coastal shelf, semi-enclosed and enclosed seas, and extreme polar waters).

- LMEs are to be characterised by their endemisms. Biodiversity audits are needed for plankton, nekton and benthos.

- Trophic complexity has to be known in terms of primary, secondary and tertiary production, P:B ratios, trophic aggregation, Cycling Index etc..

- The single and multi-species approaches to fisheries management have to take into account environmental variability for recruitment as well as the trophic complexity at all ontogenetic stages from eggs and larvae to the adult fish.

- Broad information on contamination is required. Thousands of unidentified compounds enter the North Sea by rivers and by precipitation. On an LME- wide scale we have to consider the invasion of alien species including micro-organisms with the ballast water of double hull tankers and other vessels. Obsolete oilrigs and other installations and impacts at the seabed are further sources of environmental problems.

- Political regimes in the various coastal states partly determine man's relationship with the seas. We have to study the interactions between central, regional and local government, between the public and private sector, as well as the degree of participation at the various levels and stages of decision-making. Political stability and the preparedness to transfer national rights and obligations to regional bodies are important for the success of transboundary LME management. Socio-economic information is required, as the resources of the LME should contribute to the food security of the population and to the market needs in the coastal states. Marine products become elements of a global market with self-organising networks, e.g. the electronic stock market of fish.

- Closely related to the socio-economic criteria is governance, which has been little studied so far in relation to fisheries, coastal zone management, market and societal structure in the LME and its user groups. In most North Sea countries much of such governance takes place without governmental input and interference. Hierarchical systems, market and participation amongst the various players largely govern the use of the North Sea resources.

- For the North Sea, the present day's regulatory structures and regulations on national and EU level are complex and confusing. *Nevertheless, substantial progress has been made, i.e. in precautionary regulation of North Sea herring fisheries.*

Are we better off with management on an LME scale than on single stock regulations? J. McGlade concluded: The LME approach can result in decisions based on information on relevant scales. It can provide guidelines, e.g. for the mitigation of risks caused by environmental changes, market collapses or geopolitical events and trends. The LME approach can create sets of objectives for governance with its associated risk mitigating actions to achieve sustainability.

TWO YOUNG LME PROJECTS AND TWO LARGE APPROACHES

The Benguela system and the Baltic Sea are two emerging GEF-funded LME projects only loosely linked to the North Atlantic. Their international co-ordinators provided the audience with insights in those systems and in the cumbersome process of obtaining financial support from GEF.

Mick O'Toole described the Benguela System as the world's most active upwelling system, and the only one framed by tropical waters from the North (Angola Current) and from the South (Agulhas Current). The system is highly variable from year to year in oceanographic conditions and in recruitment and accessibility of fish to fisheries. He reported on a disastrous event in 1995 independent of an El Niño, wiping out large parts of the benthos community and of many fish stocks by the entrainment of oxygen poor water.

M. O'Toole then went on to outline step-by-step the five years' process of planning the international LME project, which originated from a bottom-up initiative of scientists of South Africa, Namibia and Angola together with Norwegian and German fishery biologists. Jointly they developed the co-operative programme BENEFIT (Benguela Environment Fisheries Interaction and Training) which was formally launched in 1998. The preparation for a GEF-sponsored LME project started in parallel to BENEFIT, but with substantial interaction between the two activities. Over the past two years a tremendous amount of paperwork had to be completed to qualify for the financial support of GEF and the World Bank for a trilateral Benguela LME Project.

Only a few years ago, war struck the Benguela region and even now the situation is tense at its northern fringe. The cooperation between the marine scientists of the three countries is not only beneficial to science but has also had positive political effects as it helps to reduce the North/South gradient in scientific capacity and in self-reliance in the region. In this context, M. O'Toole stressed the importance of joint cruises as organised by Norway, Germany and South Africa and with participation by scientists and trainees of all three countries of the region. In the Benguela region there is a consensus that the time has passed when European scientists visiting the region concentrated on their own scientific interests, or when national donor agencies competed with each other rather than co-operating.

Some scientists and science administrators in industrialised countries complain about the large amounts of money given by GEF and others to developing countries for their LME projects, while in industrialised countries the direct support of LME research tends to decrease. We have to realise, however, that over many years industrialised countries were able to build up an excellent scientific infrastructure supported by high annual budgets. Much of it can be used for LME research, provided the priorities in the respective institutes are put right. Even now at times of limited resources, the total budget for marine and fisheries research and pollution monitoring in industrialised countries is much higher than the international (e.g. by World Bank) and bilateral assistance programmes for LME work in the developing and newly developed countries.

The Baltic Sea, as described by Jan Thulin, is a rather unique system: The world's largest body of brackish water, semi-enclosed, permanently stratified with episodic exchange in the deep layers. The Baltic Sea in its present configuration is only about 6000 years old and the species diversity is small compared to the neighbouring North Sea. The young system is under heavy natural and man-made stress because of the renewal of its water masses being rather sporadic and the large input from land, the drainage area being four times larger than the Baltic Sea itself. Eighty million people, and intensive agriculture and industry, cause heavy riverine and airborne pollution. The latter stems from an even larger area in the West and Southwest. Eutrophication has reduced the transparency of the water column, caused algal blooms and increased sedimentation and oxygen uptake at the seabed. All this resulted in substantial changes in the benthic and pelagic flora and fauna including fish. The input of toxic pollutants into the Baltic Sea reached its maximum in the 1980s. It affected reproduction in seals and birds, and possibly in coastal fish also. Unfavourable environmental conditions for reproduction, in combination with heavy fishing on adult and young fish, jeopardise the two populations of Baltic cod. Herring and sprat are also heavily exploited, mostly for industrial purposes.

The Baltic Sea is one of the best-monitored seas in the World, with three international organisations (ICES, HELCOM and IBSFC) advising and/or deciding on the management of the environment and its resources. Nine states border the Baltic Sea, all of them with rather different economic and scientific capacity and with different interests in the use of the sea for fishing, recreation, shipping, waste disposal etc. J. Thulin pointed to the need for increased integration among the international bodies and for long-term multinational management aiming for sustainable use of the Baltic Sea ecosystem as a whole.

A few years ago, we asked ourselves: To what extent is a GEF-funded Baltic LME project needed by the Baltic scientific and political community with its well established regional organisations and its long tradition of scientific cooperation? Now we realise that considerable amounts of money are needed "as glue and grease" for co-ordination, standardisation and rationalisation of the activities of the countries and organisations and particularly in the attempt to make the eastern coastal states full and equal partners of the research, monitoring and management system. In doing so the GEF project will not try to "re-invent wheels" or to substitute well established organisational structures of regional cooperation, but it will support HELCOM and IBSFC and the coastal states, particularly in the modules of socio-economic management and governance of fisheries and fishery statistics as well as in pollution control. Channelling much of its activities through ICES might also strengthen the LME approach to the North Atlantic in general.

Two brief contributions dealt with ocean assessments on a scale larger than LMEs. The Global International Waters Assessment (GIWA) was presented by Per Wramner. It is a four-year project executed by UNEP to provide GEF and other interested national and international agencies with a systematic and comprehensive assessment of the environmental problems in transboundary waters at sea and on land. The basic units of assessment are 66 subregions, each of them a drainage area and associated marine basin, usually an LME. This kind of subregional breakdown will help GEF in its strategic planning for further LME projects.

Jeanne L. Pagnan dealt with the Arctic Ocean and its American and Eurasian rim. There are no other regions in the World which are so sensitive to climatic changes and which have more far reaching effects in terms of Global Change. The Arctic Ocean is much larger and less uniform than a conventional LME but its subsystems are closely interlinked and have many features in common, both in terms of marine and terrestrial environments and in ethnic and societal structures. The author described the development of multinational organisations and co-operative programmes over the past decade in the wake of political opening of the region. Forerunners were the Arctic Ocean Sciences Board and other non-governmental corporations in the early and mid eighties. Presently the economic situation of the Russian Federation and the instability of its Arctic provinces call for assistance by the arctic partner countries but also by the international science community.

TELECONNECTIONS IN THE NORTH ATLANTIC

Three papers dealing with teleconnections in the North Atlantic derived their findings largely from the same long-term oceanographic data sets on the shifts in the North Atlantic Oscillation (NAO) as related to the pressure fields in the atmospheric circulation over the North Atlantic. Further time series relate to the positions of the front of the Gulf Stream/North Atlantic Drift and of the Polar Front. Reference was also made to the migration of the Great Salinity Anomaly across the Atlantic and to the recent slackening in the formation of North Atlantic Intermediate and Deep Water in the Greenland and Norwegian Sea. All those features have far reaching effects on plankton and fish in all parts of the North Atlantic and can even be traced in terrestrial and limnic communities, as reflected in the zooplankton of Lake Windermere in England.

Based on the analysis of good data sets on the position of the Gulf Stream Front, the NAO and the CPR plankton abundances in the North Sea, Arnold Taylor came to a number of meaningful conclusions on linkages between large-scale processes and local perturbations, and between physical and biological phenomena driving the North Sea system. He found the year-to-year changes in zooplankton, particularly the category "Total Copepods," around the British Isles correlated with the latitude of the Gulf Stream at the US coast, the connection being climatic and operating through the seasonal cycle of stratification with April being the key month.

Chris Reid presented in greater detail the relationship of phyto- and zooplankton parameters in the North Sea and North Atlantic to NAO, based on the same systematic and long term study of CPR data. These fully standardised time series of physical and biological data allow the detection and multivariate statistical analysis of similarities in trends and in event-like changes. One wonders to what extent NAO influences the North Sea quickly by atmospheric heating and more slowly through inflow around Scotland and through the English Channel. The comparison of CPR data sets from the North Sea with those of the Northern North Atlantic showed marked regional differences. Manmade eutrophication is presumably not the sole cause for the increase of Chlorophyll a in the North Sea as it increased west of the British Isles simultaneously.

In reviewing the synchronous ups and downs of several large-scale fisheries in various current systems of the World Ocean, Andrew Bakun assumes that the linkages between physics and fish must be simple, otherwise correlations could not be that obvious. In the search for underlying mechanisms he postulated long-term shifts in spawning and feeding grounds in reaction to (periodic) climate changes. A. Bakun assumes a certain genetic diversity in fish populations like sardines to be reflected in differences in the migratory behaviour: A small fraction of the stock moves offshore and finds rich feeding conditions at oceanic fronts while the majority of the stock keeps their traditional near-shore feeding and spawning areas. The frontal part of the stock grows rapidly until predators start concentrating at the front. High predation, fishing pressure, and possibly also shifts in the front may cause a collapse of the frontal fish population. In brief: progressive sardines go to the front, establish themselves there for a while, but then they are eaten up while the conservative party of the stock survives in their "home lands." There the detrimental effects of short-term variations in environmental conditions are buffered by indiscriminate mass spawning over all the year and over large areas. Modern fish have also a long pelagic phase for their eggs and larvae. A. Bakun considers parental care as old-fashioned - *at least in marine fish*.

There was no time at the Symposium to discuss the thought-provoking ideas of A. Bakun. They are all related to the assumption that the world-wide co-variation in fisheries is due to teleconnections caused by global climate variation. Could changes in the global market have an effect too? Over the past decades we witnessed changing demands in various commodities like fish meal and fish oil. We also noticed collapses in distant water fisheries in the wake of UNCLOS and its expansion of national fishing zones as well as for political and economic reasons. In the not too distant future, the time may be right to re-visit the concept of teleconnections on a broad scale.

LMES OFF EASTERN NORTH AMERICA

The hydrographic regime of the highly productive shelf off the **Southeastern US** is determined inshore by coastal run-off and offshore by Gulf Stream dynamics. Part of the production is exported by migratory fish species including menhaden and tuna-like fishes. John V. Merriner concentrated on the exploitation of the rich and highly diverse resources of the various systems of the region, exploited by pelagic fishing for Atlantic menhaden, inshore shrimp fisheries, and of reef and deep sea fisheries for various kinds of highly-priced demersal and pelagic fish. With the increase in human population along the coast, pressure grew on several of the stocks and on the system as a whole. Single species management has been fairly successful with the resident stocks of mackerels (Spanish and King), based on good monitoring, rigorous legislation and careful governance ensuring F's of 0.15-0.2. More difficult is the management of 300 species of reef fishes. Most fish have drastically decreased in average length. Catch statistics and stock estimates are rather unreliable. Moratoria for certain target species or for males in male-limited stocks might help. More important are the new complex effort restrictions. But in the long run, habitat management seems more promising than dealing separately with the various groups of fish. Together with a general reduction in fishing pressure, the introduction of eight fishery reserves along the coast is under consideration. Those protected areas should be strategically placed and large enough to provide the entire

LME with sufficient offspring to restore the stocks. J. Merriner assumes that it will take decades to bring the reefs back to be attractive for rich sustainable small-scale fisheries.

The experience gained by scientists and managers at the SE US Shelf should be of great assistance to other tropical and subtropical countries with degrading reef fish populations. However, those efforts of fishery administration might be in vain because of deterioration of reefs by tourists and other human activities, as well as by the threats of coral bleaching which is possibly related to global warming.

The **US Northeast Shelf** LME has been under systematic study for more than three decades. K. Sherman gave a full account of the continued monitoring of its oceanographic features (e.g. about 11 warm core rings per year) and high primary productivity (ca. 350 g $C/m^2/y$), particularly in nearshore waters where the effects of increasing human population are very noticeable and are the major concern of NOAA in its attempts to control marine environment quality. Secondary production in terms of zooplankton and zoobenthos is sufficiently high not to be a limiting factor to commercial fish stocks and their recruitment. The state of the demersal fish stocks is mainly characterised by overfishing, bringing the stocks down to less than half their pre-1960s level. Various protective measures for the reduction of fishing effort hold promise to the recovery of those stocks. The fishing industry, however, finds itself in a jungle of rules and regulations, which are difficult to follow. Fisheries for mackerel and herring by foreign fleets were closed in the mid-1970s. Since then the stocks have increased steeply to 6 mill tonnes. Surprisingly the large changes in predation pressure by the various fish stocks is not reflected in the abundance indices of plankton as recorded by CPR and other means. Although interannual variability is very noticeable, no long-term trends could be observed.

In a well-monitored LME like the US Northeast Shelf, one might look for ecological indices of stability and resilience comparable to those on fish stocks and their exploitation. In general, we may ask whether modern management should aim to reach and then keep a certain state of a system or for the potential to reach a certain state when needed.

Kees Zwanenburg described the **Scotian Shelf** LME as consisting of two parts under different oceanographic regimes and hence with different faunas, including fish populations. The western and central shelf was particularly warm after the spell of cold years in the 1960s. Only in 1998 did an intrusion of Labrador Shelf water cause a substantial cooling. The eastern shelf is generally cooler and changes are far more dramatic than on the western shelf: A cooling period started in 1983 and the bottom temperature remained low until now, while the surface temperature has fluctuated since the early 1990s. The long-term temperature drop is associated with higher chlorophyll values (CPR greenness) and abundance of copepods, particularly arctic species. In general, however, phyto- and zooplankton are rather thin on the eastern shelf. Cold water fish and crustacea invaded while the growth rate of the resident demersal fish dropped, presumably as a consequence of the low temperature. Grey seal abundance has increased greatly over the past four decades, but its impact on the fish stocks is not well understood due to poor diet information in the early part of the period.

Since the early 1980s the demersal biomass of fish, crabs, and shrimp taken together (trawlable demersal biomass) has declined all over the Scotian Shelf, but particularly in the eastern part. For pelagic fish, however, an increase was noticed. Due to decreased growth rates and heavy fishing the number of large demersal fish has dropped from 1970 to the early 1990s. Then a reduction in fishing effort and the present moratorium in the cod fishery resulted in an apparent recovery of the size structure of the stock. That seems particularly important in view of the reproduction strategy of cod: throwing as many eggs as possible in the sea, producing a successful year-class from time to time. Success of this strategy depends on a sufficient number of large, highly fecund females.

The Scotian Shelf has been exploited heavily for centuries but particularly since 1950. With the introduction of the 200 nm EEZ, the build up of the Canadian fishing fleet had priority to conservation of the stocks. Single stock management did not consider any collateral effects of the fishery. In discussing fishing policy on the Scotian Shelf, K. Zwanenburg touched on several general questions, e.g. how should management react to the invasion of arctic fish during a spell of cold years- should they be fished out quickly while they are there? In more general terms, he questioned whether the free market economic paradigm of considering fish populations as commodities ("money with fins") should be the only possible measure for fishery management. Indeed, the ultimate management goal should be a viable ecosystem. With this in mind, operational objectives have to be defined first and the many single function management agencies have to be integrated into one effective management body overlooking the entire system. In this context, K. Zwanenburg called for substantial research and monitoring on all trophic levels (see above).

While K. Sherman and K. Zwanenburg had approached "their" LMEs in a rather holistic way, the following two analyses presented by Jake Rice addressed either the fisheries aspects (Newfoundland) or the environment (West Greenland).

The waters off **Newfoundland** are very old fishing grounds with long-term changes in environment and fish catches. In the second half of this century two collapses in the fisheries occurred which were carefully documented in terms of environmental changes, trophodynamics and fishing. The first collapse in the 1960-1970s was clearly caused by overfishing but did not do basic harm to the stocks. The declining stocks responded by density-dependent increases in somatic growth rates and condition and in higher recruitment per spawner. With the introduction of Canadian jurisdiction in 1977 and the subsequent drastic reduction in international fishing effort, most stocks recovered very quickly until the mid 1980s, thanks to high growth rates coupled with high survivorship and good year-classes which have been related to the Great Salinity Anomaly. Between 1986 and 1994 most stocks showed drastic declines. Overfishing by the expanding Canadian fleet was the major cause. Other factors, however, including the large extent of the Cold Intermediate Waters in the early 1990s affected the system too, as indicated by a marked increase in natural mortality in cod, declines in non-commercial fish and squid, and an increase in shrimp. The present increase in the harp seal population from 2 million in 1972 to the present 5 million adds to the predatory pressure on the fish stocks. Although environmental conditions are now back to normal and most fisheries remain closed, the northern stocks of ground fish have not yet recovered. Therefore the income of fishery is now based on crab and shrimp, which are still very abundant, rather than on finfish.

Rice concluded that the Newfoundland stocks and fisheries were always subject to changes, but the two recent collapses were presumably the most severe ones. They were caused by overfishing. Environment and trophodynamic interactions were secondary causes for the collapses, playing major roles afterwards in the period of reduced fishing mortality. Year-class strength, growth rate, and condition as well as the abundance of non-target stocks depend on both environment and trophodynamics. Geographical distribution of the target stocks are mainly determined by environmental factors, while the high natural mortality after the closure of the fishery is the consequence of heavy predation. Altogether Rice postulated a great system inertia, which makes monitoring and precaution essential, but "documenting the need for precautionary actions is hardest when change is greatest."

LMES OF THE ARCTIC RIM

In **West Greenland** waters all major commercially exploited fish stocks live at the fringe of their distribution area, particularly for their reproduction in terms of survival and transport of the larval stages. Cod, redfish, Greenland halibut and dab (but not sandeel) all profit from a strong inflow of Irminger Current and suffer from cold Polar Current. This relationship seems mainly indirect through different provision of food, mainly copepods for larvae and early juveniles. Trends were more pronounced in cod than in long rough dab and Greenland halibut. During warm periods West Greenland is also the feeding ground for surplus cod from Iceland. Søren Pedersen and Jake Rice based their analysis of the West Greenland LME on over seventy years of observations of zooplankton, fish larvae and fisheries. Unfortunately the series of larval surveys was terminated before the hypotheses on the fishery-related effects of predation release and by-catches could be tested. The time series began in a warm period of weak Polar Current. A decline of zooplankton was noticeable from the late 1950s, i.e. long before the Great Salinity Anomaly (GSA) in the years 1969 and 1970. After 1970 the Polar Current increased and from thereon only sandeels found favourable conditions, profiting also from relatively low predation, but they still showed long-term fluctuations. Occasionally redfish larvae occurred in high numbers, even during the cold period, and the stocks of Greenland halibut increased in the 1980s and 1990s. J. Rice concluded that the "search for THE mechanism may be misguided" because of the complexity of interaction between the hydrographic regime, provision of food for the young stages, predation pressure, and fisheries.

*The waters off **East Greenland** and their fish stocks and fisheries were not discussed at the Symposium. This area is characterised by the Polar Front meridionally separating the cold Arctic East Greenland Current carrying loads of Arctic sea ice to the south and the western part of the warm Atlantic Irminger Current.*

The oceanographic conditions are extremely variable due to changes in the strength and frontal mixing of the cold and warm currents. Such meso-scale changes in the environment were found to affect significantly the growth and recruitment of the Atlantic cod stock (Rätz et al. 1999). The demersal fish community off East Greenland is composed of very few boreal species. Structures in the quantitative species composition are determined by geographical as well as depth effects, but no persistent boundaries in the demersal fish assemblage are definable. Since 1982 German annual groundfish surveys cover the continental shelves and slopes from the 3-mile offshore limit to 400 m depth contour south of 67 degrees northern latitude (Rätz 1999). Around 100-150 trawling stations are usually conducted during autumn season when ice conditions are favourable. The investigation of changes in the environmental conditions at the sample sites and oceanographic standard sections is the second main goal of the surveys. The survey results indicate fundamental shifts in species composition in coherence with dramatic changes in stock abundance and biomass, along with significant reductions in individual fish size for ecologically and economically important species. Atlantic cod and golden redfish almost disappeared, while until 1996 American plaice, Atlantic wolffish, and thorny skate displayed less pronounced declines in abundance but decreased in biomass by more than 50%. Most recently, some of these groundfish stocks and especially the deep-sea redfish showed first signs of recovery due to successful recruitment, presumably as a result of a lack of direct exploitation and a continued and significant warming in the hydrosphere.

The **Icelandic Shelf** LME as described by <u>Olafur Asthorsson</u> is much less polar than the West Greenland waters, as most of the Icelandic shelf is under the influence of the warm Irminger Current. Only the northeastern shores are affected by cold Norwegian Sea Water. The North-Atlantic Oscillation (NAO) and - to a lesser extent - local wind fields cause strong deviations in temperature and salinity within those general hydrographic regimes. Long time-series of hydrographic data and zooplankton sampling along transects normal to the Icelandic coast mirror largely the CPR data from the Northeast Atlantic. This similarity exists regardless of differences in sampling methods and seasons, *but should not be taken as an excuse to diminish ("rationalise" in) future sampling efforts.* Related to the influx of warm Atlantic water, primary production was high in the 1950s and 1960s and again in recent years. Periods of high temperature and salinity, and of high primary production north of Iceland, are also characterised by high abundance of zooplankton and capelin and by improved growth of capelin and cod. Mean weight of 6-year old cod increased with capelin stock size.

With shifts in the water mass distribution and its biological consequences, the Icelandic LME exported part of its cod population to East and West Greenland and received several immigrations of cod from there until the mid 1960s. Up to about the same time, the Icelandic herring fishery depended mainly on Atlanto-Scandian spring spawners on their feeding grounds north and east of Iceland. The cod stock decreased in 1955-1965 and has shown decadal fluctuations since then. The cod fishery is now well under control by Icelandic authorities advised by the Marine Science Institute. The capelin landings are high (around 1 mill. tons) since the early 1980s, with short periods of lower catches. The adult capelin feed mainly in cold waters off Jan Mayen while young capelin live close to the northern coast of Iceland. *Given the textbook-like food chain relationships in that area, one wonders whether the stock of capelin also profited from the disappearance of Norwegian herring and its massive zooplankton consumption.*

Pandalus is now of great economic importance for the fishing industry of Iceland, as it is all along the Arctic Rim of the North Atlantic from the Canadian coast and Greenland to Svalbard and the Barents Sea. *It seems an open question whether the stocks of Pandalus have always been of the same magnitude as today.*

Iceland has closed down its whaling industry. At the Symposium, whales and whaling were not mentioned at all nor was the harvest of seals. Twenty years ago, marine mammals would still have been topics of discussions at a Symposium of this kind. Marine ecologists should not lose sight of these top predators and their role in high latitude LMEs.

In her presentation on the **Barents Sea** LME, Padmini Dalpadado focused on zooplankton-fish interaction. Similar to the situation off northern Iceland, the hydrographic regime is characterised by the warm Atlantic waters of the southern and western Barents Sea and the cold Polar waters further to the north. The Polar Front zone meanders and is more pronounced in the western Barents Sea. Over the years an excellent database has been developed on the composition and abundance of zooplankton, particularly the various species of euphausiids and amphipods, in relation to water mass distribution. Similarly, cod and herring as well as juvenile and spawning capelin are restricted to Atlantic Water. A surprisingly strong correlation was found between the length and survival of O-group cod and the temperature along the Kola section. Adult capelin feed in summer north of the front and stay in winter south of it. So the pattern of distribution is very similar to that off Iceland.

A long-term, multidisciplinary ecosystem study Pro Mare in 1984- 1989 and subsequent investigations produced inter alia detailed information on the food composition in fish of different age groups. Capelin and herring are primarily plankton feeders, partly competing for food. Strong year-classes of juvenile herring seem to affect recruitment in capelin. Capelin is the staple food for cod, but cod can switch to amphipods when capelin abundance is low. In years of high capelin biomass zooplankton abundance was low and vice versa. *The Barents Sea seems ideal for in-depth studies of predator-prey relationships, as the number of species is low and data on trophodynamics by age groups are very good.* Capelin stocks show strong decadal changes. In the 1970s total landings of cod, herring, and capelin from the Barents Sea fluctuated around 2.5 million tons; at present only 1.0 million tons are caught, largely because of the closure of the capelin fishery which is now ready to resume.

All along the Atlantic Polar Front Zone capelin is a very important fish resource, which is mainly processed for fishmeal. Is there no way to convert capelin more directly into products for human consumption?

LARGE MARINE ECOSYSTEMS OF THE NORTHEASTERN ATLANTIC

J. Chr. Holst concentrated his presentation on the **Norwegian Sea** on stock fluctuations and growth in Norwegian spring spawning herring (Atlanto-Scandian herring). Since saga times there has been a long sequence of rich and poor herring periods of vital socio-economic importance. During rich periods herring was found all along the Norwegian coast and the stock spilled over to Iceland, while in poor periods herring was more or less restricted to the waters off central Norway between Møre and Vestfjord. After two decades of very low abundance, herring stock has recently recovered to a total biomass of over 10 millions tons. The total stock concentrates in the entrance of Vestfjord for overwintering and spreads all over the Norwegian Sea for feeding in summer. Main spawning grounds are still off Møre, but small satellite populations spawn off Iceland and southern Norway. Regular multi-ship surveys have revealed major changes from summer to summer in the vertical and horizontal distribution of herring, varying with hydrographic conditions and zooplankton distribution. The condition factor of herring was used as an indicator for growth in weight varying from year to year. The loss in condition from 1997 to 1998 amounted to more than double the total catch quota (1.3 millions tons) for the same period. So the stock biomass can shrink without changes in F and M. Year-class strength in herring seems to be related inter alia to condition of the spawners: rich year-classes occur only if maternal condition is good or at least average.

According to Eilif Gaard, the **Faroe Shelf** is a fairly closed small LME of great uniformity. It has a circular belt of shelf water separated from the offshore water by a persistent tidal front. The highly turbulent water mass is occupied by a distinct neritic fauna, quite different from the surrounding oceanic environment. Management of fisheries is relatively easy as the LME belongs to one country only and advice is provided by a single institute. Forty years of fisheries data showed rather stable landings of haddock and cod and indicated synchronous fluctuations of both stocks. A deep drop occurred in the late 1980s due to recruitment failure. The growth of cod had decreased since the 1960s and was lowest when cod stock was lowest in the early 1990s. During the same years recruitment in sandeels was also poor, resulting in poor feeding conditions for seabirds and hence high chick mortality. In 1993, simultaneous increases began in recruitment and growth in various stocks of fish and seabirds. Guillemots showed fluctuations similar to those of cod. Sandeels increased in the diet of puffins in the recent few years.

Ten years of full sets of environmental data showed very meaningful food chain relationships: nitrate uptake fluctuates from year to year together with phytoplankton production and zooplankton abundance. The period of particularly poor fish recruitment in 1989-1991 was characterised by low and late primary production and poor small-sized neritic zooplankton. Larger zooplankton, particularly *Calanus*, was dominant in those years of low nearshore production. High neritic primary production is followed by dominance of neritic zooplankton as a better food than *Calanus* for fish larvae, in spite of its relatively low biomass. E. Gaard, however, also pointed to strong winds causing offshore drifting of eggs and larvae as another potential control mechanism.

Altogether the Faroe Shelf seems ideal for modelling of trophodynamics and reproduction. The necessary information can be obtained by a relatively modest monitoring programme. But what about links of the Faroe Shelf communities to the surrounding open North Atlantic in spite of their isolation by the tidal front? It would be worthwhile to look for teleconnections with other parts of the northern North Atlantic as have been described for far more distant places, like the North Sea and the Canary Current.

Multi-disciplinary studies in the **North Sea** have continued for a hundred years and have provided very long-term series of oceanographic, biological, and fisheries data. Jacqueline McGlade concentrated mainly on recent environmental problems rather than fisheries impacts. She described an increase in large-scale climatic effects of the Atlantic on the North Sea. Since 1948 close links have been observed in the trends of North Sea plankton and sea surface temperature in the northern North Atlantic, also described by Taylor and Reid. The northern and southern North Sea are rather different in their bathymetry, current system, and fauna. The water masses are well mixed most of the time in the southern North Sea, but they are stratified in the north. There, salinity has shown a long-term increase in the deep layer. Over the past seven years, since the introduction of satellite observations by SeaWIFS, a steady increase in near surface phytoplankton abundance has been observed in the southern North Sea.

J. McGlade drew the attention of the audience to a number of recently detected environmental hazards, which might affect major parts of the system: polysaccharides produced by benthic bacteria stick at the sediments and might affect the demersal spawning of herring. Those polysaccharides have recently increased by two orders of magnitude. There is an apparent positive trend in the occurrence of algal blooms. Algal blooms in the North Sea seem to be difficult to predict in space and time and in the kind of toxins they produce.

J. McGlade's main concern for the North Sea LME lies in pollution. So far, no dangerous concentrations of single pollutants have been observed. The monitored species of pollutants, however, make up only a small fraction of the long and rather incomplete list of pollutants reaching the North Sea. As in the Baltic, carbon compounds and other matter introduced by man are kept in the system for a very long time and accumulate in sediments and in the lower trophic levels. Pollutants affect the buffering power of the system and the structure of the sediments. Studies of the lower levels of the food chain, where most of the pollutants are retained, should be intensified.

J. McGlade considers the import and release of ballast water in double hull tankers and other large vessels to be the most dangerous recent development in the North Sea. 200 millions tonnes of ballast water from all over the world are annually introduced to the North Sea, importing alien species and diseases en masse. Double hulls have been introduced as a precautionary measure against potential oil spills by tanker accidents. But ballast water seems to have become an even larger ecological hazard than oil spills.

During the Symposium there was no time to deal with the fisheries in the North Sea as another major human impact on the LME. In 1975, ICES had convened a Symposium on "North Sea Fish Stocks - Recent Changes and their Causes" (Hempel 1978a) in Århus, Denmark. The Symposium was triggered by the dramatic shift in the 1960s when pelagic fish like herring, sprat, and mackerel seriously declined and gadoids and sandeels became dominant. For the first time, the possible interaction between these species was discussed in terms of predator-prey relationships and competition for food at the various life history stages. The trophodynamic model by Andersen and Ursin (1978) tried to visualise the interaction between the key fish species and their main food as "a multispecies extension to the Beverton and Holt theory of fishing, with accounts of phosphorus circulation and primary production." Furthermore they produced a model of the biological effects of (man-made) eutrophication in the North Sea (Ursin and Andersen 1978).

In reviewing the 1975 North Sea Symposium at the Joint Oceanographic Assembly (JOA) 1976 in Edinburgh, I concluded "It seems that direct and indirect effects of changes in the fisheries as well as climatic changes in their consequences for the biotic environment caused the recent changes in the fish stocks in the North Sea. It is not possible to quantify the effects of man-made and natural factors separately because of the complexity of interactions between various fish stocks and the stages of their early life history. Obviously the old fight became pointless between those who considered fishery to be the only factor affecting each fish stock separately through growth overfishing and/or recruitment overfishing and those who tried to explain almost every change in fish production by environmental factors. A new approach to the problem has been developed which emphasises ecosystem modelling, detailed studies of early life history and the feeding relationships of the various fish stocks.

The studies of the changes in the North Sea fish stocks face a problem, which is common to many ecosystem studies. They require monitoring over large areas and decades in order to describe correlations. On the other hand the events are small-scale in space and time which determine the year-to-year differences in recruitment and hence in the built-up and decline of stocks" (Hempel 1978b). It is obvious that the "Århus Symposium" contributed substantially to the development of the LME concept.

At the JOA 1976, Lee (1978) summarised for the first time the effects of man - other than fisheries - on the fish resources of the North Sea. He dealt with pollutants and eutrophication, oil industry, extraction of sand and gravel, as well as land reclamation affecting the marine environment and its living resources.

Twenty years later, ICES re-visited Århus and the North Sea phenomena, now covering all aspects of the ecosystem. With regard to fish and fisheries as summarised by Daan et al. (1996), the gadoid outburst in the 1960s was reversed in the 1980s and the cod stock is in a poor state now. Herring had recovered in the 1980s, then decreased again, and now seems relatively well managed. North Sea mackerel did not recover and became replaced by immigrants from west of the British Isles. Total fishing pressure on North Sea fish stocks has further increased in spite of international regulatory measures and improved modelling. The quality of the fishery data has seriously deteriorated with the introduction of sophisticated quota regulations. Fishery management is still based on single species assessments but is taking the risks of by-catch into account. The majority of the commercially exploited stocks seem to be overfished and in decline, while non-commercial stocks thrive or are at least stable.

The linkages of North Sea fish stocks to environmental changes and human impacts other than fisheries are presently not much better understood than at the Århus Symposium 1975 and JOA 1976, in spite of great research efforts. Daan et al. (1996) warn against believing in highly significant correlations between environmental factors and state variables (e.g. population size) of fish stocks. Process variables like growth, mortality, and reproduction are better suited but more difficult to obtain over long periods. At the present Symposium we have seen various successful attempts to move in this direction.

Nevertheless, there is an obvious need for additional monitoring of the lower compartments of the North Sea food web. Remote sensing is of limited value because of frequent cloud cover. The excellent data sets of CPR on "greenness" require ground truthing on primary production and phytoplankton species composition – particularly in the open North Sea. We lack good long-term data sets of benthos in many parts of the North Sea.

The **Celtic Sea** and the **Gulf of Biscay** were addressed by <u>Luis Valdés</u>. They are parts of the Western European Current system but differ substantially in their productivity, which is much higher in the shallow Celtic Sea than in the Gulf of Biscay. The LME is presumably strongly affected by shifts in NAO. Tropical fish species invade the region with increasing frequency, possibly indicating effects of global warming. L. Valdés presented beautiful SeaWIFS photographs of an eddy in the Gulf of Biscay in May 1999 and asked: How does plankton cope with those meso-scale physical processes? Phytoplankton is plentiful in the eddies, but can zooplankton and fish make use of its abundance?

L. Valdés found indications of "fishing down the food chain" and applied Pauly's approach of estimating shifts in the average number of trophic levels in the system, in spite of doubts in the validity of the relatively small differences found. The Gulf of Biscay is notorious for its bad weather and very heavy traffic, rendering great risks for oil spills. But pollution does not seem a general concern for the LME.

Tim Wyatt had a fresh look at the **Iberian upwelling system** by comparing it with the Californian system from a paleoceanographic point of view. During the Ice Age no upwelling existed at the Iberian shelf. In contrast to the Californian sardine, there are two distinct sardine populations off Spain and Portugal, with one spawning inshore in spring and the other offshore in autumn. European sardines do not migrate far, while Californian sardines spread widely if the stock is large. The European sardine does not seem to be overfished. The system seems more sensitive to climate change than to pollution. In fact, Wyatt does not consider toxic pollution a major problem in the Iberian LME.

The comparisons between the Californian and Iberian upwelling systems demonstrated to my mind once more that time is ripe for a major Symposium on the upwelling systems of the world. Their physical and biological similarities and dissimilarities might provide us with further clues to their functioning, variability, global trends, and teleconnections.

Mariculture of mussels and oysters is of great importance to the economy of the Iberian coast, particularly Galicia. Bivalve culture in the Vigo area suffered recently from an introduced dinoflagellate bloom. The reasons for the increased frequency of algal blooms may be found in the increase in human coastal population and/or in the fall-out of the great number of forest fires. A sophisticated monitoring system has been developed to detect harmful algal blooms on the coast and its mariculture installations.

The **Canary Current** LME consists of one of the four major upwelling regions of the World Ocean. Its physical and biological oceanography has been studied in great detail by the international CINECA programme in the 1970s and thereafter particularly by ORSTOM. In his presentation, Claude Roy demonstrated the need for critical evaluations of long-term data sets, particularly for wind and sea-surface temperature (SST) obtained from ships of opportunity. Elimination of doubtful data let trends partly disappear: There are good reasons to believe that the 20% increase in wind speed since 1950 might be due to a systematic bias rather than to natural causes. The main findings after correction of the time series can be summarised as follows: In winter the interannual variability in upwelling and hence SST is negatively related to NAO, particularly in the northern and central part of the upwelling region. The southern part has been very dynamic over the past fifty years, and teleconnections to ENSO events were suggested. Negative SST anomalies in winter were observed in the early and mid 1970s from Spain to Senegal. In the 1980s a warming period started.

During the five decades of observation, fish and fisheries have changed drastically. The sardine populations off Morocco fluctuated widely. Several species showed temporarily great increases in abundance. The effect of upwelling intensity and SST on spatial distribution and abundance of fish was obvious. Tropical species invaded the Canary Current from the south whenever they were particularly abundant in the Guinea system. The pelagic catches follow the global pattern of long-term fluctuations with teleconnections described by Kawasaki and by Bakun (see above). The structure of fisheries has undergone drastic changes from an industrial fishery by foreign fleets to a very efficient local fishery by small vessels.

CLOSING COMMENT

Most of the LMEs under consideration in the present Symposium belong to the best-studied systems in the world. Nevertheless for all of them applies, at least partly, the statement by L. Valdés and A. Lavín for the Gulf of Biscay: "Many of the ecosystem's functional aspects remain unclear, such as connections between physical and biological processes, the parameterisation of energy flows, or the study of natural variability. In other words, we do not know how the ecosystem is regulated." The Symposium demonstrated large regional differences in the state of LME research and management. This reflects partly the specific interests of the authors and partly the actual state of general knowledge about the various LMEs.

Taking K. Sherman's (1995) five modules of any LME project as a yardstick, we can conclude:

Productivity Module. Information on primary, secondary, and tertiary production and their variability and trends was rather limited, while a wealth of data sets on zooplankton, including ichthyoplankton, was provided. Information on zoobenthos was virtually absent.

Fish and Fisheries Module. In most LMEs under consideration, information is richer on landings than on location and amount of catch including discards. Biomass data are available on many stocks, but information on changes in growth rates is limited and we know little about the actual reproduction potential of heavily exploited or poorly nourished stocks where fish are too small, too young, and too lean to produce good eggs in sufficient numbers. More information is needed on changes in the efficiency and range of operation of the fishing fleets.

Pollution and Ecosystem Health Module. Worries were expressed regarding only the North Sea and Baltic Sea. From the presentations it seems that pollution is presently of little concern to open water parts of most LMEs. This is true for the direct effects of pollution on human health and on the survival of fish, but not quite so if we consider hazards to other marine organisms and communities. The term "ecosystem health," however, has a much broader meaning. According to Costanza "to be healthy and

*sustainable an ecosystem must maintain its metabolic activity level, its internal
structure and organisation, and must be resistant to external stress over time and space
relevant to the ecosystem" (Sherman 1996). Although this definition might be too static,
it makes us realise the limitations of our knowledge about the complexity of man's
interaction with the ecosystem and where key risks exist, as demonstrated by J.
McGlade for the North Sea.*

*Socio-economic and Governance Modules. Only a few authors touched upon the
practical application of the scientific findings to the management of LMEs. Over the
past decade the interaction between fishery biologists and fishery administrators and
fishing industry has considerably improved in many places on both sides of the North
Atlantic. Management of fisheries now takes into account the results of stock
assessments and predictions. On the other hand, scientists have become more aware of
the socio-economic implications of their advice. In many cases governance, in terms of
implementation, enforcement and control of the management schemes, is still
underdeveloped. Valuing LMEs, as done for US LMEs, is a useful way to impress the
public and the politicians on the need for research and sustainable management. There
are inherent biases of the values used in economic comparisons among ecosystems.
However, there was an apparent consensus among the Symposium participants that a
strengthening of the linkages between scientifically based assessment of changing
ecosystem states, and clearer expressions of the socio-economic benefits of improved
management practices, could lead to greater global sustainability of marine resources.*

REFERENCES

Andersen, K.P. and E. Ursin. 1978. A multi-species analysis of the effects of variations
     of effort upon stock composition of eleven North Sea fish species. Rapp. P. -V.
     Réun. Perm. Int. Explor. Mer 172:286-296.
Daan, N., K. Richardson, and J. Pope. 1996. Changes in the North Sea ecosystem and
     their causes: Århus 1975 revisited. Introduction. ICES J. Mar. Sci. 53:879-883.
Hempel, G. (ed.). 1978a. North Sea fish stocks – recent changes and their causes. Rapp.
     P.-V. Réun. Cons. Int. Explor. Mer 172. 449 p.
Hempel, G. 1978b. North Sea fishery and fish stocks – a review of recent changes and
     their causes. Rapp. P.-V. Réun. Cons. Int. Explor. Mer 173:145-167.
Lee, A. 1978. Effects of man on the fish resources of the North Sea. Rapp. P. -V.
     Réun. Perm. Int. Explor. Mer 173:231-240.
Rätz, H.-J. 1999. Structures and changes of the demersal fish assemblage off Greenland,
     1982-96. NAFO Sci. Coun. Studies, 32:1-15.
Rätz, H.-J., M. Stein, and J. Lloret. 1999. Variation in growth and recruitment of
     Atlantic cod (*Gadus morhua*) off Greenland during the second half of the
     twentieth century. J. Northw. Atl. Fish. Sci. 25: 161-170.
Sherman, K. 1995. Achieving regional cooperation in the management of marine
     ecosystems: the use of the large marine ecosystem approach. Ocean and Coastal
     Management 29:165-185.

Ursin, E. and K.P. Andersen. 1978. A model of the biological effects of eutrophication in the North Sea. Rapp. P. -V. Réun. Perm. Int. Explor. Mer 172:366-377.

Index